Captain Cook
and the
Pacific

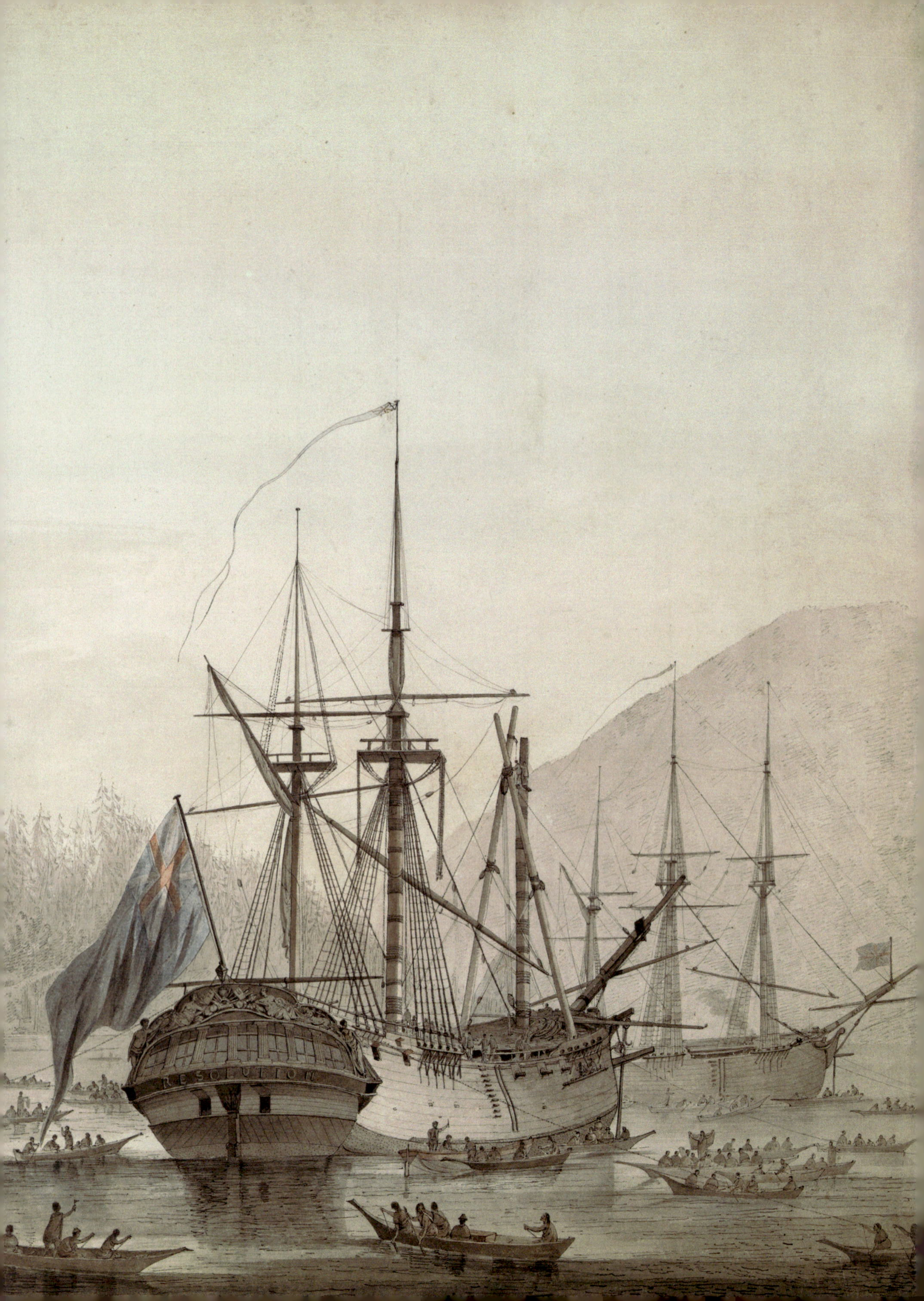
RESOLUTION

Captain Cook
and the
Pacific
Art, Exploration & Empire

JOHN McALEER and **NIGEL RIGBY**

Yale
UNIVERSITY PRESS
NEW HAVEN AND LONDON

NATIONAL
MARITIME MUSEUM
GREENWICH

First published by Yale University Press 2017
302 Temple Street, P.O. Box 209040, New Haven CT 06520-9040
47 Bedford Square, London WC1B 3DP
yalebooks.com/yalebooks.co.uk

In association with Royal Museums Greenwich, the group name for
the National Maritime Museum, Royal Observatory Greenwich,
Queen's House and *Cutty Sark*
www.rmg.co.uk

Copyright © 2017 National Maritime Museum

All rights reserved. This book may not be reproduced or transmitted
in any form or by any means, electronic or mechanical, including
photocopy, recording or any other information storage and retrieval
system (beyond that copying permitted by Sections 107 and 108 of
the US Copyright Law and except by reviewers for the public press),
without prior permission in writing from the publisher.

ISBN 978-0-300-207248 HB
Library of Congress Control Number: 2016055768

10 9 8 7 6 5 4 3 2 1
2022 2021 2020 2019 2018 2017

Designer: Will Webb
Printed in China

All images are from the National Maritime Museum collection unless
otherwise indicated, and their catalogue reference numbers are given.
Images can be ordered from the Picture Library, Royal Museums
Greenwich, London, SE10 9NF (tel: 020 8312 6704/6631)
or online at images.rmg.uk.

Front cover:
The Resolution *and*
Discovery *off Huaheine* (detail; see fig. 46)
Style of John Cleveley,
the Younger, c.1780
National Maritime Museum.
Presented by Captain A.W.F. Fuller
through The Art Fund (BHC1838)

Back cover:
Captain James Cook, 1728–79 (see fig. 1)
Nathaniel Dance, 1776
National Maritime Museum,
Greenwich Hospital Collection
(BHC2628)

Frontispiece:
Resolution *and* Discovery
in Ship Cove, Nootka Sound (detail; see fig.44)
Pen, ink and watercolour,
by John Webber, 1778
National Maritime Museum (PAJ2959)

Contents

Foreword
Dr Kevin Fewster, Director, Royal Museums Greenwich   7

1 **Introduction**
James Cook and the Pacific at the National Maritime Museum   9

2 **James Cook, the Royal Navy and His Three Voyages of Exploration**   25

3 **'The advancement of science and the increase of knowledge'**
Charting the Pacific and Enlightenment Science   63

4 **Cook's Pacific**
Exploration and Encounter   85

5 **Visualizing the Pacific**
Art, Landscape and Exploration   117

6 **Exhibiting the Pacific**
Collecting, Recording and Display   145

7 **'Men of Captain Cook'**
Pacific Voyages 1785–1803   169

8 **The Strange Afterlives of Captain Cook**
Representations and Commemorations   201

9 **Cook on Display**
The National Maritime Museum's
Cook Galleries and Exhibitions, 1937–2000   225

Notes   241
Index   249

Foreword

The links between James Cook and Greenwich stretch back to the spring and summer of 1768, when Cook was appointed to the command of HM ship *Endeavour*, then being prepared at Deptford dockyard for a voyage to 'distant parts'. Marking the 250th anniversary in 2018 of the departure of this momentous voyage, the National Maritime Museum will open *Pacific Encounters*, a new gallery that will usher in more than a decade of commemorations in Britain and overseas. Among them are Cook's landings in New Zealand and Australia, to be commemorated in 2019 and 2020 respectively; in Hawai'i and the northwest coast of America in 2028; and, of course, his death in 2029. I am grateful to Yale University Press for having worked so closely with the Museum over the last few years to anticipate the renewed public and scholarly interest in Cook that these events will bring, by publishing this handsome volume that explores the story, contexts and legacies of the man and his voyages through the Museum's unique collections.

The National Maritime Museum holds one of the finest collections of Cook material in the world. In art – largely, but far from exclusively, through William Hodges's oil paintings and sketches from the second voyage – it is pre-eminent. However, it is really the sheer diversity and depth of the collection that give it the international significance and public appeal it has enjoyed since the Museum opened its doors in Greenwich in 1937. Here one can see the four marine timekeepers developed by John Harrison to resolve the problem of finding longitude at sea; and K1, the copy of Harrison's successful design by Larcum Kendall that Cook took and tested on his second voyage. Here, too, one can reflect upon the iconic portrait of Cook, commissioned by Sir Joseph Banks and painted by the society artist Nathaniel Dance. The Museum's Caird Library and Archive holds private and public papers relating to the voyages, and a large and important collection of published journals, charts and maps. The Museum also holds a modest but important collection of ethnographic objects brought back from the three voyages. We continue to develop our Cook holdings through purchase, transfer and gift, acquiring in 2013 (with the munificent support of a number of institutional and private supporters, led by the HLF) George Stubbs's *The Kongouro from New Holland* [Kangaroo] and *Portrait of a Large Dog* [Dingo], the influential first studies of Australia's startlingly different fauna to be seen in Britain. The same year, we also acquired objects collected in the Pacific from the 1790s by the London Missionary Society, a generous gift of the Council for World Mission. No gallery can do full justice to the size, range and significance of these collections. Neither does this book set out to be a concise catalogue. It is a publication that demonstrates how the Museum's holdings can tell the story of – and, more significantly, how they shed new light on and explore the complexity of – one of the most important set of voyages of the modern era.

This book is the product of a great deal of hard work, by many different people, over a number of years. I would particularly like to thank Sally Salvesen at Yale University Press, and Rebecca Nuotio and Kara Green at the Museum, for taking the initial idea and helping to make it a reality. Over the course of the project, a number of colleagues at the Museum played crucial roles in the process. Kirsty Schaper and Kate Mason in the Publications team helped to steer the entire project, patiently and skilfully, to a successful conclusion. Authors and readers alike are indebted to Tina Warner, Senior Manager of Photography, and her colleagues in the Photographic Studio for the wonderful images of the Museum's collections. The authors are particularly grateful to the Museum's General Editor, Pieter van der Merwe, for his insightful comments and customary attention to detail. Finally, I would like to congratulate the book's authors, John McAleer and Nigel Rigby, both of whom have worked with the Museum's unrivalled collection of Cook-related material over a number of years. I hope their words and interpretations help to highlight the enduring significance of this collection, as well as the people and voyages that helped to create it.

Dr Kevin Fewster, AM, FRSA
Director, Royal Museums Greenwich

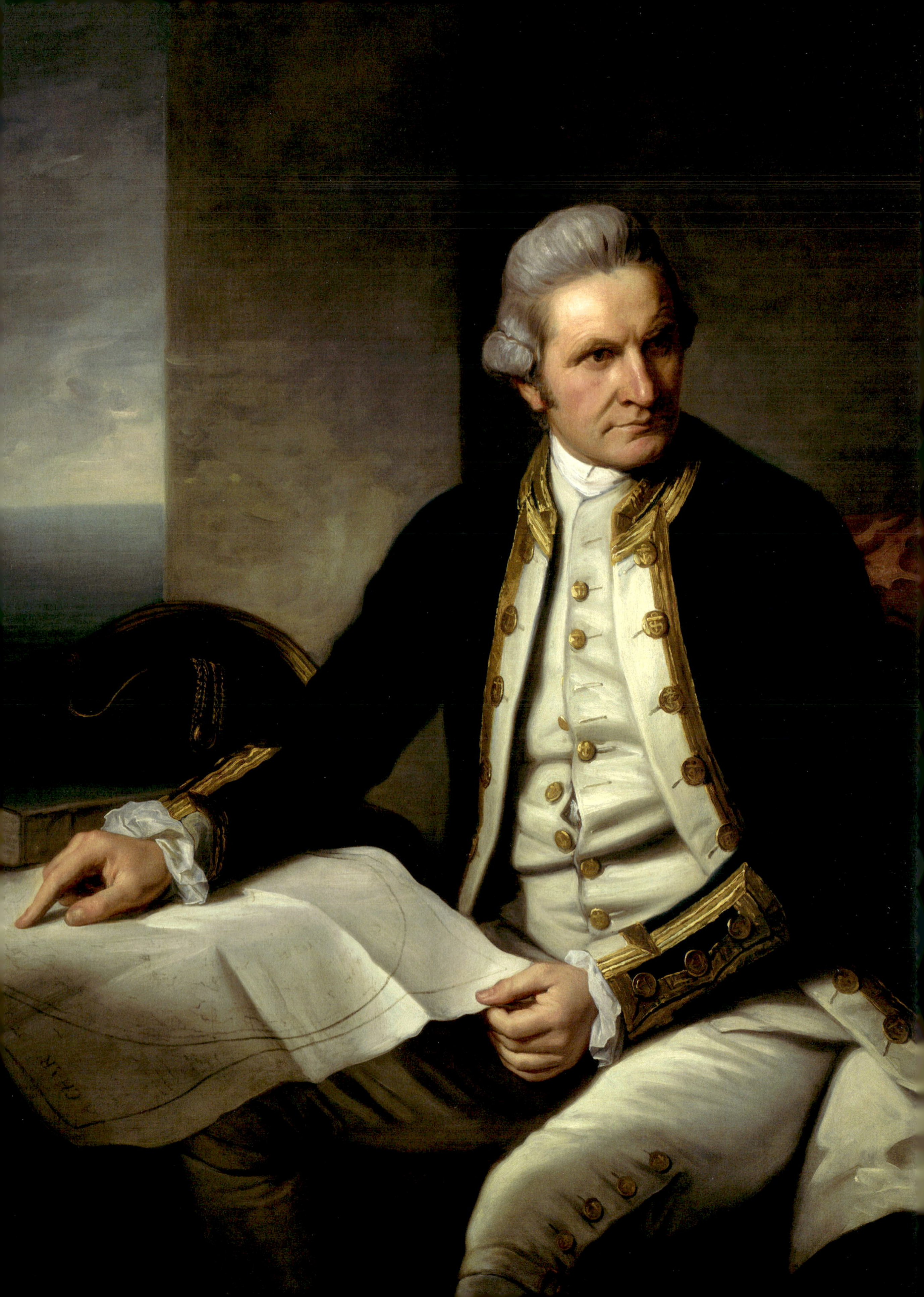

Chapter 1
Introduction
James Cook and the Pacific at the National Maritime Museum

John McAleer and Nigel Rigby

Nathaniel Dance's three-quarter-length portrait of James Cook, painted in the summer of 1776, is one of the most arresting images in the National Maritime Museum's collections (fig. 1). The viewer's attention is drawn to the confident gaze of one of the most influential and celebrated seafarers in Britain's maritime history, and confronted by some of the fruits of his extensive explorations in the vast Pacific Ocean. Cook is resplendent in the full dress uniform of a Royal Navy captain. The navy-blue jacket, white breeches and white waistcoat with gold braid and gold buttons alert us to the sitter's profession and his standing within it. A chart of the Southern Ocean, produced from the results of his voyages to the region, rests on the table, while his right hand points to the outline of the east coast of Australia. Cook's hat sits on the table behind him to the viewer's left, on top of a substantial volume, itself resting on the chart.

Originally commissioned by Sir Joseph Banks to commemorate Cook's return from his second voyage to the Pacific Ocean, this work by one of the leading portrait painters of the day was intended to hang over the fireplace in the library of Banks's Soho Square home. Banks was one of the leading lights of the British scientific establishment, but he had a personal as well as a scientific interest in Cook's exploits: he had sailed with Cook on his first voyage in the *Endeavour* and maintained a keen interest in his subsequent travels (fig. 2). By commissioning and displaying a work like this, Banks placed Cook and his travels at the heart of European Enlightenment culture and of Britain's eighteenth-century empire. In 1825, following the death of Banks's widow, his executor was easily persuaded to present the portrait to the new 'National Gallery of Naval Art' in the Painted Hall of Greenwich Hospital, and it thereby passed with the rest of that collection into the care of the National Maritime Museum (NMM) in 1936. Its early and continuing presence in the national pantheon of naval and maritime memory at Greenwich demonstrates a consistent fascination with Cook's story since his own time, and his ongoing relevance to our understanding of Britain's place in the world.

The voyages of exploration across and around the Pacific Ocean commanded by James Cook, undertaken between 1768 and 1780, are among the most significant events in the history of European engagement with the wider world. In their day, they represented maritime achievement and scientific progress on a global scale, bringing Enlightenment ideals to bear on the huge expanse of the Pacific. Their work was based on reliable instruments, mathematical methods and verifiable observations. The ships and their crews tested scientific theories, collected data and speculated on the people and places they encountered in the course of their travels. They inspired some of the most influential images of the Pacific made by Europeans, and encouraged a host of future expeditions and exploration schemes. Following his death on a Hawai'ian beach, James Cook – the 'British Columbus' – became a powerful and enduring icon of British maritime prowess.[1]

Cook's expeditions ushered in a new era in world history, with the Pacific Ocean presenting a novel and challenging field of experience for Europeans. In the words of Joseph Conrad, these voyages offered potential solutions to 'the problems of our earth's shape, its size, its character, its products, its inhabitants'.[2] The Pacific is the largest of the world's oceans, covering an area greater than the Earth's land area combined. For thousands of years it supported extensive exploration, migration and settlement, yet it was the last to be navigated and explored by Europeans, with Britain coming late to the endeavour. Cook's voyages in the 1770s built on foundations laid by previous explorers, but they did more than any other to

Fig. 1
Captain James Cook, 1728–79
Nathaniel Dance, 1776
National Maritime Museum,
Greenwich Hospital Collection
(BHC2628)

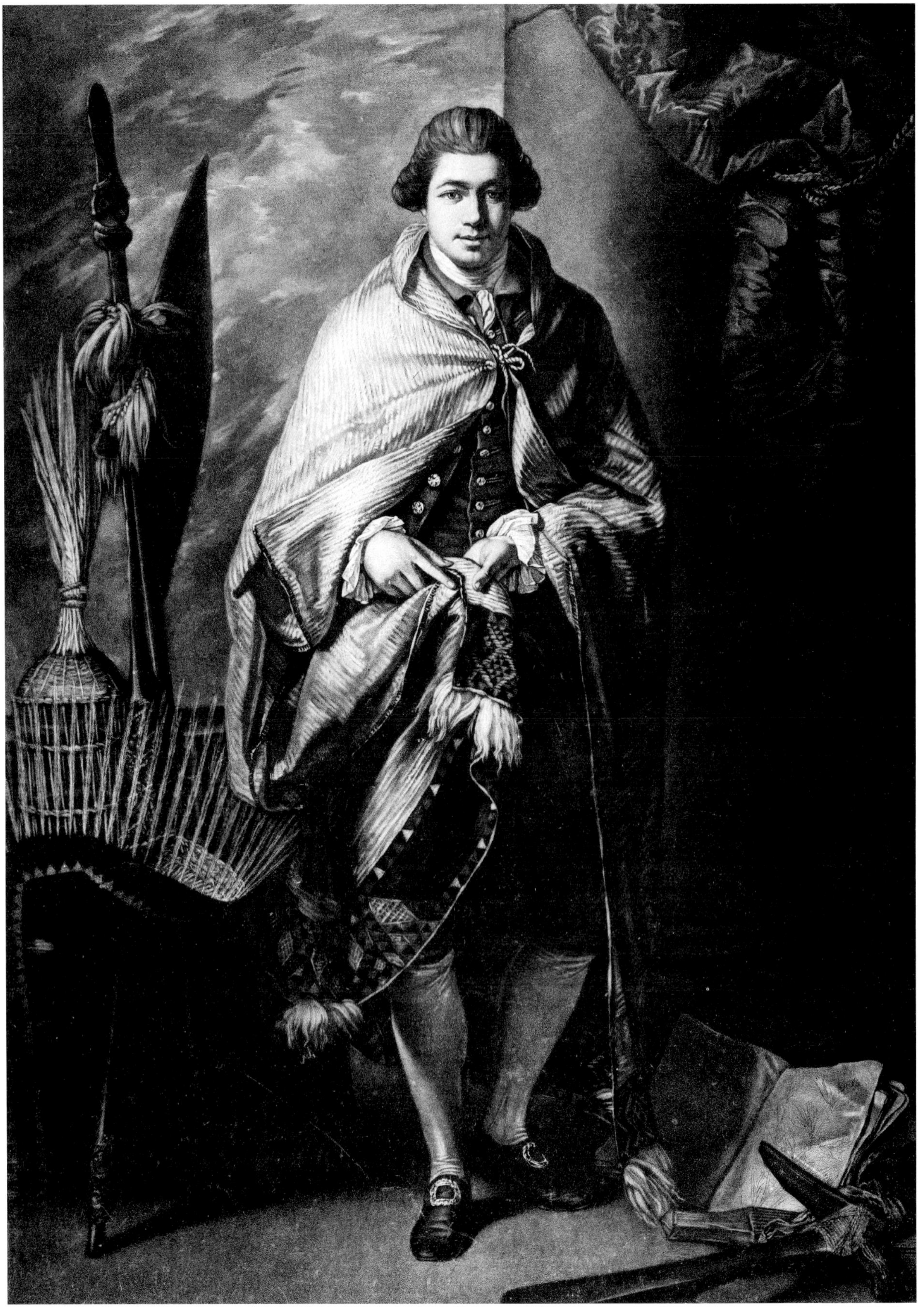

change the way Europeans viewed the wider world, as well as themselves and their own cultures. The expeditions returned with a wealth of knowledge in the form of raw scientific data and tangible objects. They inspired a host of artistic and philosophical responses whose reverberations were felt in every branch of the arts. During his voyages Cook not only encountered many Pacific cultures for the first time, but also assembled the first large-scale collections of Pacific objects brought back to Europe. He returned with the first reliable maps of the region and some of the earliest images and surveys of the flora and fauna of many Pacific islands. By the end of the eighteenth century – thanks in large measure to Cook, his companions and his successors – the myth of a vast 'Southern Continent' had been dispelled, virtually the entire Pacific basin had been charted, and its diverse cultures had been brought to the attention of the West. Cook, in particular, 'left nothing to those who might follow in his track to describe or fill up', as was lamented by the great French explorer Jean-François de La Pérouse (1785–88).[3] This navigational success also opened the way for commercial and colonial developments in the nineteenth century. If Europeans came late to its exploration, the Pacific was the first section of the world to be included in the imperial archive of the nineteenth century.

Of course, an interest in the world beyond Britain existed long before Cook embarked to plot the transit of Venus in 1768. William Dampier's voyage in HM ship *Roebuck* at the very beginning of the eighteenth century, for example, prefigured expeditions later in the century. Dampier was given instructions to make careful observations, collect specimens and bring back 'some of the natives, provided they shall be willing to come along'. He was also ordered to keep an 'exact journall', which was to be handed in to the Admiralty 'and no other' on his return.[4] A few decades later, John Campbell remarked that Arthur Dobbs's expedition to discover the North-West Passage in the 1740s was 'the topic of common discourse, and of almost universal expectation'.[5] In 1764, Commodore John Byron sailed with instructions to carry out explorations in both the North and South Pacific, while two years later Captain Samuel Wallis and Lieutenant Philip Carteret also left for the Pacific. Byron's instructions urged that 'nothing can redound more to the honour of this nation as a maritime power, and to the advancement of trade and navigation thereof, than to make discoveries of countries hitherto unknown'.[6] Cook's voyages took this interest to new heights. By 1772, the year in which he set off on his second voyage to the South Seas, Johann Reinhold Forster commented that 'circumnavigations of the globe have been of late the universal topics of all companies'.[7]

James Cook was, in the words of Marshall Sahlins, 'responsible for the shape of the world as we know it', but he did not live to see the full impact of his voyages.[8] His death in Hawai'i in 1779 on his third voyage, in an altercation with the local people, offers a powerful illustration of the forces unleashed by the European exploration of the Pacific. Encounters between indigenous populations and European sailors were frequently charged with confrontation, confusion and easy misunderstandings. Cook's voyages, their results and their widespread and enduring impact on European society also demonstrate that these expeditions shed a great deal of light on British society at the time. The lionization of Cook following his death enrolled him in the pantheon of British scientific, naval and imperial achievement – as shown by the early desire to have his portrait in the new Naval Gallery at Greenwich. He represented a new kind of hero for an industrial, proto-democratic age. This image of Captain Cook – described by Joseph Conrad, in language reminiscent of medieval chivalric romance, as 'the navigator without fear and without reproach' – sustained British views of its maritime history and its place in the world for generations, as well as providing potent founding myths for colonial settler societies in the Pacific.[9]

Biographies of Cook have appeared regularly since his death, and interest in him has supported a healthy industry in biographical and hagiographical writing since Andrew Kippis published his account of Cook's *Voyages* in 1788. Although this book, *Captain Cook and the Pacific*, concentrates on James Cook and the voyages he commanded, it is neither a comprehensive biography of the man nor an exhaustive account of his voyages. These have already been documented in great detail by others, and all students of the subject are particularly indebted to J.C. Beaglehole, who made it his life's work to document the voyages comprehensively and wrote an extensive and still-standard biography. Rather, it attempts to chart the history of Britain's engagement with the people and places of the Pacific – with the three voyages of Captain Cook at its core – through a specific collection of objects and images.

British voyages of exploration to the Pacific in the eighteenth century tested geographical theories, solved scientific problems, broadened artistic and academic horizons, and brought cultures into contact with each other for the first

Fig. 2
Sir Joseph Banks Bt.
President, F.A.S.
John Raphael Smith, after
Benjamin West, 1 May 1788
National Maritime Museum
(PAH5497)

This engraving from Benjamin West's portrait of Sir Joseph Banks shows him surrounded by 'curiosities' brought back from the Pacific.

INTRODUCTION 11

time. This book touches on all of these themes and attempts to elucidate them through the objects and collections that form one of the most tangible and compelling legacies of Cook's voyages. Held largely by the NMM at Greenwich, London, these collections offer a unique window on this crucial moment in the history of cultural interaction and encounter. Lisant Bolton argues that Cook is 'more trope than history' – a man whose 'popular fame today is partly constructed through and by museums, and through the collections made on his voyages'.[10] In using these objects, artefacts and images to tell the history of Cook's voyages to the Pacific Ocean, the chapters that follow demonstrate the importance of the material legacies and public histories of exploration as crucial ways of understanding (and writing about) it and its consequences in terms of encounter and empire. Each chapter introduces readers to different aspects of, and examples from, the unrivalled collection of Cook-related material held at Greenwich: oil paintings, prints and drawings, navigational instruments, globes, charts and maps, ship plans, rare books and manuscripts, coins and medals, ethnographic material and personal effects. Taken together, these artefacts present a compelling picture of Cook, the voyages he commanded and their impact on world history, bringing the dramatic stories of Britain's maritime engagement with the Pacific Ocean, its people and its places vividly to life.

Fig. 3
Plans showing works completed to convert the collier Earl of Pembroke *for exploration as HM ship* Endeavour
Signed by Adam Hayes, Master Shipwright, Deptford Dockyard, 11 July 1768
National Maritime Museum (ZAZ6587)

The Museum holds the original plans for Cook's ships, and for many of those that followed in their wake.

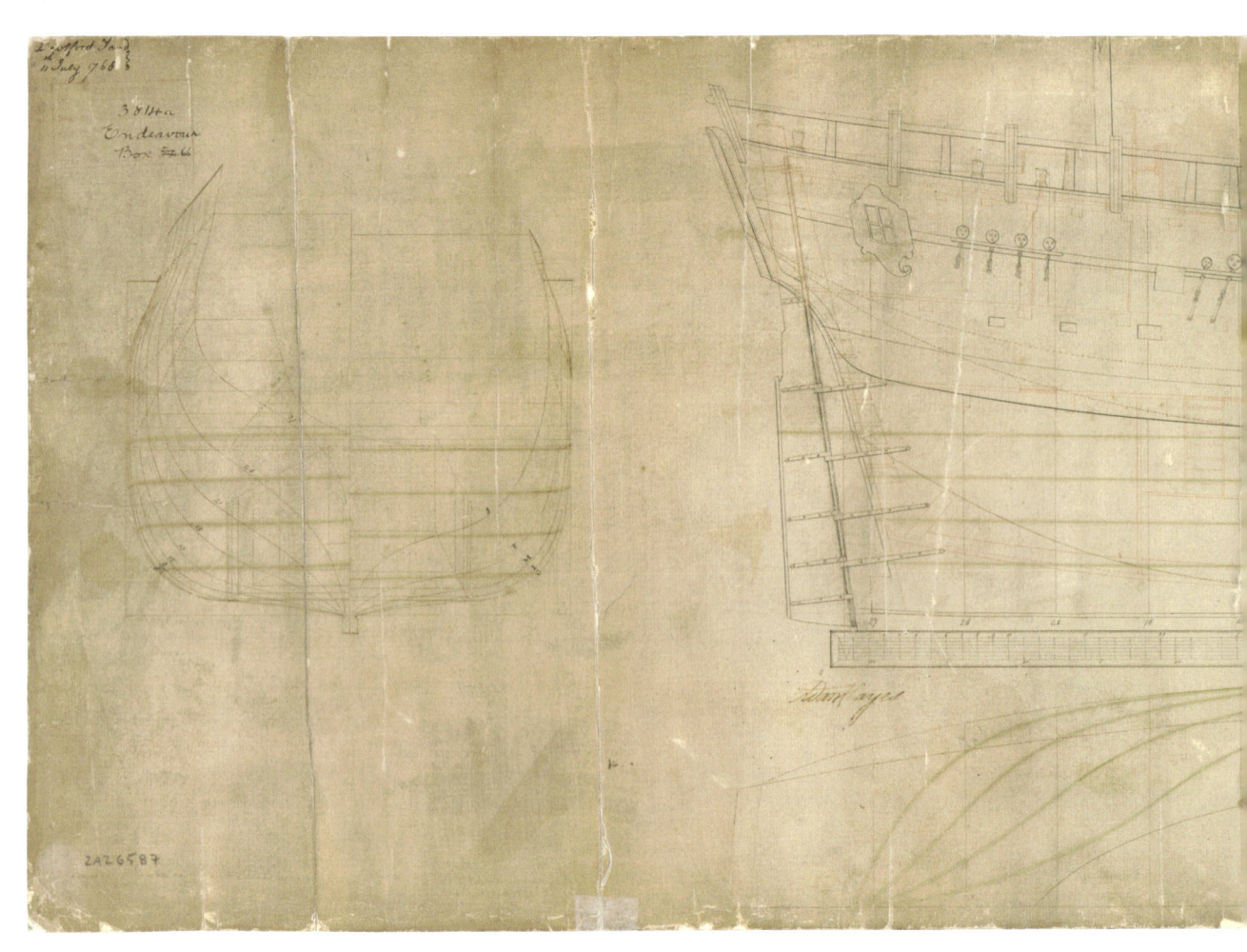

IMAGES, OBJECTS AND COOK'S VOYAGES

The portrait by Nathaniel Dance, like many of the objects associated with him in the Greenwich collections, reminds us that Cook's voyages were about the acquisition of knowledge. However, they were also about the acquisition of things and the ways in which objects could convey information and ideas. In the preface to the official account of the third voyage, John Douglas remarked on the curiosities derived from Cook's voyages, which were then on display to the public in the museum of Sir Ashton Lever – his so-called 'Holophusicon' on London's Leicester Square. Lever himself had named the room in which the objects were housed after the 4th Earl of Sandwich, First Lord of the Admiralty and a key mover behind the Cook voyages. Douglas was clearly impressed: 'If the curiosities of Sir Ashton's Sandwich-room alone, were the only acquisition gained by our visits to the Pacific Ocean, who that has the taste to admire, or even the eyes to behold, could hesitate to pronounce, that Captain Cook had not sailed in vain?'[11] Objects and their interpretation have always played a crucial role in assessing the significance of Cook's voyages.[12]

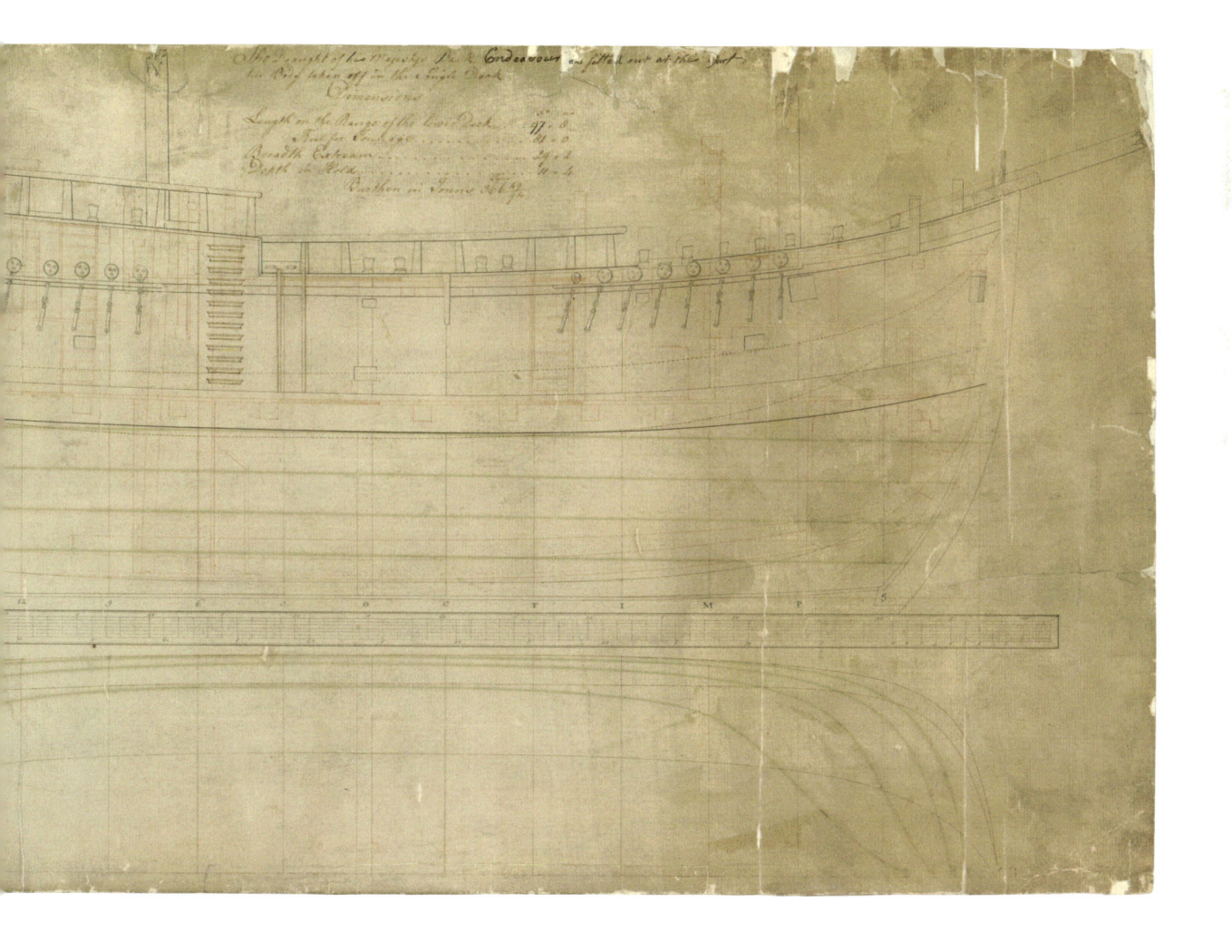

Although some early viewers of Dance's portrait, like David Samwell, surgeon's mate in *Resolution* on the second voyage and surgeon in *Discovery* on the third, thought it 'a most excellent likeness', Mrs Cook was less convinced. Apparently she thought that her husband looked too stern.[13] In the confident pose of its sitter, the painting seemed to be the visual embodiment of John Hawkesworth's view that Cook was 'universally allowed . . . to be as good an officer and as able a navigator as the world has ever seen'.[14] Historians, by contrast, have tended to focus less on the pose and physiognomy of Cook than on the attributes with which the artist has surrounded him. Indeed, the portrait tells us much about the important role played by such material culture in representing exploration to the wider public. Brian Richardson, for example, has drawn attention to the way in which map, text and image come together in the portrait. Cook points to the map, on which rests a book or journal (figs 4, 5). For Richardson, 'the painting is a theoretical statement concerning the relationship between knowledge, authority and the world'.[15] We might add that it is a statement expressed through physical artefacts and material culture. The painting exhibits exploration in a very literal sense: in the two-dimensional representation of the map and the three-dimensional form of the book (as well as the canvas itself, of course).

This book similarly aims to place other material evidence of Cook's voyages, now housed at the NMM, in the wider historical contexts in which it was collected, created or displayed. Objects tell us something about the voyages and the voyagers, but they also give us insights into the ways in which these are commemorated, the priorities of their audiences and the importance they attached to different aspects of this episode in Britain's maritime, naval and imperial history. In taking this approach, the book draws on well-established anthropological ideas about the cultural biographies of things, in which the identity, interpretation and relevance of an artefact changes, depending on the context in which it is being displayed and viewed.[16]

Interpretations of objects do not remain static. Indeed, the Dance portrait of Cook – just like the reputation and legacy of

Figs 4, 5
Journal of Captain Cook, *Endeavour*, 1768–71
National Maritime Museum
(JOD/19)

Presented to the Museum by George VI in 1937, along with Cook's journal of the second voyage.

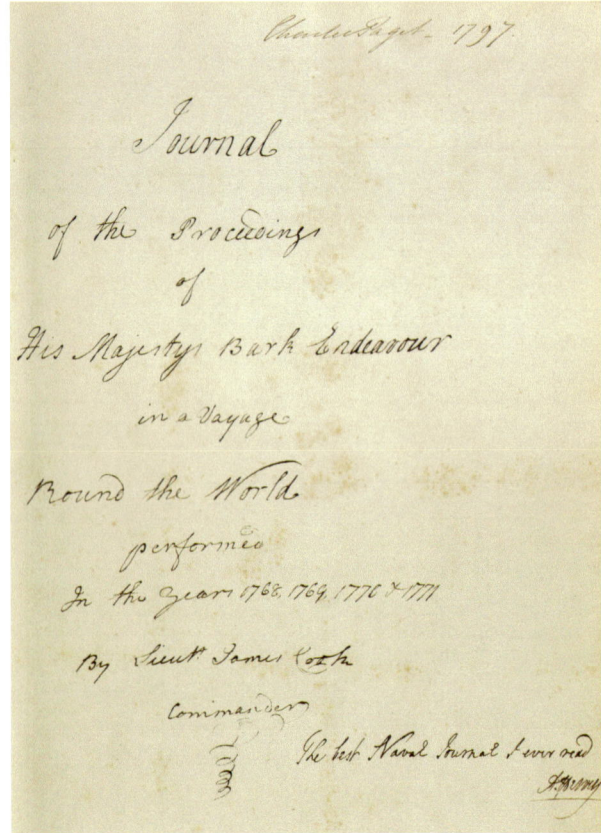

14   CAPTAIN COOK AND THE PACIFIC

the sitter – has undergone many re-evaluations over the years. Those viewing the painting in Banks's library, in the inner sanctum of one of the architects of the Enlightenment, were clearly intended to associate the portrait's owner with the sitter and recall their respective positions in British scientific, governmental and naval circles. Banks was an influential figure in the corridors of Whitehall, at the heart of government. Cook, on the other hand, acted out his career largely on the peripheries of the known world, bringing European knowledge and curiosity, Enlightenment science and British naval power into contact with the people and places of the Pacific. This context for viewing the painting changed, however, with Sir Joseph's death in 1820. After the portrait was presented to the Naval Gallery at Greenwich Hospital by Banks's executor, Sir Edward Knatchbull, its exhibition there presented an opportunity to foreground other interpretations of the work. As Geoff Quilley has noted, the establishment of the gallery at Greenwich was an attempt to address issues of nationhood and national identity.[17] The creator and *de facto* first curator of the gallery, Edward Hawke Locker (Secretary and later senior Commissioner of the Hospital), regarded the portrait as embodying the essence of 'native genius' and the kind of meritocracy that made Britain great. Cook was the exemplar of someone who rose through the ranks from humble origins to greatness by virtue of hard work, dedication and talent.[18] In a similar way, many of the objects discussed below would have been regarded and understood in different ways by different people at different times over the intervening two-and-a-half centuries.

The collections of the NMM have dictated the strands explored in the chapters that follow. However, while the focus is necessarily on Cook, we have tried to do this while simultaneously retaining in view other, sometimes competing narrative strands. For example, within these collections and our interpretations of them we have tried to reflect the critical role played by Pacific Islanders as voyagers themselves, as well as in facilitating Cook's voyages and in resisting some of their legacies. As Cook himself acknowledged, Polynesians were extraordinarily skilled mariners and people like Tupaia and Omai (Ma'i) played crucial roles in aiding the travels of these European ships around the ocean and its islands.

The presence of the Dance portrait in a public gallery from the early nineteenth century reinforces a point made by Christian Feest, who reminds us that the objects and artefacts collected by Cook and his crew in the Pacific have spent much more time in museums than in their original source communities.[19] One might say something similar about the images and objects created back in Britain, also associated with Cook's voyages. The collections at Greenwich illuminate every aspect of these voyages and represent one of the richest resources of Cook-related material in the world. Of course, their presence in a museum underlines the need to be sensitive to the ways in which objects can be interpreted and presented, and the wider narratives that they are used to support. Like all museum collections, the one at Greenwich is a partial one, built up in what often appears to be a random and piecemeal fashion. Nevertheless, the images and objects there – and in other notable Cook collections, of which there are several – bear powerful testimony to Britain's engagement with the rest of the world. From the vibrant paintings of William Hodges, the official artist on Cook's second expedition, to the ethnographic items collected by the crews of his ships, to maps, charts and accounts published after the return of the voyages, the Museum's array of two- and three-dimensional objects provides an intriguing window on this historical episode.

THE STRUCTURE OF THE BOOK

The themes of individual chapters in the book follow the contours of the collection at Greenwich. They weave together the histories of individual objects, placing them in the wider context of Pacific voyaging to tell the story of Cook, his successors and their various encounters with the region, through the material legacy of their travels.

The trajectory of Cook's career intersected with one of the most tumultuous and dramatic periods in British history, as the country developed into a major power on the world stage. The Royal Navy played a central role in this process as it helped to win, protect and sustain the burgeoning British Empire. The story of Cook's life, career and expeditions needs to be evaluated and considered in the wider panorama of Britain's commercial interests, naval power and expanding imperial ambitions and responsibilities. The book begins by looking at Cook's early life and career, as well as his formative naval experiences, in these social and political contexts. Chapter 2 also provides an overview of the three Pacific expeditions that he commanded and the motives that inspired them: their key personalities, events and locations, as well as the mechanics and logistics associated with their preparation, their course and their results. These expeditions were important for the sheer scale and diversity of the scientific work performed by Cook and his crew. They engaged in comprehensive surveying and charting work; encountered unfamiliar cultures and named new lands; collected and gathered specimens and objects; and recorded, in both word and image, the people, places and cultural practices they witnessed. As well as focusing on the voyages themselves, this chapter situates them in the broad sweep of earlier European exploration of the Pacific, demonstrating how they fit into a larger pattern of British interest in the maritime discovery and commercial exploitation of the region.

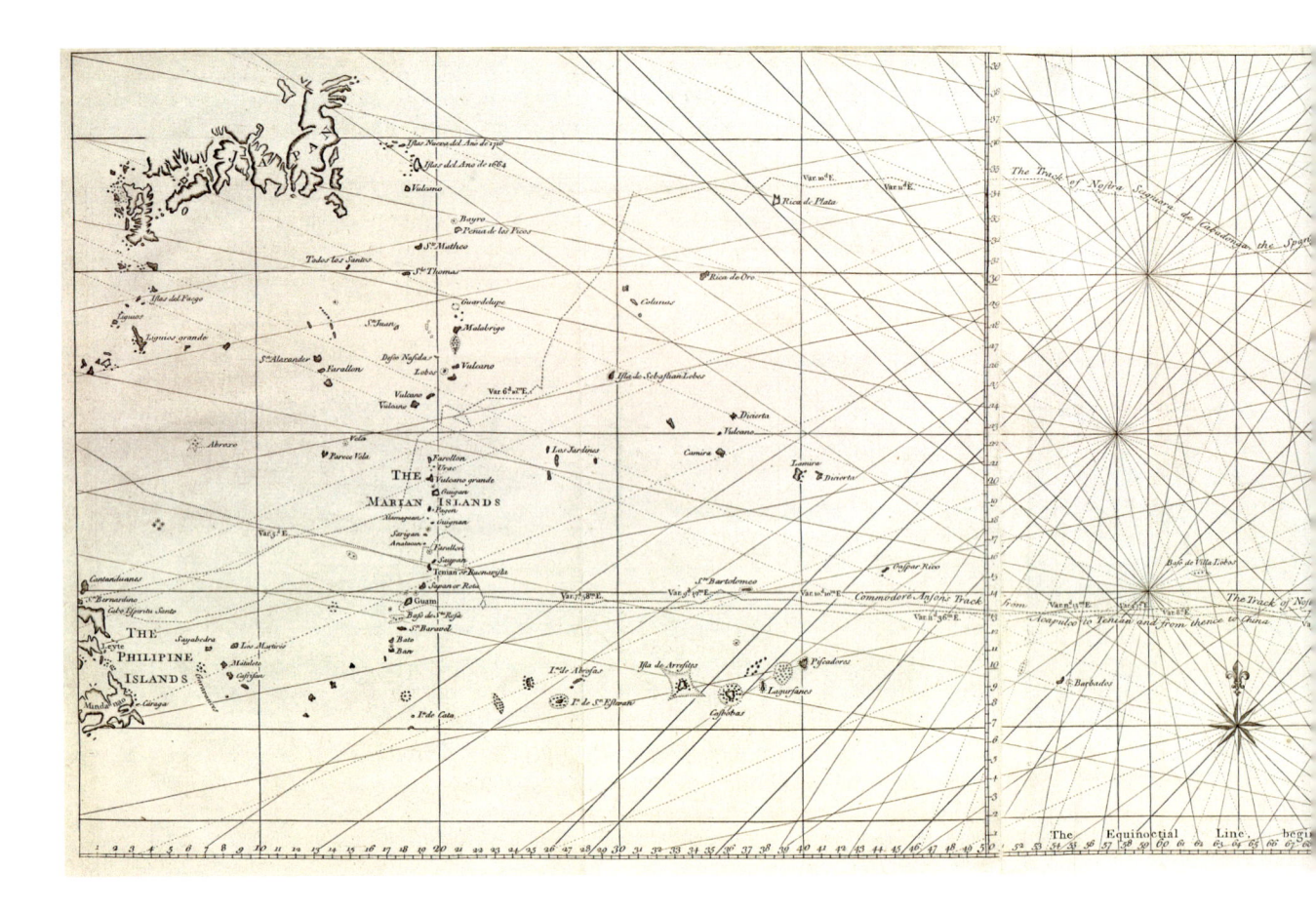

Fig. 6
A chart of the Pacific Ocean from the equinoctial to the latitude of 39½° No.
R.W. Seale, c.1744
National Maritime Museum
(G266:2/8)

The chart was reproduced in George Anson's *Voyage Round the World* (1748).

William Bligh, one of the most famous (or perhaps notorious) figures to follow in Cook's wake, maintained that the voyages to the Pacific undertaken during the reign of George III had 'the advancement of science, and the increase of knowledge' as their principal objectives.[20] Cook's voyages were at the heart of this scientific impulse. Chapter 3 assesses their role in surveying, charting and mapping the globe and defining our concepts of the modern world, as well as the material results these efforts produced. In bringing Enlightenment ideals to bear on the huge expanse of the Pacific Ocean, the voyages of Cook and others spawned an array of published material, including maps, charts (fig. 6), narratives and images. The production of these items not only added to the store of knowledge, but also brought it to the attention of a rapidly expanding audience. When awarding him the Royal Society's Copley Medal in 1776, Sir John Pringle remarked that Cook had 'fixed the bounds of the habitable earth, as well as those of the navigable ocean'.[21] The scientific achievements of the voyages were extraordinary and were largely a result of the combined efforts and creative interaction of the sailors and savants on board. These 'floating academies' of science produced great advances in European knowledge of the vast and relatively uncharted waters of the Pacific: they proved that New Zealand was composed of two islands; they surveyed the coast of New Holland (Australia) for 2000 miles and established that it was separate from New Guinea; they traversed the Antarctic Ocean on three successive voyages, sailing completely around the globe in its high latitudes and proving that the dream of a habitable great Southern Continent was a chimera; and they traced 3500 miles of the North American coastline. These enquiries produced detailed sea charts, maps and accounts of the people and places encountered, and meant that – although it was the last oceanic region to be explored by Europeans – the Pacific was among the first to be added to the archives of Enlightenment knowledge.

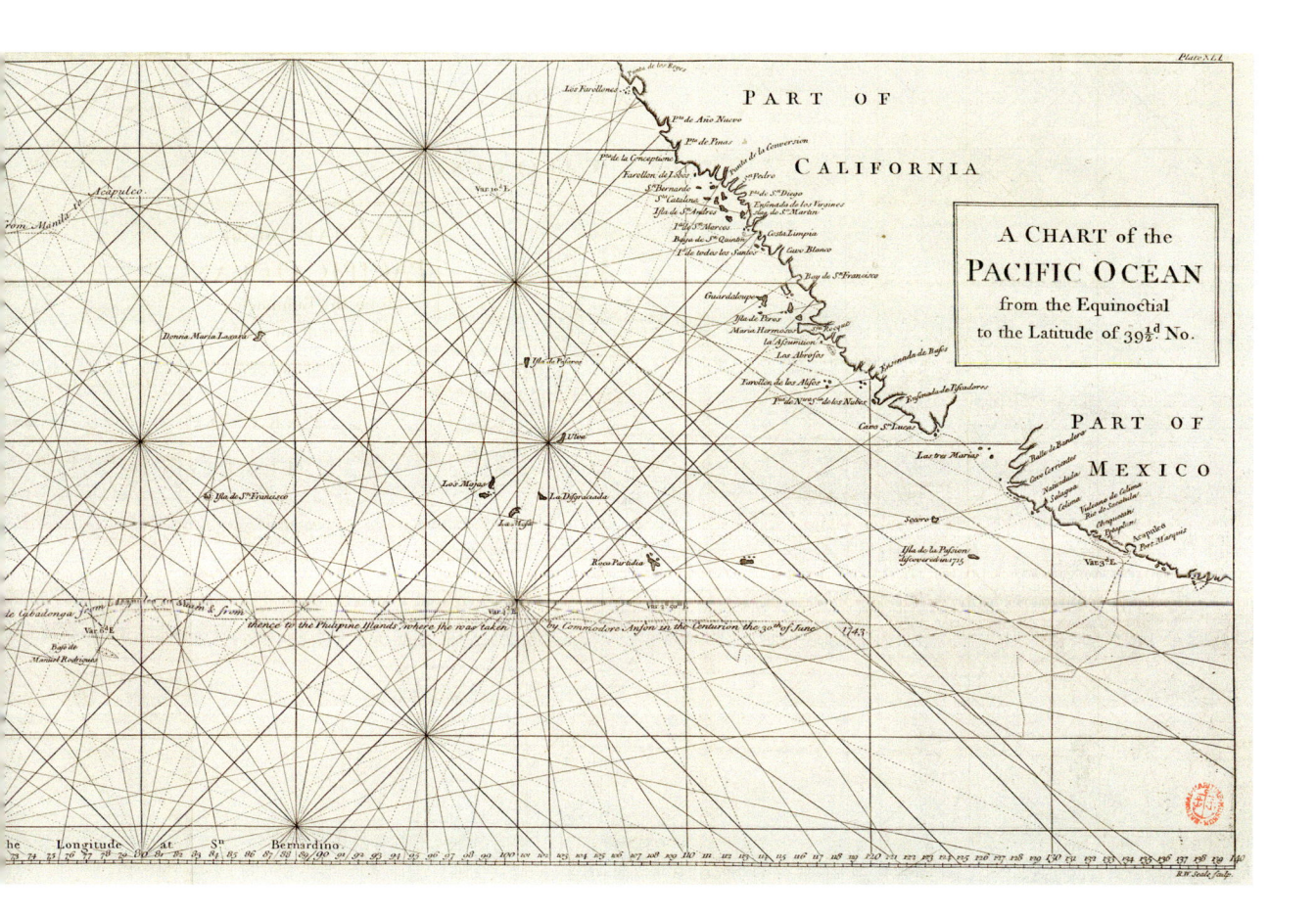

In a similar way to their geographical and navigational achievements, Cook's voyages brought the rich tapestry of the many cultures of the Pacific into view for Europeans. All manner of information – from religious beliefs and ritual ceremonies, to language and everyday household routines – was recorded. This aspect of the voyages forms the basis of chapter 4, which explores the engagement of European sailors and travellers with the various Pacific cultures they encountered. It uses the objects they collected, as well as the descriptions and images they created, to discuss the ways in which European voyages (and voyagers) interacted and engaged with Pacific people and cultures (fig. 7). The expeditions of Cook and his successors were instrumental in acquiring large quantities of art and artefacts from the societies of various Pacific islands, America and Australasia. They also helped to establish a taste for, and an interest in, these objects in Europe. Material culture from the Pacific – the 'numerous specimens of the ingenuity of our newly discovered friends', as John Douglas termed them – offered 'real matter for important reflection' on the nature of Britain's place in the world, the status of the societies encountered and the relative merits of each.[22] Today, they also provide us with an unparalleled resource for understanding the encounters and exchanges that characterized the relationship between European sailors and Pacific Islanders in this momentous period in the history of the region.

In addition to returning with a host of natural history specimens, scientific results and three-dimensional objects, a significant number of visual images were created on, or as a result of, these voyages. Chapter 5 focuses, therefore, on the role of art, image-making and the voyage artists who accompanied the

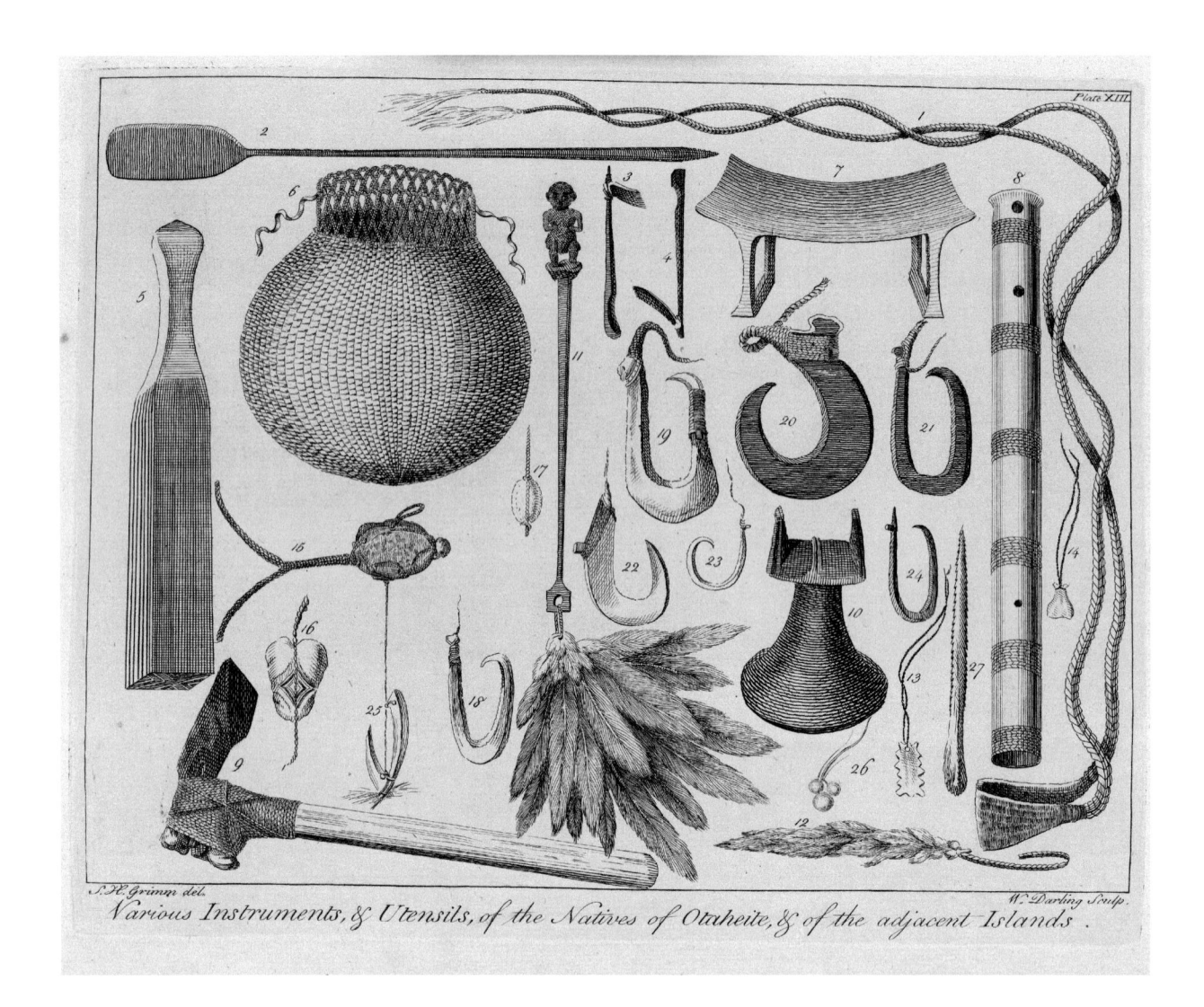

Fig. 7
A Journal of a Voyage to the South Seas, in His Majesty's Ship the Endeavour
Sydney Parkinson, *c.*1773
National Maritime Museum
(PBC4680)

The journal of the artist on the first voyage, Sydney Parkinson, reveals the wide-ranging interests of the scientific party and their fascination with 'new' and exotic cultures.

18 CAPTAIN COOK AND THE PACIFIC

expeditions. Even before Cook set sail in 1768, the practice of taking dedicated artists on maritime expeditions was a well-established one. William Dampier, for example, was provided with the services of 'a person skill'd in drawing' to help him, at the beginning of the eighteenth century.[23] In an era before photography or clearly defined scientific 'expertise', artists played a crucial role in recording scientific data on voyages of exploration. Voyage artists were entrusted with making visual records that complemented written accounts. They supplied what, as Cook himself admitted, 'can only be described by the pencle [pencil] of an able painter'.[24] While these artists brought their European artistic ideas, classical training and aesthetic preferences with them to the Pacific, they returned with eyewitness experience of landscapes that had rarely been seen before by Europeans. The NMM's unique collection of oil paintings, watercolours, prints and drawings, which vividly evoke this tension, spans the three Cook voyages as well as the records of those voyages that went to the Pacific in their wake. These works of art not only offer a visual testimony to the respective voyages, but also provide clues to the ways in which contemporary European audiences engaged with these regions and their exploration by naval vessels. For example, John Webber exhibited some 29 paintings based on the third Cook voyage and utilized his 'storehouse' of Cook images for the rest of his career. In 1788, he issued a series of prints based on drawings and studies not included in the official account, which were later published as *Views in the South Seas*.[25] The 'image' of the Pacific was one that many artists – including those who had never travelled there – were willing to represent to a public eager for visual information about these places. Often they owe more to European ideals of aesthetic composition than to empirical knowledge derived from the voyages themselves. The channels by which information about Cook's voyages entered the wider public consciousness in Britain, as well as the ways in which they were presented and interpreted to fit with contemporary preoccupations and concerns, are themes that run through many of the subsequent chapters in the book.

Writing in 1777 – a year after Cook had embarked on his third and final Pacific voyage – it appeared to Edmund Burke as if all the regions and peoples of the world were 'at the same instant under our view'. For those living at the heart of Britain's empire, it seemed that 'the great map of mankind is unroll'd at once'.[26] Voyages of exploration – or, more precisely, their various results in the form of objects, images and the texts and cultural products they inspired – were crucial conduits for presenting the wider world to people in late-eighteenth-century Britain. Published exploration narratives became wildly popular and were avidly consumed by armchair travellers in Europe. The first edition of Cook's *Voyage to the South Pole*, for example, sold out in a single day in May 1777. Cook and King's *Voyage to the Pacific Ocean* appeared in June 1784 and, despite retailing at the relatively expensive price of 4½ guineas, sold out within three days. More than 100 editions and impressions of Cook's voyages were published by the end of the eighteenth century.[27] The results of these voyages appeared in a variety of other contexts, too: theatre, museum displays, art exhibitions and even pocket globes. Chapter 6 assesses the various channels through which information about the Pacific percolated into British and European cultural consciousness. The paintings of voyage artists, such as William Hodges (figs 8, 9) and John Webber, adorned the walls of influential art exhibitions (albeit with mixed success); lavishly illustrated official travel narratives and cheaper, more popular accounts were eagerly anticipated by the reading public; audiences were exposed to narratives of exploration in pantomimes, panoramas and music halls. The presence of indigenous people in the imperial metropolis powerfully confirmed that exploration literally brought the world back to Britain. With objects such as the hand-held terrestrial globes that were then so fashionable, it was even possible to carry in one's pocket the story of exploration and Britain's maritime engagement with the wider world. The ways in which material culture shaped the public understanding of Cook's voyages in his own day was mirrored in later periods. Before considering this, however, we return to the Pacific Ocean to assess how Cook's exploits inspired further expeditions there.

Cook's voyages were part of a broader context of European engagement with the Pacific in the eighteenth century. Although Europeans had been traversing the ocean for centuries, the six voyages undertaken by Wallis, Byron, Cook and Vancouver from the 1760s to the 1790s, for example, together with those of their French and Spanish contemporaries, transformed the character and pace of European engagement with the Pacific in a manner that was to have far-reaching consequences.[28] Although his expeditions were among many undertaken, Cook's voyages seemed to capture the collective imagination in Britain and laid the groundwork for much of the later British exploration of the region. The feats of endurance and exploration performed by Cook, his crew and their ships encouraged emulation and aroused the interest of contemporaries. Indeed, many of the people who sailed in Cook's wake had previously served under him, such as William Bligh of *Bounty* mutiny fame but of much greater success in his subsequent *Providence* voyage. Other British expeditions, such as those commanded by George Vancouver and Matthew Flinders, made significant contributions to scientific knowledge in their own right; and, of course, Cook's influence extended beyond the boundaries of Britain, inspiring a whole host of European voyages to the Pacific. Still others were involved in exploiting the commercial possibilities it offered. For example, Nathaniel Portlock, who had served on

CAPTAIN COOK AND THE PACIFIC

Cook's third Pacific voyage, was appointed by the King George's Sound Company in 1785 to lead a fur-trading expedition to the north-west coast of America. James Strange decided to fit out a voyage from India to sail to the Columbia River on the Pacific coast of North America by virtue of 'an attentive perusal of Captain Cook's last voyage'.[29] John Ledyard's account of Cook's final voyage – published in 1783 – was the first book to be copyrighted in America.[30] A history of Cook's last voyage, Ledyard concluded, 'may be essentially useful to America in general but particularly to the northern states by opening a most valuable trade across the North Pacific Ocean to China and the East Indies'.[31] Chapter 7 explores some of these late-eighteenth- and early-nineteenth-century voyages by the 'men of Captain Cook'. The NMM also holds important artefacts and relics relating to them, emphasizing the widespread and profound impact made by the Pacific and its exploration on British naval and national consciousness at the time and subsequently. Ultimately, the chapter shows that the story of Cook and his influence over Pacific and maritime exploration did not end with his death.

The final two chapters take up this theme of Cook's 'afterlife'. They assess his variegated legacies in terms of his continued importance and enduring fascination for more than two centuries after his death, through the plethora of material culture inspired by his career. Chapter 8 begins by analyzing the representation of Cook and his voyages in the period following his death. Texts, images and objects were swiftly deployed in the service of presenting and commemorating the various facets of his personality and reputation in the era of Nelson, Napoleon and post-1815 change: the intrepid explorer, the resolute naval officer and the man of science. The discussion then considers some of the ways in which Cook's memory was harnessed in the nineteenth- and twentieth-century heyday of empire, when – as an eighteenth-century explorer – he seemed to herald Britain's burgeoning imperial reach and ambitions for later generations. Walter Besant, writing at the end of the Victorian period, offers a good example. For him and many of his contemporaries, Cook's legacy was simple: he had 'given to Great Britain Greater Britain'.[32] The interweaving of his life, death and afterlife also forms the basis of the final chapter in the book, where we explore the representation of Cook and his voyages at the NMM since it first opened to the public in 1937. Decolonization and the post-colonial world of the late twentieth century undoubtedly affected the ways in which Britain's maritime and imperial history was presented over that time. Nevertheless, where many of the traditional naval heroes of the Victorian age, such as George Brydges Rodney, Edward Vernon and George Anson, all but disappeared from the Museum's exhibitions and galleries, Cook's story largely survived and prospered in the first 80 years of its history. Despite his endurance as a subject, however, the interpretation of Cook has undoubtedly altered over this period to take account of changing social, cultural and political attitudes. Chapter 9 shows the interplay and tensions between the various forces – institutional, scholarly, personal, museological and financial – that kept Cook and related exploration of the Pacific at the forefront of the NMM's public displays. The enduring appeal of objects, exhibitions and displays relating to maritime exploration, throughout the nineteenth century and up to the present day, demonstrates the long-lasting impact of eighteenth-century Pacific exploration.

EMPIRE, SCIENCE AND CELEBRITY

The story of Cook's voyages illustrates the complex intertwining of science and empire in eighteenth-century Britain. It also illuminates some of the ways in which an individual, and the cult of celebrity surrounding him, were harnessed to wider themes of national identity and an emerging imperial consciousness.

One of the most important motives for embarking on the expeditions undertaken in the second half of the eighteenth century was their potential scientific benefit. The voyages and their results contributed to new knowledge in many areas and even inspired the creation of new academic disciplines. From anthropology to zoology, art history to taxonomy, voyages of exploration expanded intellectual as much as geographical horizons.[33] Although their promoters and sponsors in the Royal Society, in the Admiralty and in government more generally

Fig. 8 (opposite, above)
Landing at Mallicolo [Malakula], *one of the New Hebrides*
William Hodges, *c.*1776
National Maritime Museum, Caird Fund (BHC1904)

Hodges produced a series of 'encounter' paintings to be engraved for the official publication of Cook's second voyage.

Fig. 9 (opposite, below)
The Landing at Tanna [Tana], *one of the New Hebrides*
William Hodges, 1775–76
National Maritime Museum, Caird Fund (BHC1905)

Another of the four small 'encounter' paintings that Hodges made for engraving in the official publication of Cook's second voyage.

Fig. 10 (left)
The Kongouro from New Holland [Kangaroo]
George Stubbs, 1772
National Maritime Museum (ZBA5754). Acquired with the assistance of the Heritage Lottery Fund; the Eyal and Marilyn Ofer Foundation; The Art Fund (with a contribution from the Wolfson Foundation); and other donors.

The Museum's collections of world-class art related to Cook's voyages continue to grow. Stubbs's 'Kangaroo' and 'Dingo' (fig. 152) were acquired in 2013.

Fig. 11 (opposite)
'The Apotheosis of Cook', from *Outlines of the Globe*
Thomas Pennant, 1787–92
National Maritime Museum (P/16/4)

were not uninterested in the economic and political possibilities of exploration, the three state-sponsored expeditions led by Cook were undertaken with the improvement of navigation and geographical knowledge in mind. They were justified by the desire to add to the store of human knowledge about uncharted lands, seas, peoples, plants and animals (fig. 10). Emmanuel Mendes da Costa, a London-based shell collector, wrote to the Swedish naturalist Daniel Solander before he left on Cook's first expedition. He thought the voyage that Solander was about to embark upon would be 'the most scientific and illustrious event that has ever happened to that useful and enchanting study. It may be justly called the Argonautic Expedition for the study of nature.'[34] Cook's ships were, in effect, floating laboratories to achieve this, and their voyages both reflected and refined the ideals of the Enlightenment and served as inspiration for the equally ambitious British, Spanish, French and Russian expeditions that followed.

Cook's voyages were never simply disinterested scientific missions, however. The Admiralty's instructions to their appointed commander made it clear that scientific exploration went hand-in-hand with commercial opportunities and national aggrandisement. Their Lordships advised that 'the making of Discoverys [sic] of Countries hitherto unknown . . . will redound greatly to the Honour of this nation as a maritime power, as well as to the dignity of the Crown of Great Britain, and may tend greatly to the advancement of the Trade and Navigation thereof'.[35] P.J. Marshall and Glyndwr Williams caution that 'the motives for the Pacific expeditions after 1763 were not simply, or even primarily, scientific':

> *This second New World promised resources of such potential that its discovery and control might tip the commercial balance of power in Europe – for Britain confirm the overseas superiority brought by the wartime conquest, for France redress the humiliations of an unsuccessful war.*[36]

Exploration of the Pacific was necessarily bound up in complex ways with Britain's expanding global role in the late eighteenth century.

Perhaps most intriguingly of all, and something that continues to shape the ways in which societies and individuals interpret the story of Cook, is his status as a 'celebrity'. Kathleen Wilson has demonstrated how late-eighteenth-century thinking about national identity was articulated and refined through representations of Cook and the South Pacific.[37] In the years immediately after his death, Cook's name became a cipher for empire. His voyages became a way of representing the eighteenth-century British Empire and of understanding the country's place in the world. Even at the time, Cook's near-contemporaries interpreted the voyages and their commander as symbols of British maritime expertise and its successful commercial exploitation of the world's oceans. This can be seen in an account of the island of Madagascar, penned anonymously by Sir George Young in the first decade of the nineteenth century. The fact that Cook and his actions in the Pacific are employed in this relatively obscure document about an island far distant from it demonstrates the kinds of cultural cachet that his memory held. In this prospectus, Cook's name is used to justify Britain's imperial destiny, as the author wonders: 'To what end are all the discoveries of our great fore-fathers and lately those of the wonderful Cook?' For Young, the scientific discoveries of the voyages and the commercial possibilities they implied were of a piece. They were entirely consistent with 'the spirit of discovery, the genius for trade and skill in navigation, for which we always have been and still are famous'. Neither the opportunities opened up by the voyages nor the sacrifice of Cook himself should be given up lightly: 'For certain it is, Cooke's [sic] discoveries were not given in vain and a very small beginning on this plan is sufficient to promote the general good of the human race.'[38]

A similar, if slightly more disinterested, impulse is in evidence in *Outlines of the Globe*, a manuscript written by the naturalist and armchair traveller Thomas Pennant (fig. 11). In it, he offers a touching testimony to the departed Cook. At first, he attempts to use language to deliver 'an eulogium on our great circumnavigator', but is defeated by the enormity of the task: 'my powers cannot attain the extent of his merits'. Giving up on the written word, Pennant turns to images instead: 'My fancy has endeavoured to sketch a monument worthy of so illustrious a character whenever his country thinks proper to pay him that respect'. The naturalist proceeds to explain the composition and iconography:

> A Globe is represented, and the great scene of his actions the Pacific ocean is turn'd towards the admiring spectator. His passing the Arctic and Antarctic circles is delineated and the fatal spot on which he fell by savage hands is expressed. Above is his Apotheosis. An angelic form is conveying our hero to heaven and pointing towards the immortal abodes with a naval coronet which he bears in the other. Strong rays of glory beam on both. The inscription above tells the illustrious shade Now receive your regards you have explored every part of the mortal coasts.[39]

The harnessing of Cook's voyages and his memory by both Young and Pennant serves to illustrate just how far the voyages, their clear scientific objectives and their commander had penetrated the general cultural consciousness of late-Georgian Britain, to an extent that could not be disentangled from the nation's newly found global role. As we have seen already, however, Cook himself, as leader and commander, has inspired a great deal more commentary than this. His life, his voyages and his afterlife – especially as represented through objects and images – have come to epitomize a formative passage both in the history of British seafaring and in Britain's more general cultural encounter with the wider world.

The Sailor's Return.

Just on the Beach arriv'd, with great Surprize,
Jack sees his Molly; Him too Molly Spies:
What! is it Thou? with open Arms She cries.
Then drops the brittle Goods She sells for Bread,
While all aghast beside stands Messmate Ned,
And points where flows the Bowl, & Gen'rous Red.

But Molly's Mother, more sagacious, opes
The wealthy Chest, on which She plac'd her hopes,
And for the richest Prizes careful gropes.
The settled Crew gay Mirth and Love proclaim:
One leads aloft the mercenary Dame,
Who drunk, returns her Load from whence it came.

Contemning Wealth, which they with Risk obtain,
Thus Sailors live, and then to Sea again.

Printed for Carington Bowles in St. Pauls Church Yard, London.

Chapter 2
James Cook, the Royal Navy and His Three Voyages of Exploration

Nigel Rigby

NAVY DAYS

In June of 1755, James Cook left a secure, if unexciting, job as mate of a Whitby-based collier to enlist at Wapping, East London, as an able seaman in the Royal Navy. Exactly why, at the age of 26, he decided to exchange the imminent prospect of commanding one of his employer's ships for the uncertainties of life on the lower deck of a man-of-war, just weeks before the start of the Seven Years War in Europe (1756–63), has long been a matter of speculation. The eighteenth-century Navy's popular reputation for brutal discipline, sadistic officers, dreadful working conditions and inedible food – not to mention, as Dr Johnson put it, the chance of drowning – makes Cook's decision appear singular. N.A.M. Rodger, however, has exposed the falsity of this myth, showing in his masterly studies *The Wooden World* and *The Command of the Ocean* that, by the standards of the time, the Royal Navy was a good employer, and one under which it was possible for a seaman to advance in a respected profession and even, on occasion, cross what was not always a clear line between the 'tarpaulins' who worked the ships and the gentlemen who commanded them. It is doubtful that Cook's sights were set so high at this point in his career, but his action was certainly not remarkable, especially as the imminent war held the strong possibility of prize money and promotion. Daniel Baugh has recently shown that the number of sailors in the Royal Navy increased from around 58,000 at the beginning of the war to about 75,000 at its height, some through the unpopular press gangs, but many more as volunteers.[1]

Cook was posted to HM ship *Eagle*, a 60-gun fourth-rate commanded by Joseph Hamar and undergoing refit in Portsmouth. She was 147 feet overall, with a beam of 42 feet and a crew of around 400. The 60-gun ships were considered just big enough to take their place in the line of battle – the Navy's traditional fighting formation – although these were gradually giving way to the heavier 74-gun ships that would come to form the backbone of the line. Within a month Cook was rated master's mate, a rapid but scarcely surprising promotion given his experience, and one that some biographers suspect had already been agreed when he signed on. A master's mate was a petty officer who assisted the sailing master – the senior warrant officer (generally referred to as the 'master') who, beneath the captain, held responsibility for the navigation and pilotage of the ship. A master's mate could be an experienced, professional seaman hoping in time for promotion to master, but within the growing professionalism of the mid-century Navy he could just as easily be a young aspirant officer training towards his lieutenant's examination, which required candidates to have served at least two years at sea as either a midshipman or a master's mate. Either way, master and master's mate were respected positions and, as Rodger points out, even though the holders of these roles were not given an official uniform, as commissioned officers had been in 1748, they had an occupational dress, great responsibility and were expected to look respectable and behave like officers.[2]

Soon after Cook's posting to *Eagle*, Hamar was replaced by Captain Hugh Palliser, a well-connected, able and respected officer (fig. 13). Palliser's patronage would play an influential role in Cook's progress. By the time Cook left *Eagle* in 1757, he had been given the responsibility of taking back to English ports two French prizes captured by *Eagle*. Crucially for his career, he had also passed the oral examination conducted by two captains at Trinity House in London, which qualified him to 'take charge as Master of any of His Majesty's Ships from the Downs thro' the Channel to the Westward and to Lisbon'.[3]

On passing the examination, Cook was appointed master of *Solebay*, a sixth-rate, 20-gun ship engaged in patrolling the seas around Scotland. He did not stay long, moving before the

Fig. 12
The Sailor's Return
Charles Mosley, c.1750
National Maritime Museum
(PAF3801)

The print shows a sailor's occupational dress of a short blue jacket and loose trousers.

25

Fig. 13 (left)
Captain Hugh Palliser, 1723–96
George Dance, the Younger,
c.1775
National Maritime Museum,
Greenwich Hospital Collection
(BHC2928)

Palliser became one of Cook's greatest patrons. After the explorer's death, Palliser built a monument to Cook at his house in Chalfont St Giles, dedicated to 'The ablest and most renowned Navigator this or any country hath produced'.

Figs 14, 15 (opposite)
*Two Views of
Cape Breton Island*
J.F.W. Des Barres, 1777
National Maritime Museum
(PAH5777 and HNS64A)

The two vignettes show the everyday work of a surveyor: taking sights using a plane table and theodolite, and taking soundings and bearings from a small boat.

end of the year to the newly built 60-gun fourth-rate *Pembroke* under the command of Captain John Simcoe. In 1758, *Pembroke* crossed the Atlantic to take part in the capture of Louisbourg, arriving in Halifax, Nova Scotia, after a long and tedious voyage with almost half the crew incapacitated by scurvy. Cook would be based on the North American coast for the rest of the war, on *Pembroke* and later *Northumberland*. Some days after the fall of Louisbourg in 1758, he happened to meet the army engineer Samuel Holland, who was making a detailed plan of the town and its defences. Cook was intrigued by his use of a plane table, a portable instrument mounted on a tripod that was used in the field to sight and map topographical detail (figs 14, 15).

Although well established in land surveys, the plane table had been little used in maritime contexts, if at all.

The two men met again on board *Pembroke* with Captain Simcoe – 'a truly scientific gentleman', as Holland would describe him many years later.[4] This was a pivotal moment for Cook who, with the backing of Simcoe and the advice of Holland, began to develop a serious professional interest in surveying, apparently reading Euclid and applying himself to the related study of mathematics and astronomy for the first time.[5] This went well beyond what would have been expected of a master in the Royal Navy, and Cook's willingness to learn and apply the new techniques of land surveying,

26   CAPTAIN COOK AND THE PACIFIC

A View of Richmond Isle, near the Entrance of the Gut of Canso.

JAMES COOK, THE ROYAL NAVY AND HIS THREE VOYAGES OF EXPLORATION

and to appreciate the increasing advances in the manufacture of instruments, would have brought an impressive accuracy to his coastal surveys.[6] Cook would play an important part in Holland's survey of the St Lawrence River, which made possible the risky but successful attack on Quebec the following year. More importantly, he began to build a reputation both within and outside the Navy. After the end of the Seven Years War, Holland persuaded the British government of the need to chart its now-extended North American possessions, a project that essentially became part of a larger one to map America's entire eastern seaboard from Labrador to the Gulf of Mexico.[7] Palliser's patronage brought Cook the job of surveying the strategically and commercially important but inadequately charted island of Newfoundland, acquired by Britain in the Treaty of Paris in 1763. The detailed survey of a large part of Newfoundland's 8400 miles of heavily indented coastline was to occupy his next three years. Cook spent the summers surveying and the winters back in London drawing up his charts. He observed an eclipse of the Sun in Newfoundland in 1766, and his observations were published by the Royal Society. He also renewed his professional relationship with the commander-in-chief and governor of Newfoundland, Hugh Palliser.[8] By 1768, 'Cook was arguably the most experienced marine surveyor in the navy and the only one able to command a ship'.[9] With these qualifications, together with the patronage of Palliser, he was certainly a strong candidate for the command of HM ship *Endeavour* on the proposed voyage of scientific exploration to the Pacific Ocean.

Fig. 16 (below)
A representation of the surrender of the Island of Otaheite to Captain Wallis by the supposed Queen Oberea
John Hall, *c.*1770s
National Maritime Museum (PAI3953)

Fig. 17 (opposite)
The attack of Captain Wallis in the Dolphin *by the natives of Otaheite*
Edward Rooker, *c.*1770s
National Maritime Museum (PAJ2145)

The two engravings are taken from John Hawkesworth's *An Account of Voyages Undertaken...for Making Discoveries in the South Seas* (1773). Its many inaccuracies persuaded Cook to play a far greater part in publishing an official account of the second voyage.

HM SHIP *ENDEAVOUR*: THE FIRST VOYAGE, 1768–71

The origins of the *Endeavour* voyage lay in the Royal Society's decision to support an international project to observe the transit of Venus across the Sun. These rare astronomical events happened in pairs, the last transit having been in 1761, with the second due on 3 June 1769, but not then repeated until 1874. It was hoped that bringing together data from the observations at a number of known points around the world would enable that basic astronomical measurement, the distance of the Sun from Earth, to be determined, from which it would be possible to calculate the size of the solar system. That was the theory, but when it was applied to the 1761 transit, it became evident that at least one of the observations needed to be made from the southern hemisphere. In the plan drawn up for the 1769 transit, the Pacific observations were going to be made from Tonga, which the Dutch navigator Abel Tasman had briefly visited more than a hundred years before, in 1643. However, there was little confidence in Tonga's reported location, so when Captain Samuel Wallis's frigate *Dolphin* (fig. 18) returned from its circumnavigation early in 1768 with an accurate position of a previously unknown but suitably placed island with a good anchorage and plentiful supplies of fresh food and water – albeit with inhabitants that Wallis reported to be 'rather treacherous than otherwise' – *Endeavour*'s destination was quickly changed to Tahiti (figs 16, 17).[10]

The Royal Society realized that funding such a voyage, with Britain's scientific reputation at stake, would be beyond its means, and successfully petitioned George III for £4000. For its part, the Admiralty agreed to supply a suitable ship, but there was disagreement about who would lead the expedition. The Society wished to appoint Alexander Dalrymple, an East India Company official and gentleman geographer who had led some exploratory voyages through the East Indies with a professional sailing master at his side, and who had proposed himself as commander for the voyage. Although the Society was agreeable to the suggestion, the Navy, having perhaps learnt from the unhappy outcome when it agreed a similar arrangement that had put the astronomer Edmund Halley in command of the *Paramour* in 1698, saw things rather differently. The First Lord of the Admiralty reportedly declared that he would rather have his right hand cut off than allow a civilian to command a naval ship.[11] Dalrymple was offered a place on the voyage, but not its command, and he promptly withdrew. In that sense, Cook was a compromise: he satisfied the Navy's requirements for a proven seaman, surveyor and commander, and the Royal Society was comfortable with his scientific credentials. James Cook the second – for there were two James Cooks in the Navy – was therefore commissioned lieutenant and took command of *Endeavour* on 25 May 1768.

Dalrymple's objectives had been fundamentally geographical rather than astronomical, for he was a strong and persuasive advocate of the existence of a still-undiscovered Southern Continent, a theory that went back to an ancient belief that the Earth's northern continents had to be balanced by a similar land mass in the south. Samuel Wallis's circumnavigation (1766–68) in the *Dolphin* had been directed to search for the elusive continent, but after rounding Cape Horn, which was the favoured route into the Pacific for English ships, it crossed the ocean at too low a latitude either to prove or disprove the theory. Crossing it from east to west, in the teeth of strong eastbound currents and prevailing westerly winds of legendary force, also required some determination and tended to push ships too far to the north to have any realistic chance of finding a continent in its higher, southern latitudes. Nevertheless, Wallis's cheerily optimistic sailing master, George Robertson, reported having seen mountain peaks many miles to the south of Tahiti, which he thought could have been the continent. Wallis appears to have been sceptical – if Robertson reported it to him at all – for he made no effort to examine the sighting more closely.

While Cook's 'Instructions' from the Admiralty focused on the all-important transit of Venus, he was also given sealed 'Additional Instructions' to follow up Robertson's sighting and search for the continent as far as latitude 40° south, once the transit was completed. In the event of the continent not being found, Cook was then free to return to England via New Zealand, charting as much of it as the condition of *Endeavour* would then allow, and 'with the Consent of the Natives to take possession of Convenient Situations in the Country in the Name of the King'.[12] Effectively, the transit of Venus acted as a screen for a commercial and imperial assessment of the South Pacific. This was politically sensitive, for quite apart from Spain's jealously guarded and centuries-long claim on the Pacific Ocean, France was also seeking to explore the possibilities of a new empire of the south, after having lost the vast majority of its American colonies after the Seven Years War. To that end, France had already despatched Louis-Antoine de Bougainville to the Falklands and the Pacific for that very purpose. Bougainville's

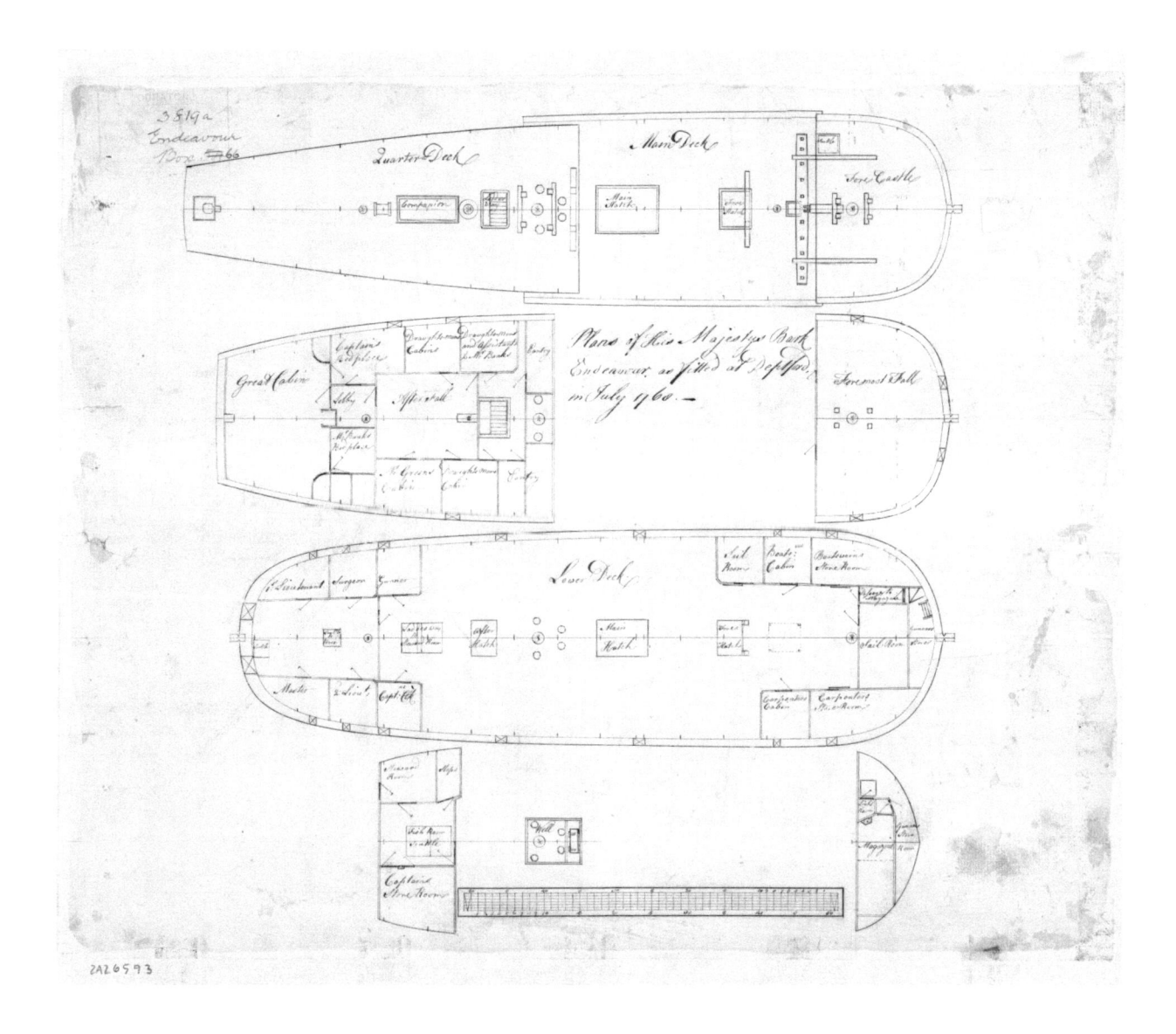

Fig. 18 (opposite, above)
Model of a 24-gun, sixth-rate Dolphin-class frigate
Maker unknown, c.1745
National Maritime Museum, Caird Collection (SLR0475)

Dolphin circumnavigated the world twice: under Captain John Byron, 1764–66, and Captain Samuel Wallis, 1766–68. On the second voyage it became the first European ship to visit Tahiti.

Fig. 19 (opposite, below)
Model of *Endeavour* (1768)
Robert A. Lightley, c.1973
National Maritime Museum (SLR0353)
Reproduced with kind permission of the late Robert Lightley.

Fig 20 (above)
Plan showing the works to convert the Earl of Pembroke *for exploration as HM ship* Endeavour, 1768
National Maritime Museum (ZAZ6593)

La Boudeuse became the second European ship to reach Tahiti, less than a year after Wallis and only a few months before Cook's arrival. Bougainville took possession, a formal act that Wallis had also made, and brought back to France the first Polynesian to visit Europe, Ahu-toru.

The vessel selected for Cook's voyage was the *Earl of Pembroke*, a collier typical of the small but capacious ships that plied the English east-coast coal trade. Cook himself had no hand in its selection, however, for it had been bought by the Admiralty before he was given command. The ship was renamed *Endeavour* and adapted in Deptford Dockyard to carry the stores and people needed for what was expected to be a three-year voyage (figs 19, 20). As *Earl of Pembroke*, a crew of 12 or 15 would have been normal, but *Endeavour* had to find

Fig. 21 (left)
Portable reflecting telescope
James Short, mid-18th century
National Maritime Museum,
Caird Fund (AST0941)

Fig. 22 (opposite)
Chart of the Island Otaheite
Lieutenant James Cook;
engraved by J. Cheevers, 1769
National Maritime Museum
(G267:47/1)

Cook's chart of Tahiti was first published in John Hawkesworth's *An Account of Voyages Undertaken* (1773).

room for nearly 100 men, including a private scientific party of astronomers, naturalists and artists: the 'scientific gentlemen' whose work would come to define the voyage, influence subsequent ones and generate intense public interest in the Pacific. They were led by Sir Joseph Banks, a young man with a passion for botany and the means to indulge it. The *Endeavour*'s was very much an experimental voyage, as were the later ones of *Resolution* and *Adventure* (1772–75) and of *Resolution* and *Discovery* (1776–80), providing an opportunity for an extended trial of the lunar method of finding longitude at sea (the chronometric method would not be tested until the second voyage); carrying the latest Gregorian reflecting telescopes, astronomical quadrants, sextants, clocks and watches provided by the Royal Society and the Royal Observatory at Greenwich (fig. 21);[13] and stowing away all manner of preventatives for scurvy, including portable soup, wort of barley, carrot marmalade and sauerkraut.

Endeavour left Plymouth on 25 August 1768, reaching the Pacific via Rio de Janeiro and Cape Horn in January 1769, and arriving eight months later at *Dolphin*'s old anchorage in Matavai Bay, Tahiti (which Wallis had named Port Royal Harbour and King George's Island respectively, names that did not survive for long) six weeks before the transit of Venus.

Cook lost four men on the voyage out: two to accidental drownings and two more to hypothermia in Tierra del Fuego; another, Alexander Buchan, the landscape artist employed by Banks, died after an epileptic fit on Tahiti. By the standards of the times and the distances involved, this was an acceptable level of loss, similar to that of Wallis, who lost but five men during his entire voyage: two to accidents and three to fevers caught in Batavia (Jakarta), a notoriously unhealthy port where Cook would lose over a quarter of his men in 1770. Although scurvy had clearly been present on both Wallis's and Cook's ships, particularly on the former, no one died from the disease.

After two previous visits from European ships, the Tahitians had developed a strategy for dealing with the visitors: the attack with which they greeted *Dolphin* had demonstrated only too clearly that, if it came to force, only one side would win. *Endeavour* reaped the benefits of this lesson, and trading relations were quickly established. A fort and observatory were built on a spur of the bay named Point Venus and were sketched by the botanical artist Sydney Parkinson, who now had to take on some of his dead colleague's work. Despite its forbidding appearance (more the engraver's imagination than an accurate representation) and its real and heavy defences, which included two four-pounder cannon, a number of swivel guns and the space to house 45 armed men, the fort was less secure than it appeared. Overnight, an astronomical quadrant needed for the observation was stolen by an enterprising islander and, though Cook and Banks acted swiftly to get it back, which they did with difficulty, this was one of a number of thefts that, together with Cook's increasingly draconian responses across his three voyages, would cast a dark shadow over Polynesian–European encounters. Theft led to the first killing of a Tahitian on the *Endeavour* voyage; to the vicious and gratuitous destruction of houses and canoes (in pursuit of a stolen goat on the third voyage); and eventually, and almost inevitably, to the sad circumstances precipitating Cook's death.

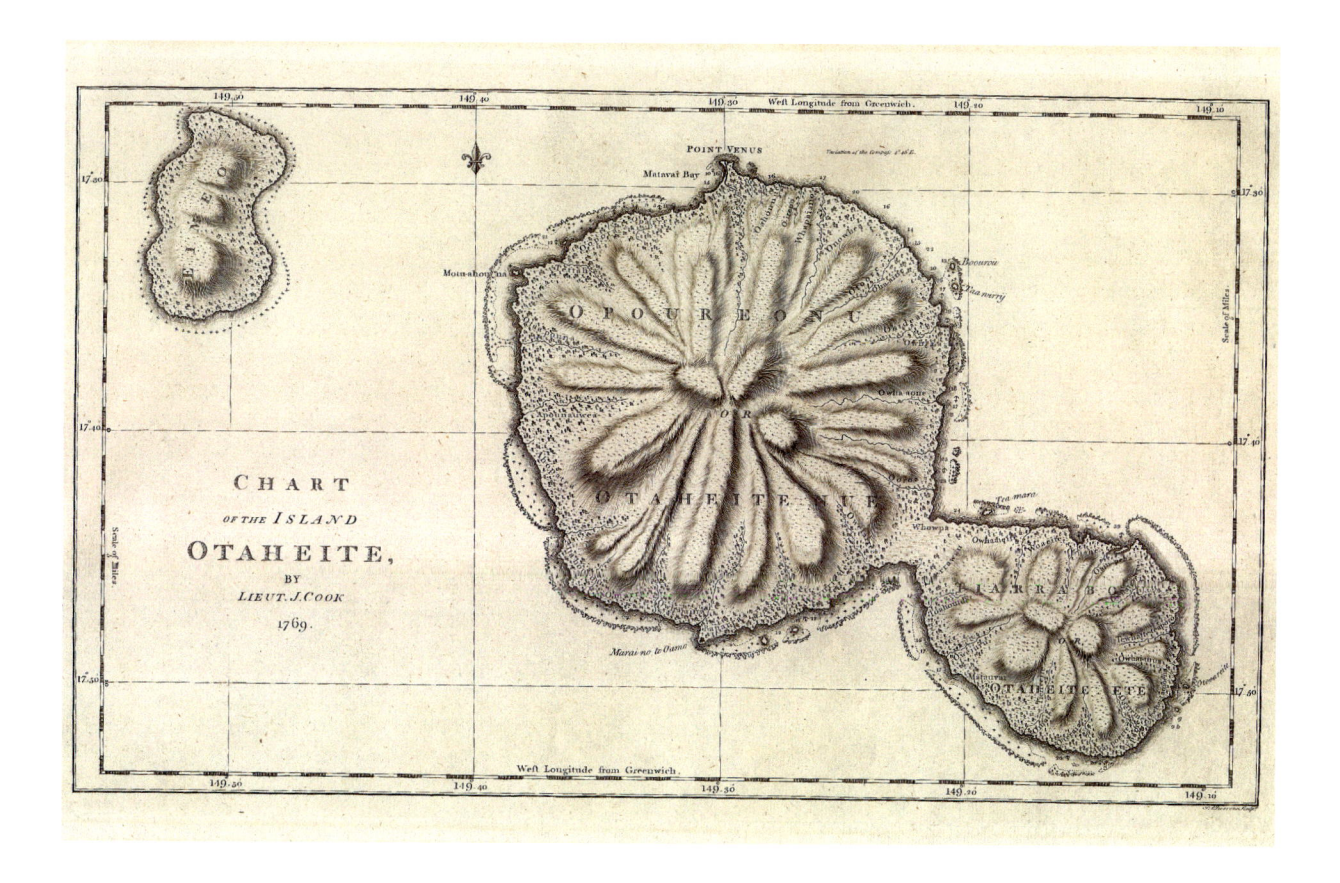

The transit of Venus was duly observed on 3 June 1769, the lengthy and precise process receiving little more than a few anticlimactic lines in Cook's journal, where he noted the unusual heat of the day (119°F) and an unexpectedly large variance between his observations and those of the astronomer, Charles Green, for both of them found it difficult to judge the precise moment when Venus began and ended its transit of the Sun. This was also true of the observations taken at Cook's two back-up points, as it was of the others taking place globally; indeed, the problem recurred in the two nineteenth-century transits, which used photography to help avoid human error.

The transit completed, and with the ship in good order and loaded with fresh water and supplies, Cook was free to leave Tahiti and visit some of its near neighbours, which he named the Society Islands ('as they lay contiguous to one another,' said Cook, rather than as an acknowledgement of the voyage's scientific sponsors). The ship now had two additional supernumeraries, a priest and navigator called Tupaia, together with his servant boy, Taita – for both of whom Banks took responsibility, Cook being reluctant to take them to England with no prospect of being returned. Tupaia proved an immediate and trusted asset on the island sweep, and Banks remarked that:

> We have now a very good opinion of Tupias [sic] pilotage, especially since we observed him at Huahine send a man to dive down to the heel of the ship's rudder; this the man did several times & reported to him the depth of water the ship drew, after which he has never sufferd her to go in less than 5 fathoms water without being much alarmd.[14]

It had been obvious to Cook's professional eye that the boat-building skills and seamanship of the Polynesians was of a high level – Tupaia's expertise and knowledge of the waters

JAMES COOK, THE ROYAL NAVY AND HIS THREE VOYAGES OF EXPLORATION

convincing him, accurately, that 'these people sail in those seas from Island to Island for several hundred leagues'. Their navigational techniques, 'the Sun serving them for a compass by day and the Moon and Stars by night', together with their observations of currents, winds and seabirds, used the same basic principles as European navigators, although without the sophisticated technology provided by the London instrument-manufacturers.[15] Tupaia drew up a chart of the known Polynesian world, which stretched for thousands of miles from east to west, although scholars and seamen from that time to this have not found it easy to align it with European conceptions of distance and direction. Tupaia was a strong force in *Endeavour*'s exploration of the Society Islands and New Zealand, and Cook later described him as 'a Shrewd, Sensible, Ingenious Man, but Proud and Obstinate' – an epitaph that could just as easily have been applied to Cook himself.[16]

Obedient to his sealed orders, Cook then headed south in search of the supposed 'Southern Continent'. Finding nothing but cold, strong and contrary winds, and duly reaching the latitude of 40° south that his orders required, he then turned west towards New Zealand. This had been a squiggle on maps of the world since Tasman's brief and unhappy encounter with the Maori at Murderers' Bay (now Golden Bay) in 1642 (fig. 23). *Endeavour*'s first meeting with the Maori was at Tuuranga-Nui (which Cook named Poverty Bay) on New Zealand's north-east coast. It was no less tense than Tasman's encounter, with a number of Maori shot and killed and others wounded. While the aggressive stance taken by the Maori on the beach was probably a ritual challenge to the landing, it was equally likely that the line of Europeans facing them was seen in the same light. Tupaia, who discovered he could understand and make himself understood by fellow Polynesians (fig. 24), however distant from the Society Islands, helped defuse the situation on the beach, but there was more bloodshed yet to come.[17] 'Thus ended the most disagreeable day My life has yet seen black be the mark for it,' wrote Banks, with Cook similarly depressed at having ordered his men to open fire and aware that his excuse – that he could not 'stand still and suffer either my self or those

Fig. 23 (left)
Terrestrial table globe
Guillaume Delisle, 1700
National Maritime Museum,
Caird Collection (GLB0146)

The Dutch chart on this globe shows the extent of Europe's knowledge of New Zealand in 1700.

Fig. 24 (opposite)
Table of linguistic differences, from J.R. Forster's *Observations made during a voyage round the world, on physical geography, natural history, and ethic philosophy*, 1778
National Maritime Museum
(PBG3249)

The Forsters were keenly interested in studying the differences as well as the similarities between the languages, cultures and implements of the islands of the South Pacific.

34 CAPTAIN COOK AND THE PACIFIC

that were with me to be knocked on the head' – was at best a partial truth.[18] An encounter further down the coast brought more Maori deaths, but eventually a cautious respect was established and trade ensued.

Cook followed the coast along as far as 40° south to a point he named Cape Turnagain, where he began to retrace his steps, completing an anticlockwise circumnavigation and survey of North Island by February 1770. He promptly began his running survey of South Island, discovering what would become a favourite anchorage in Ship Cove, Queen Charlotte Sound, and finally leaving the coast of New Zealand in April 1770. The running survey was a method refined by Cook that enabled a coastline to be laid down quickly and with surprising accuracy, through taking a constant stream of bearings of points on the shoreline as the ship sailed along the coast, while simultaneously taking into account the ship's position, course, speed and the direction and strength of the current.

Once the survey of New Zealand was done, the Admiralty had given Cook considerable latitude in choosing his route home. Aware of the parlous state of the ship after so long at sea, he decided that the route would have to include a call at a European port for repairs and refitting, and so he decided to cross the Tasman Sea to Van Diemen's Land (Tasmania) to establish whether it was an island or part of the mainland, to survey the as-yet-uncharted east coast of New Holland (Australia) and finally return to England via the Torres Strait and Batavia in the Dutch East Indies. Contrary winds prevented Cook from reaching Van Diemen's Land, so his first landing in

New Holland was in what he initially named Stingray's Harbour, and later Botany Bay, a name that has echoed down the centuries, even though its first settlement by Europeans (as a penal colony in 1788) lasted only days (fig. 25). The ship then headed north, following the coast and unwittingly driving further and further into an ever-narrowing gap between the mainland and the outlying Great Barrier Reef. The inevitable happened on the night of 10 June 1770, when the ship struck what is now called Endeavour Reef and quickly began to take in water. Efforts to float her off at the next high tide failed because she had run aground near high water and was firmly wedged on the reef. The ship was lightened by some 50 tons, by throwing cannons and stores overboard, and was eventually kedged off, only to reveal that the pumps had been able to cope with the original ingress of water only because the reef on which the ship was sitting had partially blocked the hole it had cut in the timbers. The water flooded in with redoubled force and *Endeavour* faced 'an alarming and I may say terrible Circumstance', as the captain put it, until a sail was fothered over the hole to reduce the flow. Slowly they managed to nurse the ship to the coast some 20 miles distant, where she was run ashore in a serendipitously located river, now Endeavour River, to effect repairs (fig. 26). Here it was discovered that the coral reef that had so nearly destroyed the ship had also been its saviour, for a large lump of coral was still stuck in the hole and had kept the inflow of water to a (just) manageable volume.

The repairs took nearly two months and *Endeavour* did not leave the river until 6 August, picking a careful way through the

Fig. 25 (below)
A Chart of New South Wales or the East Coast of New Holland Discover'd and Explored by Lieutenant J Cook, Commander of his Majesty's Bark Endeavour ... *MDCCLXX*
William Whitchurch, 1772–73
National Maritime Museum (PAI4016)

Fig. 26 (right)
A view of Endeavour River, on the coast of New Holland, where the ship was laid on shore, in order to repair the damage which she received on the rock
William Byrne (engraver), after Sydney Parkinson, 1773
National Maritime Museum (PAI3988)

The enforced stay in Endeavour River gave the scientific party their first chance for an extended study of Australia's flora and fauna. It was here they first saw kangaroos and dingos – and found the former good eating.

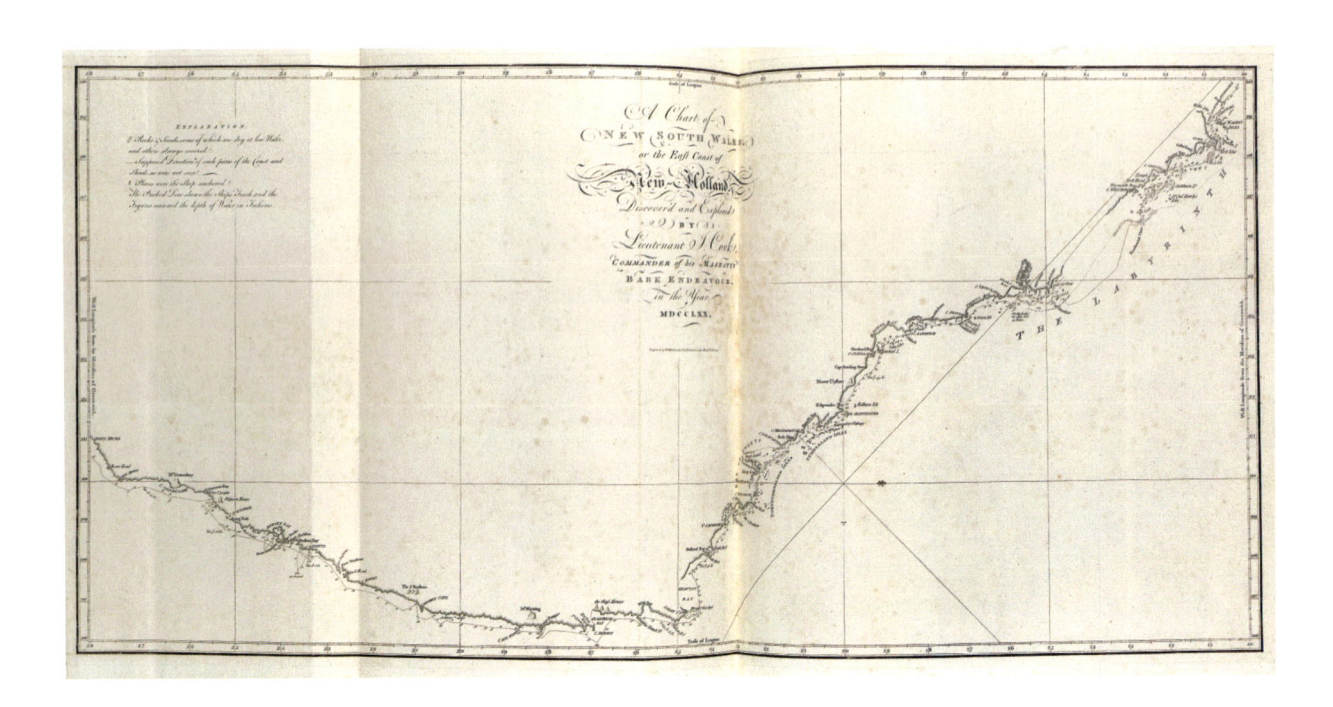

intertwined reefs until she was able to carry on northwards, although not without incident, as the ship was nearly lost again, and its fortunate escape was immortalized in the name Providential Channel. The weeks in Endeavour River had given Banks and his scientific party the opportunity to collect more specimens of Australia's flora and fauna, among which were a kangaroo and a dingo (or, rather, their skins). Relations with the inhabitants, however, were tentative and fleeting, as they had been in other landings along the east coast.

Endeavour reached Batavia on 10 October 1770, forced there by circumstance but an unwise choice as it turned out, for it was notoriously unhealthy for Europeans, as the crew were to discover for themselves. Crucially, though, it was a port able to carry out the necessary repairs in the small Dutch dockyard there (the only European dock in the East, other than Bombay, far off Cook's track). While Endeavour was being refitted, the crew, probably already weakened by months at sea – for Charles Green, Lieutenant Hicks and Tupaia had certainly suffered from scurvy – were exposed to dysentery, malaria and dengue fever. Banks came down with a malarial disease and recovered, but nearly one-third of the crew and supernumeraries, including Tupaia, Taita, Sydney Parkinson, the astronomer Charles Green and Banks's secretary Herman Spöring, died of fevers caught at Batavia. A further nine died during the voyage from other, unrelated causes.[19]

Endeavour finally dropped anchor at the Downs, the busy anchorage just south of the Thames Estuary, on 13 July 1771, confounding a rumour heard from a passing ship some weeks earlier that the ship had been lost, so long had they been absent. This must have been received with sardonic smiles, for the ship was by this time in a very fragile state, having just lost two sails, the main topmast and two of the supporting stays in a fresh gale. 'Our Rigging and sails are now so bad that some thing or another is giving way every day,' noted Cook. As soon as they anchored, he left the ship and travelled overland to London with his reports and papers.[20]

Cook was well received by the Admiralty, as might be expected, and was presented to the king, promoted to the rank of commander and had all his own recommendations for promotion confirmed – which in the Royal Navy was as much an honour to the captain as to his subordinates. Cook had laid down, or charted, more than 5000 miles of coastline and had clarified the positions of many islands – an extraordinary achievement, setting professional standards that the great European maritime powers would envy and soon emulate. Outside the Navy, though, it was largely seen as Mr Banks's voyage and he undoubtedly basked in the social and scientific limelight – as well he might, for the botanical results alone of the Endeavour voyage were considerable, Banks having collected 1300 plant specimens previously unknown to European science.

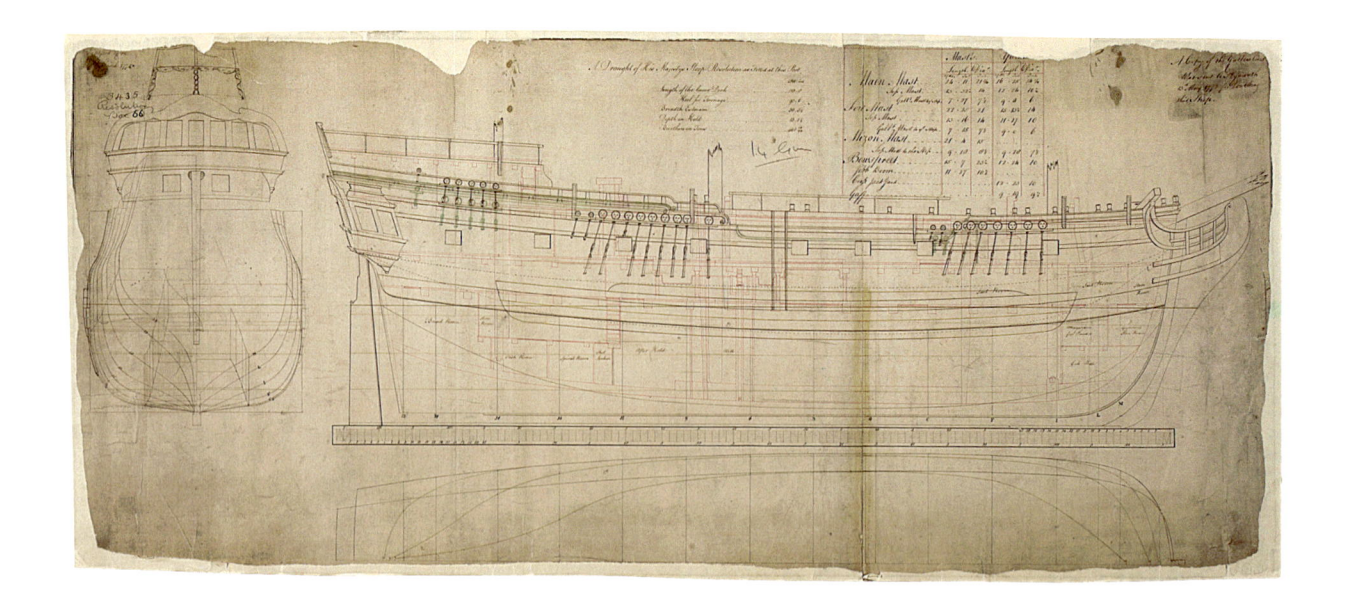

He also commissioned the fashionable animal artist George Stubbs to paint the dingo and kangaroo killed at Endeavour River, and these iconic representations of two of the characteristic exotic species of 'New South Wales' – as Cook called the Australian east coast – were displayed to great acclaim in London in 1773, as indeed they still are today. The professional writer John Hawkesworth was soon commissioned by the Admiralty to bring together the stories of Byron's, Wallis's, Carteret's and Cook's circumnavigations. The resulting book, *An Account of the Voyages Undertaken by the Order of His Present Majesty for Making Discoveries in the Southern Hemisphere*, was not published until after Cook's departure on a second voyage, but swiftly became a best-seller, if a controversial one.

Endeavour was refitted as a store ship and was detailed to take a cargo of supplies to the Falkland Islands. It was eventually sold out of the Navy and renamed once again, this time as the *Lord Sandwich*. Its service to its country was not yet ended, however, for during the American Revolutionary War it was one of a fleet of transport ships that delivered badly needed stores to Newport, Rhode Island, where it was scuttled to protect the town from a French seaborne attack. Its precise location in Newport Harbour has never been established, for it was one of 12 transports scuttled at the time. However, a document in the archives of the National Maritime Museum proves conclusively that it was one of five sunk in the north channel between Goat Island and the mainland, although at this distance in time, and bearing in mind the similarity of the ships, we may never know precisely which one is *Endeavour*. Ironically, the French attack never materialized, as their fleet was dispersed in a gale.

Fig. 27 (above)
Plan of Resolution
1772
National Maritime Museum
(ZAZ6560)

The plans show the original alterations proposed to convert the collier *Marquis of Granby* into *Resolution*.

Fig. 28 (opposite)
K1, Kendall's copy of Harrison's prizewinning chronometer, H4
Larcum Kendall, 1769
National Maritime Museum
(ZAA0038)

RESOLUTION AND *ADVENTURE*: THE SECOND VOYAGE, 1772–75

The big question, theoretically still left unanswered after the *Endeavour* voyage, was the whereabouts of the Southern Continent. For that reason alone, and leaving aside the fact that the Pacific still promised to provide further discoveries of importance to science, a second voyage was almost inevitable. Cook had already given the matter thought and drew up a plan to establish once and for all whether a continent did exist in habitable latitudes. This proposed to take advantage of prevailing winds and currents rather than fighting them, and to circumnavigate the world in high latitudes from west to east, heading north in the southern winter to refresh and recuperate in warmer waters, and south again in the summer to continue the search. With the backing of Lord Sandwich, First Lord of the Admiralty and a close friend of Banks, the plan received royal assent and a second voyage was announced. Learning from the near-loss of *Endeavour*, two ships were purchased and, acting on Cook's advice, they were both sturdy east-coast colliers: the *Marquis of Granby* (462 tons) was somewhat bigger than *Endeavour*, and the *Marquis of Rockingham* (340 tons) somewhat smaller. They were initially renamed *Drake* and *Raleigh*, after two famous Tudor explorers. However, one person's explorer is another's pirate and so, in deference to Spanish sensitivities on the matter, the ships were renamed once more, as *Resolution* (fig. 27) and *Adventure*. Command of the smaller ship was given to Tobias Furneaux, who had sailed as second lieutenant on Samuel Wallis's *Dolphin*.

If a second voyage was inevitable, its leader, in Banks's view, had to be Banks himself, with Cook effectively in a subordinate role as sailing master – ironically, the same arrangement that Dalrymple and the Royal Society had imagined for the first voyage, and which the Admiralty had firmly rejected. Banks's assumption suggests that the relationship between science and the sea was not consistently harmonious during the *Endeavour* voyage. In the event (and if, indeed, this was Banks's plan) it was never tested, for Banks withdrew in dudgeon when he discovered that the additional deck built on *Resolution* to accommodate his much-expanded scientific party had been demolished, after sea trials showed that it made the ship dangerously unstable.

To take the place of Banks and suite, the Admiralty appointed the father-and-son team of natural historians, Johann Reinhold Forster and his son George (properly Georg). The older Forster was a man of penetrating intelligence who would raise the standard of scientific enquiry and debate on the voyage to new levels, but this came at a cost, for he was not always a congenial travelling companion and it was a long voyage. His son, however, was both popular and a good scientist.

A third naturalist, Anders Sparrman, was hired by Forster senior when *Resolution* reached Cape Town on the voyage out. After the withdrawal of Banks's chosen artist, Johan Zoffany, the Admiralty engaged the landscape artist William Hodges to chronicle the voyage. Hodges was a happy choice, as were the appointments of William Wales and William Bayly, the astronomers on *Resolution* and *Adventure* respectively. As on the *Endeavour* voyage, the ship sailed with a range of experimental equipment and stores for extended trials at sea, not the least of which was 'our trusty friend the watch', as Cook would refer to K1, Larcum Kendall's first copy of John Harrison's successful 'H4' chronometer (fig. 28). A full range of antiscorbutics was again taken, for the second voyage presented the opportunity to double-test the measures and diets introduced on the first.

The success of the first voyage and the high-profile preparations for the second encouraged many an aspiring officer to seek a place as a 'young gentleman'. Successful applicants

(54)

of the most inchanting little Harbours I ever saw, it was surrounded with high Lands intirely cover'd with tall shady trees riseing like an Amphitheatre, and with the sweet swelling Notes of a number of Birds made the finest Harmony —

We hauld the Ship into a little cove that was just big enough to admit her and had 6 fathoms water, where we fastned her to the Trees, the Branches of which in many places hung over her —

Next Day we spent in Clearing away the woods, for to Errect Tents &ca and makeing a Platform from the shore to the Ship and likewise sent a Boat fishing which Practice we fallow'd during the whole time we lay in this sound, and in consequ of theis catching enough for every Body

included Alexander Hood, the young cousin of Admiral Lord Hood; George Vancouver, who would also serve on the third voyage and eventually command his own survey of the north-west coast of America in 1791–95; James Burney, the son of the musical scholar Charles Burney and brother of the novelist Fanny Burney; and John Elliott, whose uncle had used 'Intrest' to get him a position on the voyage, and who left a memoir of it with some brief but sharp character sketches of his colleagues (Forster *père*, for example, being described as 'A clever but a litigious quarelsom fellow', and Charles Burney as 'Clever & Excentric'). A number of officers and crew who had previously served on *Endeavour* were also selected for the second voyage: Richard Pickersgill, whose journal was acquired by the NMM in 1957 and who had served as a master's mate on the first voyage (fig. 29), joined *Resolution* as third lieutenant; and John Ramsey, an able seaman in *Endeavour*, who was promoted to gunner's mate on *Resolution*. The surgeon's mate, William Anderson (described as 'a Steady clever man' by Elliot), was appointed for his interests in botany and his facility with languages as much as for his medical expertise.[21] Cook's appointments demonstrate the mix of patronage, skills and experience that would have been found on most Royal Navy vessels of the time.

The ships left England on 13 July 1772, sailing in company and replenishing their stores at Cape Town, where Hodges painted *Adventure* at anchor in Table Bay, before setting out on the first of the great west-to-east sweeps through the high latitudes of the Southern Ocean. In 1739 the French navigator Jean-Baptiste Charles Bouvet de Lozier had reported sighting land that he called Cape Circumcision, which was suspected to have been a part of the Southern Continent: confirming or denying the sighting was Cook's first objective, but although tiny Bouvet Island did exist, its position remained unconfirmed and would stay so for decades. On 17 January 1773, *Resolution* and *Adventure* became the first ships to cross the Antarctic Circle, as far as it is known. The frost and cold were so intense, wrote John Elliott, that they covered the rigging with ice 'like compleat christal ropes, from one end of [*Resolution*] to the other'.[22] Richard Pickersgill remarked that the freezing conditions and the length of time at sea were having their effect on health, with scurvy making its inevitable appearance: 'even the Catts and Doggs were affected so much as to cause their teeth to become loose'.[23] Cook later noted that 'we have only one man on board that can be called ill of this disease and two or three more on the Sick list of slight complaints'.[24] The two ships became separated in fog, so Tobias Furneaux's *Adventure* made for the agreed rendezvous in Queen Charlotte Sound, and Cook decided to head instead towards Dusky Sound at the south-western tip of New Zealand. He had glimpsed it one evening on the last leg of *Endeavour*'s circumnavigation of New Zealand – hence the name – but would not risk entering a strange harbour in the dark and a rising onshore wind. Three years later, he reasoned that the advantages of finding a 'good Port in the Southern part of this Country' outweighed the disadvantages of *Resolution* keeping *Adventure* waiting at the rendezvous for several weeks, and so after five gruelling months at sea, *Resolution* cautiously entered Dusky Sound.[25]

It was to stay there for nearly six weeks, much of the time moored in a small, sheltered creek that Cook named Pickersgill Harbour. In truth they were spoilt for choice, for wherever they looked they found 'convenient and well-sheltered harbours, with plenty of wood and water; and wherever they went they met with such an abundance of fish and water-fowl that they entertained hopes of a constant supply of refreshments'.[26] All were taken by the fecund beauty of the place (less so by the sandflies and the frequent heavy rain), although their heightened responses to the landscape were clearly affected by the harshness of the icy and forbidding seas they had lately quitted. More recently it has been suggested that their extreme sensitivity to light and colour might also have been an early sign of scurvy, which was more widely present than realized or acknowledged.[27] During conservation work in 2004 it was discovered that the voyage artist, William Hodges, was moved to overpaint the first-ever oil depiction of towering and threatening icebergs with what may be a *plein air* oil on canvas of *Resolution*'s 'rural' mooring, *View in Pickersgill Harbour, Dusky Bay* (figs 30, 31).[28] In the eyes of the seamen and supernumeraries, the bay appeared a pastoral idyll, a place of healing and one that *Resolution*'s young midshipman, George Vancouver, would remember and return to 20 years later when in command of his own voyage of Pacific exploration. The ailing sheep and goats intended for the Society Islands did not, however, recover in the lush vegetation as well as was hoped, and Cook considered it unlikely they would survive the voyage.

The expedition met few inhabitants, but Hodges's monumental *Dusky Bay*, which was painted (like the overwhelming majority of his oils) after the voyage, from sketches taken on the spot, brought together a striking view of the waterfall in Cascade Cove with a romanticized depiction of a small group of Maori with whom the *Resolution*'s crew established a brief trading relationship. Although the stop in Dusky Sound was welcome, and produced some excellent charts and striking paintings, the ecstatic European reaction to its landscape could

Fig. 29
Narrative account by Lt Richard Pickersgill of the voyage of HM bark Resolution, *1772–73*
National Maritime Museum (JOD/56)

Pickersgill describes his discovery of a snug berth for *Resolution*, in what Cook would name Pickersgill Harbour (fig. 30).

Figs 30, 31 (above and opposite)
View in Pickersgill Harbour, Dusky Bay, New Zealand and X-ray of same showing underpainting of icebergs
William Hodges, 1773–76
National Maritime Museum (BHC2370)

Hodges's view of Antarctic icebergs, which he overpainted either in Pickersgill Harbour or later, was probably done at sea and is the first attempt ever to paint polar ice in oils. Why he obliterated it is uncertain. In the visible view, the observatory tent is in the centre and *Resolution* moored to shoreline trees on the right.

Fig. 32 (overleaf)
A View of Maitavie Bay, in the island of Otaheite [Tahiti]
William Hodges, 1776
National Maritime Museum (BHC1932)

First used by Samuel Wallis's *Dolphin*, Matavai Bay became the favourite anchorage for Cook's ships.

CAPTAIN COOK AND THE PACIFIC

not hide the fact that, apart from fish, it offered little in the way of the provisions needed by the ship. *Resolution* eventually made its way to the richer pastures of the rendezvous in Queen Charlotte Sound, where on 17 May 1773 she met *Adventure*, which had arrived there a month ahead, after a brief visit to Van Diemen's Land.

With both ships reunited at the beginning of the southern winter, Cook made an unusual decision for the time of year: to sail south and east towards latitude 40° south, with the intention of making sure that no continent existed to the west of his earlier foray in *Endeavour*. The contingency was remote and the weather not ideal, but Cook felt that this was 'too important a point to be left to conjector [sic], facts must determine it'.[29]

Finding no trace of the continent, the two ships then headed north for the Society Islands, where they could be sure of a warm welcome and fresh food, reaching Tahiti on 16 August 1773 (fig. 32). While fruit and vegetables were readily available, hogs were in short supply. Cook noted that he gave presents of shirts and axes and 'in return they promised to bring me Hogs and Fowls, a promise they neither did nor never intended to perform'.[30] Pickersgill complained that 'the King' would not allow any livestock to be sold until Cook agreed to sell him the 'best boat' for ten hogs, a price that Cook declined to pay. While this was evidence of the increasing price of a pig, which rose from a nail to an axe in the ten years between the arrival of Wallis and the last visit of Cook, the politics of the island

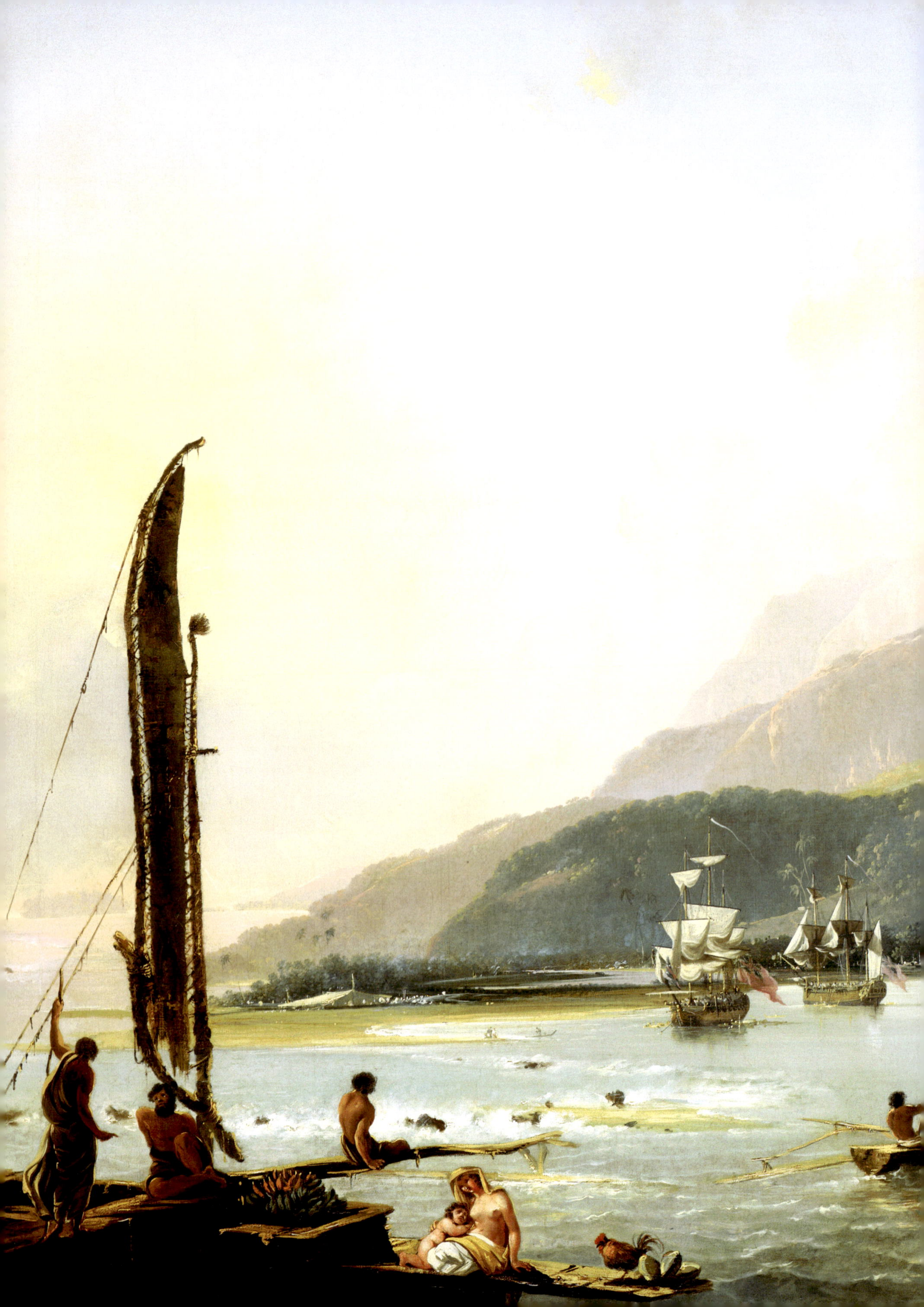

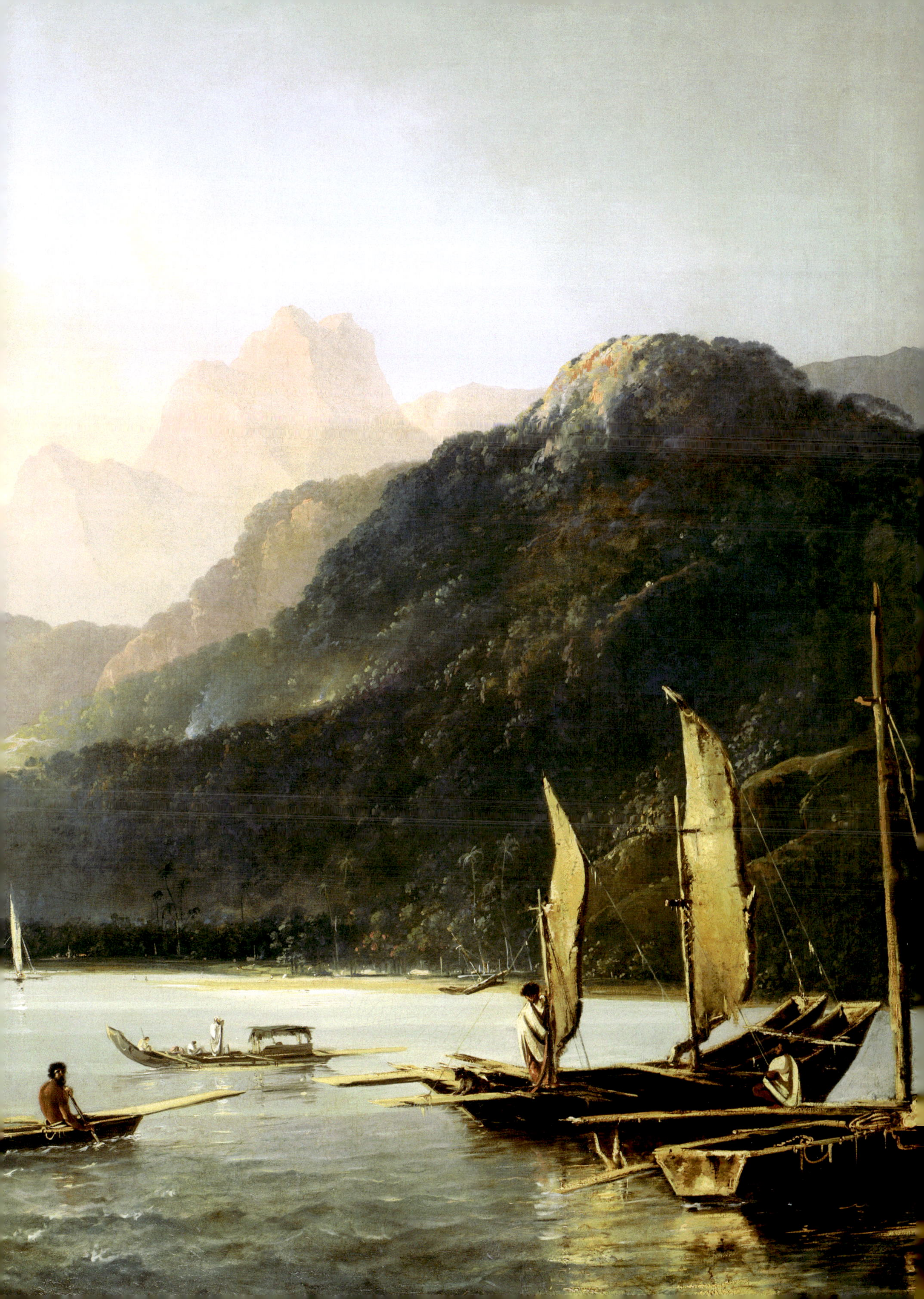

Fig. 33 (left)
The Resolution *in the Marquesas, 1774*
Pen and wash, by William Hodges, 1774
National Maritime Museum (PAF5791)

Fig. 34 (opposite)
O-Hedidee (native from the Pacific)
James Caldwell (engraver), after William Hodges, 1776
National Maritime Museum (PAI2098)

The Raiatean who sailed briefly with Cook and whose name is now generally spelt 'Hitihiti'.

had also altered and a Spanish ship had called in 1772: both of these factors had an impact on trade, prices and the voyagers' reception.[31]

After staying in Tahiti for a fortnight, the English ships left on a sweep westwards through the Society Islands, revisiting Raiatea, where Cook took on board a young man named Hitihiti (O-Hedidee) (fig. 34), returning him the following year; and Huahine, where they were able to get more provisions in four days than they had in 14 at Tahiti, and where Furneaux embarked the islander Omai (Ma'i) with the intention of taking him back to London for Sir Joseph Banks. Omai was no Tupaia, being of little use as an interpreter and even less as a navigator. He was, however, a great success in London, where he was lionized by high society and introduced to George III, before being taken back to Huahine on Cook's third voyage.

The ships continued west towards the Tongan group, sighting Eua on 2 October 1773, which Tasman had named Middleburgh. Here they stayed for a day, before sailing the short distance to the larger island of Tongatapu (Tasman's Amsterdam). Cook named the group the Friendly Islands, to reflect the welcome they received. 'Here', wrote Pickersgill, echoing the reactions of many of the journalists, 'we were presented with one of the most delightfull views I ever beheld the land . . . planted with regular rows of cocoa nut, plantain another [*sic*] kind of Fruit trees'. George Forster approvingly described the islanders' arts, manufactures and music as 'more cultivated,

46 CAPTAIN COOK AND THE PACIFIC

complicated, and elegant than at the Society Islands'.[32] This was a different sort of beauty from the lush, natural excess of Dusky Sound: here nature was ordered and improved by civilization, creating a land that 'put us in mind of the finest spots in England' (a comparison made more than once).[33] The Forsters were trying to understand and record different Pacific societies, establishing a hierarchy of what they saw as different stages of development among the many islands they visited, seeking a theoretical base to accommodate both their diversity and their common roots. George Forster particularly noticed the more 'advanced' designs on the Tongan *tapa* cloth. Naturally, the naval men engaged with the Tongans' boat-building skills, and Pickersgill described their canoes as 'the Neatest I ever saw'.[34] Cook managed the trading with a firm hand, although once he finally lifted his ordinance against private trade shortly before the ships left, the exchanges quickly became a free-for-all, and the sailors proved such avid collectors that they became 'the ridicule of the Natives by offering pieces of sticks stones and what not to exchange'.[35] Inevitably, theft was as rife here as on other islands, although conducted with an élan that even Cook seemed to admire.

They stayed on Tongatapu for a week before embarking for Queen Charlotte Sound, the last stop before the next Antarctic sweep. The two ships became separated in a storm off New Zealand: *Adventure* finally arrived at the Sound five days after *Resolution* had departed, Cook being anxious to leave as quickly as possible to make the most of the short southern exploring season. Furneaux found the voyage plan that Cook had left for him buried under a tree with the words 'Dig here' carved on its trunk, but did not attempt to follow him, as he quickly had problems of his own when a boat's crew was killed and eaten by Maori. Now seriously undermanned by the loss of ten of his complement of 81 crew, and probably in shock, Furneaux left New Zealand the next day and took *Adventure* back across the Pacific in high latitudes, entering the Atlantic via Cape Horn. There he searched yet again, and unsuccessfully, for Bouvet's Cape Circumcision, refitted at Cape Town and sailed home to England, arriving on 14 July 1774, a year before Cook's return.

Resolution's second sweep was an extraordinary achievement – one that proved beyond reasonable doubt that if there was a continent in the South Pacific, it was so deep within the ice fields as to be uninhabitable and inaccessible. The ship sailed south from New Zealand until it reached the antipodes of London on Christmas Day 1773, the celebrations bringing a brief respite from the intense cold. Sea ice prevented them from going much further south and they turned east, 'sometimes 2° to the North of the Antarctic Circle, and sometimes 2° to the South', as John Elliott remembered it. Elliott provides some insight into Cook's methods of command, for he would neither confirm nor deny where they were actually headed: 'Many hints were thrown out to Capt. Cook to this effect, but he only smiled and said nothing, for he was close and secret in his intentions at all times.'[36] Cook's chart of the Southern Ocean shows that he was progressing eastwards in a series of enormous zigzags: a turn to the north greeted with great joy by the ship's company, and a return to the freezing south by equally great despair. The numerous ice packs made this dangerous work, with *Resolution* inches away from disaster on more than one occasion. They reached their furthest south on 30 January 1774, prevented from going deeper by a barrier of ice, and here Cook made his famous statement that 'I whose ambition leads me not only farther than any other man has been before me, but as far as I think it possible for man to go, was not sorry at meeting this interruption.'[37] They turned north for the last time after Cook evolved a final voyage intention of conducting a second tropical sweep that would take them back through the Society Islands towards New Guinea and New Zealand, before returning home via Cape Horn and making one last attempt to find the Southern Continent in the South Atlantic.

They reached Easter Island on 12 March, but stayed only for a few days, since it was unable to offer much in the way of supplies. These days saw Hodges's pen busy sketching the striking monuments, from which he would develop one of his finest and most iconic paintings (see fig. 132). Cook, who had been so ill that George Forster thought 'his life was entirely despaired of', landed with a small party, but did not feel well enough to move much beyond the beach, delegating the exploration of the island to the reliable Pickersgill. They left the next day, steering north and east to the Marquesas Islands (fig. 33), which had been sighted and visited bloodily by Alvaro de Mendaña during one of Spain's early crossings of the Pacific in 1595. Cook noted in his journal that, apart from the brief visit to Easter Island, they had been at sea for 19 weeks continuously, but that 'on our arrival here, it could hardly be said that we had one Sick Man on board, and not above two or three who had the least complaint, this was undoubtedly due to the many antiscorbutic articles we had on board'.[38] The elder Forster disagreed, pointing to a number of the crew who 'crawled about the deck too weak to take enough wort'.[39]

And so, with one or two small landfalls in between, they returned to Tahiti, sighted on 21 April. After the poor pickings of his last visit to the island, Cook intended to stay only for a few days, but favourable signs of 'a riseing state', and plenty of hogs in consequence, persuaded him to refit in Matavai Bay. The visit also gave him the opportunity to see a fleet of more than 150 Tahitian war canoes gathered for an impending attack on nearby Moorea (Eimeo). Hodges's sketches of this were the source of his largest surviving canvas, the enormous *War-Boats of the Island of Otaheite* (figs 66, 105), which was exhibited at the Royal Academy in 1777. He also did a smaller variant on panel, for engraving to accompany the official publication of the voyage (fig. 35).

From the Society Islands (where Hitihiti disembarked), *Resolution* headed west, as it had done the year before, towards a group that Cook would name the New Hebrides (today Vanuatu). This extended from the north-west to the south-east for a distance of approximately 400 nautical miles. Working from north to south, Cook began to chart their positions, landing or attempting to land on Malekula, Erromango and Tanna – the Melanesian populations of all proving unfriendly – and Hodges also later produced a series of small paintings of these fraught encounters for the official publication (fig. 36).

Leaving these islands, they made for the familiar haven of Queen Charlotte Sound, 'discovering' en route Noumea, which Cook named New Caledonia and of which he managed to chart just the south-east side. In New Zealand they heard disturbing stories of a shipwreck and people being killed, but also confirmation that *Adventure* had left Queen Charlotte Sound safely. *Resolution* then headed back across the Pacific Ocean, made its landfall at Desolation Island in Tierra del Fuego, rounded Cape Horn and continued the search for the Southern Continent in the southern reaches of the Atlantic. While this still proved elusive, they sighted and claimed South Georgia – an action that would have political consequences 200 years later – and the South Sandwich Islands, before turning north to Cape Town. Here they at last learnt what had happened to *Adventure's* boat crew, before sailing back to England via St Helena, and finally dropped anchor in Portsmouth Harbour on 30 July 1775, after a voyage that had lasted three years and 18 days.

Fig. 35 (opposite)
Review of the War Galleys at Tahiti
William Hodges, 1776
National Maritime Museum (BHC2395)

Fig. 36 (above)
Landing at Erramanga (Erromango), one of the New Hebrides
William Hodges, c.1776
National Maritime Museum (BHC1903)

Cook's cautious landing here on 4 August 1774 met immediate hostility and forced him back to the ship. It was the most violent of his Melanesian encounters, here recorded in another of Hodges's small paintings to illustrate the voyage account.

Figs 37, 38 (right)
Medal commemorating the second voyage of Captain James Cook, 1772
W. Barnett, c.1772
National Maritime Museum (MEC1384)

Joseph Banks had more than 2000 such medals made in silver and bronze for distribution during Cook's second voyage, from which Banks finally withdrew.

JAMES COOK, THE ROYAL NAVY AND HIS THREE VOYAGES OF EXPLORATION 49

RESOLUTION AND *DISCOVERY*: THE THIRD VOYAGE, 1776–80

And that should have been that. After two gruelling but outstandingly successful voyages, Cook more than deserved to enjoy his public fame, promotion to post-captain (which, provided he lived, would eventually see him become an admiral) and the Admiralty's reward of a retirement position as one of the four resident captains at Greenwich Seamen's Hospital. Friends once more with Banks, he sat for the definitive portrait of 'Cook-the-great-navigator' by the society artist Nathaniel Dance. His success in keeping scurvy at bay was recognized by the presentation of the Royal Society's Copley Medal, establishing for generations his reputation as the man who had defeated the disease. Cook's fury at John Hawkesworth's sensationalist and imaginative rewriting of his journals – and those of Banks, Wallis, Carteret and Byron, which had been brought together as a single narrative – was heeded, and it was agreed that the narrative of the second voyage would be written by Cook, with the assistance and advice of the literary scholar Dr John Douglas and illustrated by William Hodges. Although this was a victory for Cook, it did not amuse J.R. Forster, who had understood that the original terms of his involvement in the voyage included the responsibility (and reward) for writing the official publication.

Cook's retirement was short-lived, for another voyage was already being contemplated. He was soon involved in its planning and was reappointed commander of *Resolution* a mere six months after his return to England. Charles Clerke had originally been so appointed, as well as being chosen as overall commander of the voyage, but if he felt any anger at his demotion he kept it to himself. Exactly how, why or when Cook took the decision to lead the expedition is not really known, but pause for thought or the brief experience seems to have given him some doubts about his appointment to Greenwich Hospital: 'a fine retreat and a pretty income', but 'whether I can bring myself to like ease and retirement, time will shew'.[40]

The main purpose of the voyage was to search the northwest American coast for one of Europe's last great geographical mysteries: the North-West Passage. To this was added a search for islands reported by the French navigators Marion du Fresne and Yves-Joseph de Kerguelen to be deep in the Southern Ocean, midway between South Africa and Australia. A third purpose was to take Omai home: he had become something of a celebrity under Banks's protection since his arrival in the *Adventure* and had been painted by Sir Joshua Reynolds and William Parry, but after two years was probably overstaying his welcome.

Joining the refitted *Resolution* (and none-too-well refitted, as Cook would bluntly claim) was the *Discovery*, the smallest

enced navigator. There was little room for scientists on this voyage, with King and Cook acting as astronomers on *Resolution* and a professional astronomer, William Bayly, appointed to *Discovery*. The large party of scientific gentlemen on the second voyage was replaced by the assistant surgeon William Anderson, a veteran of the second voyage and a keen naturalist, thereby settling science firmly within the naval command structure. A talented young landscape artist and portraitist of Anglo-Swiss birth and upbringing, John Webber, was appointed to provide the visual record of the voyage.

Resolution sailed from Plymouth on 12 July 1776, with *Discovery* following two weeks later: on the eve of departure, the unfortunate Clerke had been gaoled in London as guarantor for his defaulting brother's debts, and this is probably where he contracted the tuberculosis that later saw him buried on the coast of Siberia. By the end of November both ships were ready to depart from Cape Town: a good-humoured Cook wrote to Sandwich that they had provisions for two years and that 'Nothing is wanting but a few females of our own species to make the Resolution a Compleate ark, for I have taken the liberty to add considerably to the number of animals your Lordship was pleased to order to be put on board in England.'[42] Although Cook had taken animals out before, with the humanitarian and pragmatic intention of leaving breeding stock on the islands, keeping them alive on a long voyage was difficult. For this last voyage he supplemented the animals loaded in England with a number bought in Cape Town to try to minimize the losses, and they included two horses for Omai.

Kerguelen Land was found on Christmas Day 1776: an isolated and forbidding place that offered needy mariners little more than fresh water, seals and penguins (fig. 39). The whalers and sealers who followed Cook's voyages would decimate the islands' fauna within a few years. A sealed bottle was found with a note inside, claiming the island for France. Cook raised the Union flag, added George III's name to the French piece of paper, dated it and put a coin in the bottle before resealing it, a procedure that William Anderson thought 'not only unjust but truly ridiculous, and perhaps fitter to excite laughter than indignation'.[43] Continuing east in the Southern Ocean, the ships made for New Zealand, although diverting to *Adventure*'s old anchorage in Adventure Bay, Van Diemen's Land, for a brief visit to replace a topmast lost in a storm. Here Webber would make a small number of sketches of Australian Aborigines, despite the general distaste for their habits felt by most of the crew (fig. 40). Curiously, Cook no longer felt the need to establish whether Van Diemen's Land was an island or part of the main; so once repairs were completed, he sailed for his old base of Queen Charlotte Sound, leaving the question of its insularity open for Matthew Flinders and George Bass to answer 20 years later.

Fig. 39 (opposite)
A view of Christmas Harbour, in Kerguelen's Land
Newton (engraver),
after John Webber, 1774
National Maritime Museum
(PAI1518)

Fig. 40 (above)
A Man of Van Diemen's Land
James Caldwall,
after John Webber, 1777
National Maritime Museum
(PAI3894)

of the four ships used on the three voyages, now commanded by Charles Clerke, who had been deservedly promoted to commander on *Resolution*'s return. John Gore was first lieutenant of *Resolution*, and James King – 'the intellectual of the voyage', as Beaglehole describes him – was the second, and was destined to take over command of *Discovery* after the deaths of Cook and Clerke.[41] A now-infamous addition to its crew was the new master William Bligh, a capable seaman and experi-

Fig. 41 (above)
Frontispiece and title page to
A Voyage towards the South Pole, and round the World
James Cook, 1779
National Maritime Museum
(PBC4700)

The official narrative of Cook's second voyage, written by John Douglas and Cook, was published while he was at sea.

Fig. 42 (opposite)
A night dance by women, in Hapaee
William Sharp (engraver), after John Webber, 1784
National Maritime Museum
(PAF6438)

It was in the Sound, as everybody on board the ships knew very well, that ten of *Adventure*'s crew had been killed and eaten by Maori. Cook's men wanted him to take decisive action against the perpetrator, Kahura, who had helpfully been pointed out by other Maori expecting his prompt execution, as did Omai. Cook, however, declined to seek revenge, and his inaction pleased neither the Maori, who saw it as contemptibly weak, nor his men, who believed he was being provocatively perverse.[44] Cook was knowingly putting his public reputation for humane, enlightened exploration above his standing on board ship. The decision, right or wrong, was not taken lightly or hurriedly, for when he first heard the story of the massacre two years earlier in Cape Town, he wrote that he would withhold judgement until he got the full story, but with the caveat that 'I must however observe in favour of the New Zeland[er]s that I have allways [sic] found them of a Brave, Noble, Open and benevolent disposition, but they are a people that will never put up an insult if they have an oppertunity [sic] to resent it.'[45] Some now see the incident as a pivotal point on the voyage, after which Cook's good judgement, inspiring leadership and carefully measured actions could no longer be relied upon. Certainly it contributed to a growing distance between Cook and his men, which some have seen as being directly, although not wholly, responsible for the manner of his death.

The ships then headed north and east to revisit the Tongan islands. These had not been on Cook's original itinerary, but they stayed and cruised there for nearly three months, stocking up on fresh provisions and observing the culture and society more closely, this time seeing a darkness and oppression behind the ordered plenty. Charles Clerke's description of a rich Tongan soil covering a harsh coral bed could have stood as a metaphor for what now seemed a thin and vulnerable veneer of 'civilization'. 'tho' productive of such abundance of good things [the soil] is no where deep, go where you will upon the utmost elevation of them [the islands], you very frequently find the points of the Coral projecting themselves above the surface'.[46] Years later it would emerge that there had been a plot to murder the European party during one of the alluring night entertainments arranged by their hosts, drawn by John Webber (fig. 42). It failed to materialize after disagreements among the Tongans, but was a sign that their greater familiarity with the voyagers was certainly not gaining the latter any increased respect.

And so to the Society Islands, where Omai – less than enthusiastic about reverting from being a celebrity in British society to a relatively low-status indigenous curiosity among his own people – was settled on Huahine in a house built for him by the ships' carpenters. This was at least large enough to hold the many gifts he had received from the great and good in England, which included a full suit of armour and two horses (fig. 43). The planks for the house had been taken from a number

Fig. 43 (left)
Engraving taken from *Journal of Captain Cook's last voyage to the Pacific Ocean, on Discovery*, by John Rickman, 1781
Engraver and artist unknown
National Maritime Museum (PBC4715)

Rickman was a lieutenant in *Discovery* and his unauthorized journal was published three years before the official account. The illustration of Cook and Omai riding on the beach at Tahiti shows some of the gifts the latter had received during his time in England, including a suit of armour.

Fig. 44 (opposite)
Resolution and Discovery in Ship Cove, Nootka Sound
Pen, ink and watercolour, by John Webber, 1778
National Maritime Museum (PAJ2959)

Nootka Sound became the main base for the sea-otter trade that followed Cook's visit, and for George Vancouver's survey of the north-west coast of America.

of native craft broken up by Cook on Moorea, during a vengeful and violent rampage through the island after the theft of a goat, burning canoes and houses and inflicting brutal punishments that even his men thought excessive. As Bougainville had done in 1768, Cook had the dubious pleasure of witnessing a human sacrifice, albeit the victim was (as usual) already dead, and did so primarily from a sense of obligation. He was happier at landing his remaining stock of animals, thereby finding himself 'lightened of a very heavy burden, the trouble and vexation that attended the bringing these Animals thus far is hardly to be conceived'.[47] The final act in the islands was the desertion and eventual recapture of two men, one of whom was the son of Captain Mouat, who had served on Byron's circumnavigation of 1764–66. The incident would produce one of the defining images of the South Pacific: the portrait of Poedua (Poetua; see fig. 82) – the daughter of the chief, Orio – who was taken hostage against their return and of whom Webber made a preliminary study in the great cabin of Clerke's *Discovery*.

Finally leaving the Society Islands on 8 December, the two ships stood north to begin their search for the North-West Passage. Much to their surprise, on 18 January 1778 they sighted an island that was clearly inhabited by the same peoples they had but recently left. 'How shall we account for this Nation spreading it self so far over this Vast ocean?' Cook pondered. 'We find them from New Zealand to the South, to these islands to the North and from Easter Island to the Hebrides; an extent of 60° of latitude or twelve hundred leagues [approximately 4000 miles] north and south and 83° of longitude or sixteen hundred and sixty leagues [5700 miles] east and west.'[48] They were approaching Kauai, one of the Hawai'ian islands. A safe anchorage was hard to find and, although they were in the islands for a fortnight, they spent only three days ashore, long enough to understand that it was a large and populous group and its peoples enthusiastic traders; but it was also long enough for them to plant the scourge of venereal disease there. Another ominous portent was the death of one of the Hawai'ians, shot and killed by the unimpressive and unpopular Lieutenant Williamson, who became rattled by a crowd gathering round his shore party.

The ships made a landfall on the north-west coast of America on 7 March. Strong onshore winds kept them from approaching too closely and they followed it north, looking for a suitable place to repair spars and rigging damaged by storms. This they found at Ship Cove in Nootka Sound on the oceanic shore of Vancouver Island, which they believed to be part of the mainland. It received its name of Vancouver and Quadra Island years later when George Vancouver surveyed the area, still vainly looking for the North-West Passage, and shared the rights of discovering its insularity with the Spanish navigator Francisco de la Bodega y Quadra, who was on the coast at the same time. The officers and men of *Resolution* and *Discovery* traded eagerly with the local people, the Nuu Chuh Chah, largely for animal skins and particularly for the fine and dense furs of the. An observatory was erected on the point and work progressed on the repairs, including the replacement of two masts. John Webber's watercolour of both Cook's vessels in Ship Cove shows the busy scene, as well as *Resolution*'s missing foretopmast, drawn from the ship like a bad tooth (fig. 44).

Again, foul weather kept the ships further off the coast than was ideal at times, but they worked their way north and west and then south and west along the coast of modern-day Alaska and the Aleutian Islands before returning to the mainland,

JAMES COOK, THE ROYAL NAVY AND HIS THREE VOYAGES OF EXPLORATION

Fig. 45 (right)
The Resolution *beating through the Ice, with the* Discovery *in the most imminent Danger in the distance*
John Boydell (engraver), after John Webber, 1 April 1809
National Maritime Museum (PAJ1500)

Cook's ships were prevented from proceeding further north by pack ice on two occasions, in August 1778 and again after Cook's death in July 1779. It is not known to which incident Webber's engraving refers.

Fig. 46 (overleaf)
The Resolution *and* Discovery *off Huaheine*
Style of John Cleveley, the Younger, *c.*1780
National Maritime Museum. Presented by Captain A.W.F. Fuller through The Art Fund (BHC1838)

56 CAPTAIN COOK AND THE PACIFIC

JAMES COOK, THE ROYAL NAVY AND HIS THREE VOYAGES OF EXPLORATION

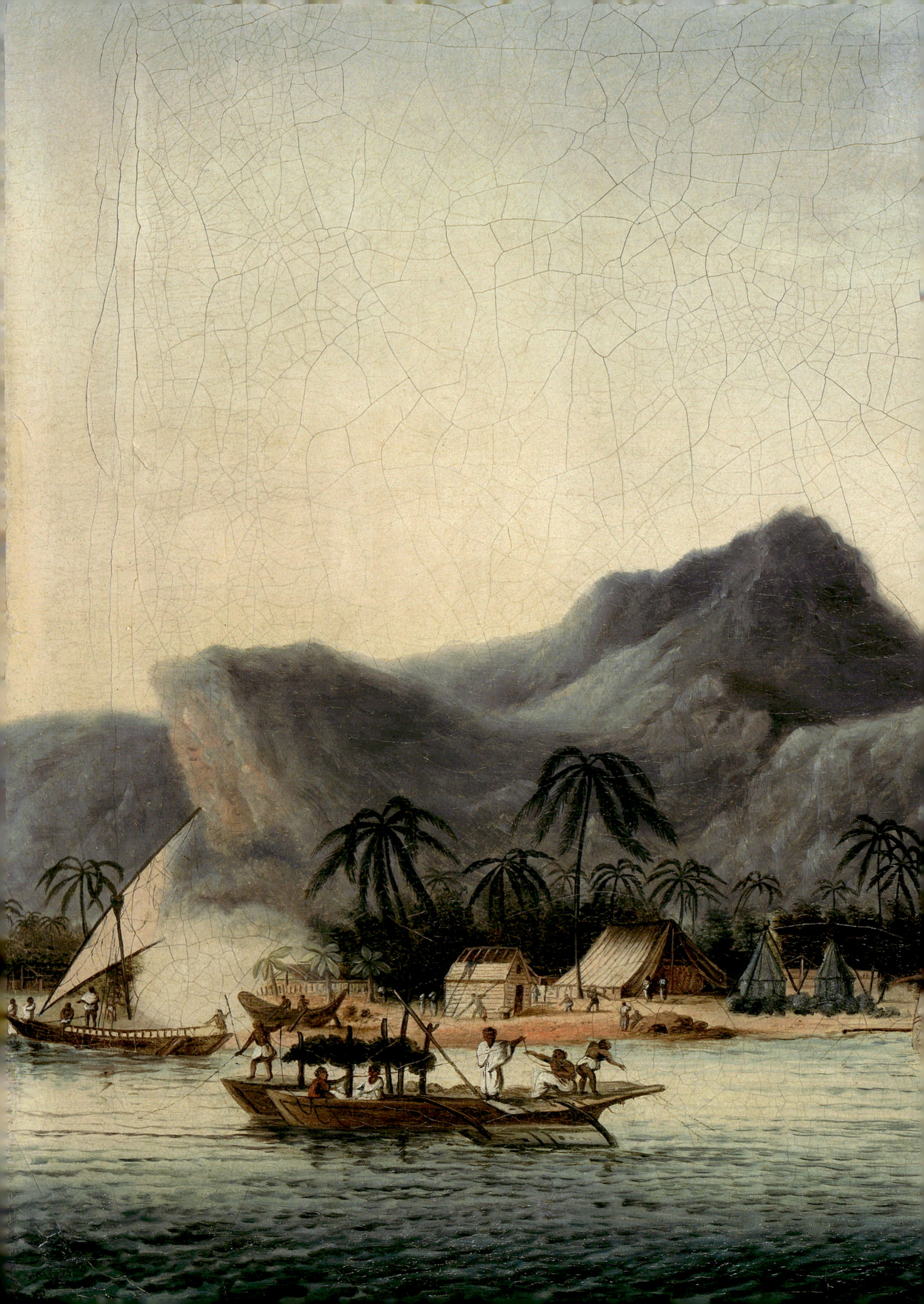

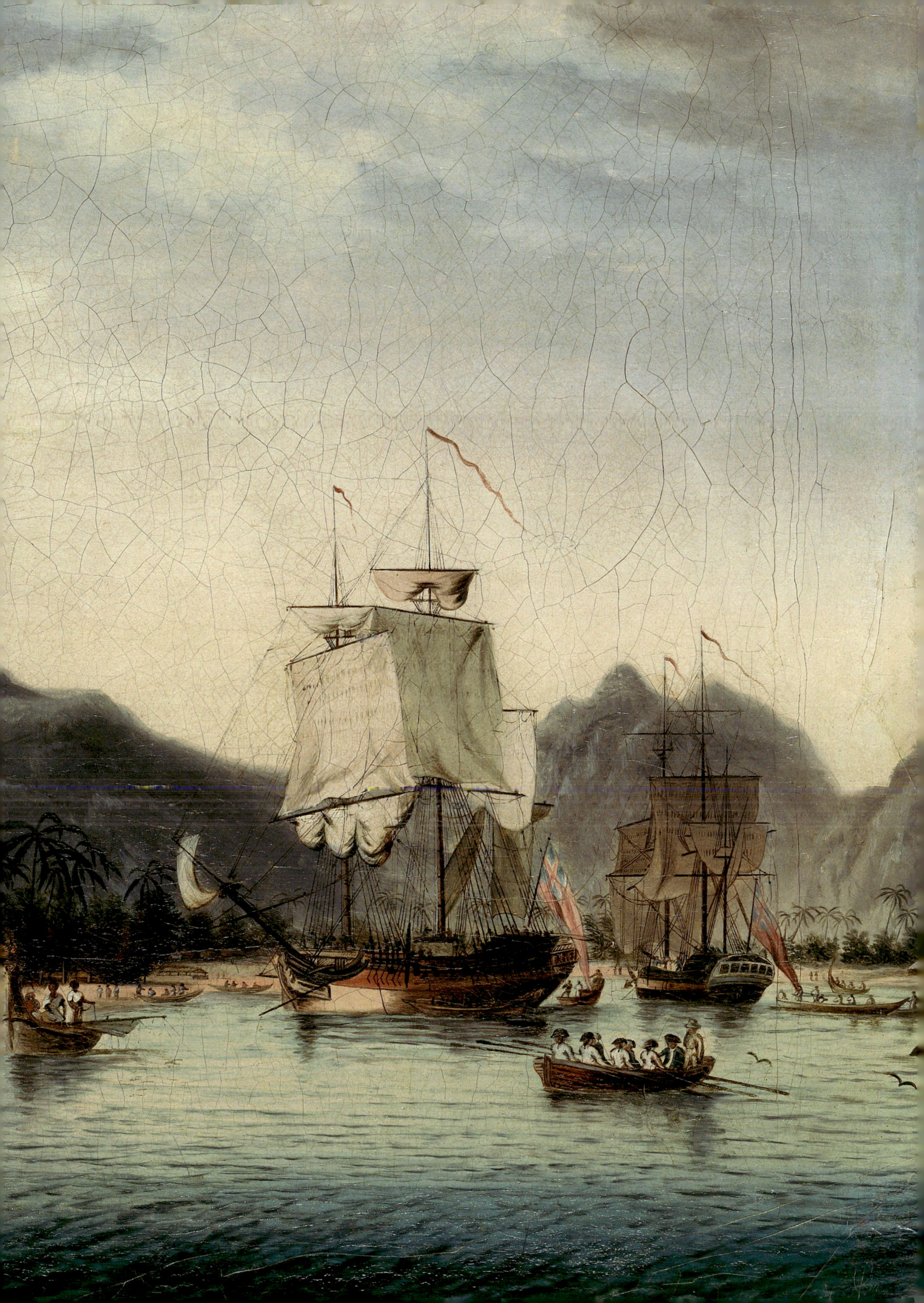

passing through the Bering Strait and almost due north into the Arctic Ocean. After first hugging the American side, they were soon defeated by the increase of ice and some near-disasters in consequence (fig. 45). They therefore crossed the Strait and followed the Siberian coast south and east, returning to the Aleutian Islands on 2 October 1778. The crews continued to acquire furs whenever the opportunity arose and their potential for a formal trade was clearly discussed on board, although Cook perceptively noted that 'unless a northern passage is found it seems rather too remote for Great Britain to receive an emolument from it'.[49]

Cook sent a letter back to the Admiralty through the good offices of Russian traders on the Aleutians, outlining his revised plans for the completion of the voyage. They would winter in the Hawai'ian islands rather than returning much further south to Tahiti, as originally intended, then make a final search in 1779 for the North-West Passage before returning to England. The ships reached the Hawai'ian group after a month and sailed slowly round the islands, with women banned from coming aboard (much to the frustration and anger of the crews). They finally found the anchorage they sought in Kealakekua Bay on the big island of Hawai'i on 17 January. Over the years, the tragedy that followed has been examined and interpreted by many distinguished writers: J.C. Beaglehole, Bernard Smith, Alistair MacLean, Marshall Sahlins, Gananath Obeyesekere, Nicholas Thomas, Glyn Williams and Frank McLynn. This list is not exhaustive, but in the context of this brief survey of Cook's voyages, there is only room for a short account of the chain of events.

The two ships stayed in the bay for two weeks, during which the voyagers were welcomed, provided for and entertained far more lavishly than they had been anywhere else in the Pacific. Cook was made the focus of a ritual at a nearby *heiau* (temple), after which the Hawai'ians prostrated themselves when he came ashore, chanting the name Orono (Lono).[50] 'It was', wrote David Samwell, surgeon's mate on *Resolution*, 'a long, & rather tiresome ceremony, of which we could only guess at its Object & Meaning, only that it was highly respectful on their parts.'[51] The numbers of canoes around the ships began to dwindle towards the end of the month and Samwell recorded that, on the last day of January, the few canoes present were offering no hogs for sale. The ships left on 4 February, only to return unexpectedly a week later to replace a badly damaged foremast. They took up their old moorings in the lee of the promontory at the north end of the bay, with *Resolution* about half a mile from the village of Kawaloa – where Kalani'op'u, the high chief of the island of Hawai'i – was staying, and *Discovery* anchored about a cable's length (600 feet) to the north.

Early in the morning of 14 February the alarm was raised when it was found that *Discovery*'s cutter had been stolen overnight: the cutter was the largest and most important of the ships' boats. Cook quickly put a tried-and-tested plan into operation, by sending off armed boats to the north and south ends of the bay to prevent movement in or out of it. He then took command of 35 armed men in three boats, including nine marines and their officer, Molesworth Phillips, and made for the village, with the intention of bringing Kalani'op'u back to the ship as a hostage, pending the return of the stolen cutter. The boats remained just offshore under the command of Lieutenant Williamson to cover Cook's shore party of marines, which landed 100 yards or so from the village and found the chief's house, where Cook persuaded him to come with him to the boats. One of Kalani'op'u's wives told him to remain where he was, and two of his chiefs joined in. The three men then sat down on the ground, not moving. The situation was not improved by a rumour that a boat on the far side of the bay had shot a chief. Armed warriors began appearing in greater numbers and the marines started to fall back towards the boats. The crowd pressed in and Cook fired a barrel of his shotgun at a warrior who came uncomfortably close, but it was loaded with small-shot and the man was completely unhurt. The emboldened crowd then rushed forward: Cook fired again, with ball this time, then turned and gestured to the boats, but was knocked to the ground and beaten and stabbed to death in the shallows. When the smoke cleared, Cook, four marines and 17 Hawai'ians were dead.

Even before the numbing shock had worn off, naval process – for the Navy has a process for everything – restored some order: Captain Clerke was rowed over to *Resolution* and took formal command of both the ship and the expedition, and John Gore was rowed in the other direction to command *Discovery*. Neither showed much inclination to prevent the desire for vengeance that saw more islanders killed in an attack on a village near their watering place. The next day some of Cook's body parts were returned, showing signs of the flesh having been burnt off some of the bones, and these were buried at sea. The rest of his body, they were told, had been cut up and distributed to the chiefs, as was the Hawai'ian custom. A week later, the two ships left Kealakekua Bay, circled and charted the islands and then sailed north once more to resume the search for the North-West Passage. They broke the journey at Kamchatka, where the Russian authorities made them welcome with much-needed supplies and repairs, before continuing to the Arctic Sea, but with no better results than the last time. Clerke died of tuberculosis just before the ships arrived back on the Kamchatka Peninsula and was buried at Petropavlovsk. Gore took over command of *Resolution* and the expedition, and James King took over *Discovery*. And so the ships slowly made their way home via Macau, the Cape of Good Hope and Stromness in the Orkney Islands, reaching Deptford on 7 October 1780 after, said one of the midshipmen, a 'long, tedious and disagreeable voyage of four years and three months'.[52]

Fig. 47
Captain James Cook
Engraving after John Webber,
date unknown
National Maritime Museum
(PAD4633)

JAMES COOK, THE ROYAL NAVY AND HIS THREE VOYAGES OF EXPLORATION

North Sea

Albemark Isl[e]
Cape Sethbauco
Cape
The Strait[s]
Port

Cape S.t Ives
Cape of Rocks
Entry of S.t Cebastian
Point of Arima
C. Holy Ghost
C. Virgin Mary
R. Gallegos

Jesus Bay / Bay of S.t Phillip

Terra del Fogua

Great Bay
Roquerom
S.t Valentine
Hills of Ishes
F. Nebada
Kings Citty
Bay of 5 Capes
Bay of
Himosa

Cape Victoria
4 Evangelists
Straights of Magellan
C. Desire
12: Apostles

Chapter 3
'The advancement of science and the increase of knowledge'
Charting the Pacific and Enlightenment Science[1]

Nigel Rigby

Nathaniel Dance's magnificent portrait of Cook (see fig. 1) is referred to frequently in this volume. It remains the outstanding and instantly recognizable likeness of the 'great navigator', one that demonstrates the pride Cook had in his chart-making. His right hand, resting on his chart of the southern hemisphere, points towards the east coast of New Holland (Australia), the survey of which was so nearly his nemesis in 1770, but which finally resolved the geographical question of whether New Holland and New Guinea formed a single land mass or two. Cook's finger points firmly to the chart (fig. 64), as if to emphasize the reality of these achievements. The power of the image is that it has come to stand for all the coasts that Cook surveyed, all the geographical myths he exploded and all the new information he, his men and his ships brought back to Britain over all three voyages. It is a painting that celebrates a new form of geographical knowledge, one that values absence as much as presence, one that is built on personal experience, precise measurement and verifiable fact. The painting celebrates the ethos and achievements of a new type of exploration, which, as Charles Withers has argued in his important study of geography in the Age of Reason, saw the Pacific becoming the principal testing ground for Enlightenment science.[2] This chapter's aim is to explore Europe's growing knowledge of the Pacific through the National Maritime Museum's collections, from early Spanish forays to Cook's voyages, which were so representative of the new scientific climate.

Fig. 48 (opposite)
Spanish chart of the Straits of Magellan and Le Maire
William Hack (detail), 1685
National Maritime Museum
(P/33(147))

Fig. 49 (left)
Americae Retectio: Ferdinand Magellan
Phillips Galle and Adrianus Collaert (engravers), after Johannes Stradanus, c.1589
National Maritime Museum
(PAD6823)

63

Fig. 50 (above)
Map of Magellan's voyage showing the global wind directions
Battista Agnese, 1554
National Maritime Museum (P/24(29))

Fig. 51 (opposite)
Pre-Magellan world map
Francesco Rosselli, 1508
National Maritime Museum, Caird Fund (G201:1/53)

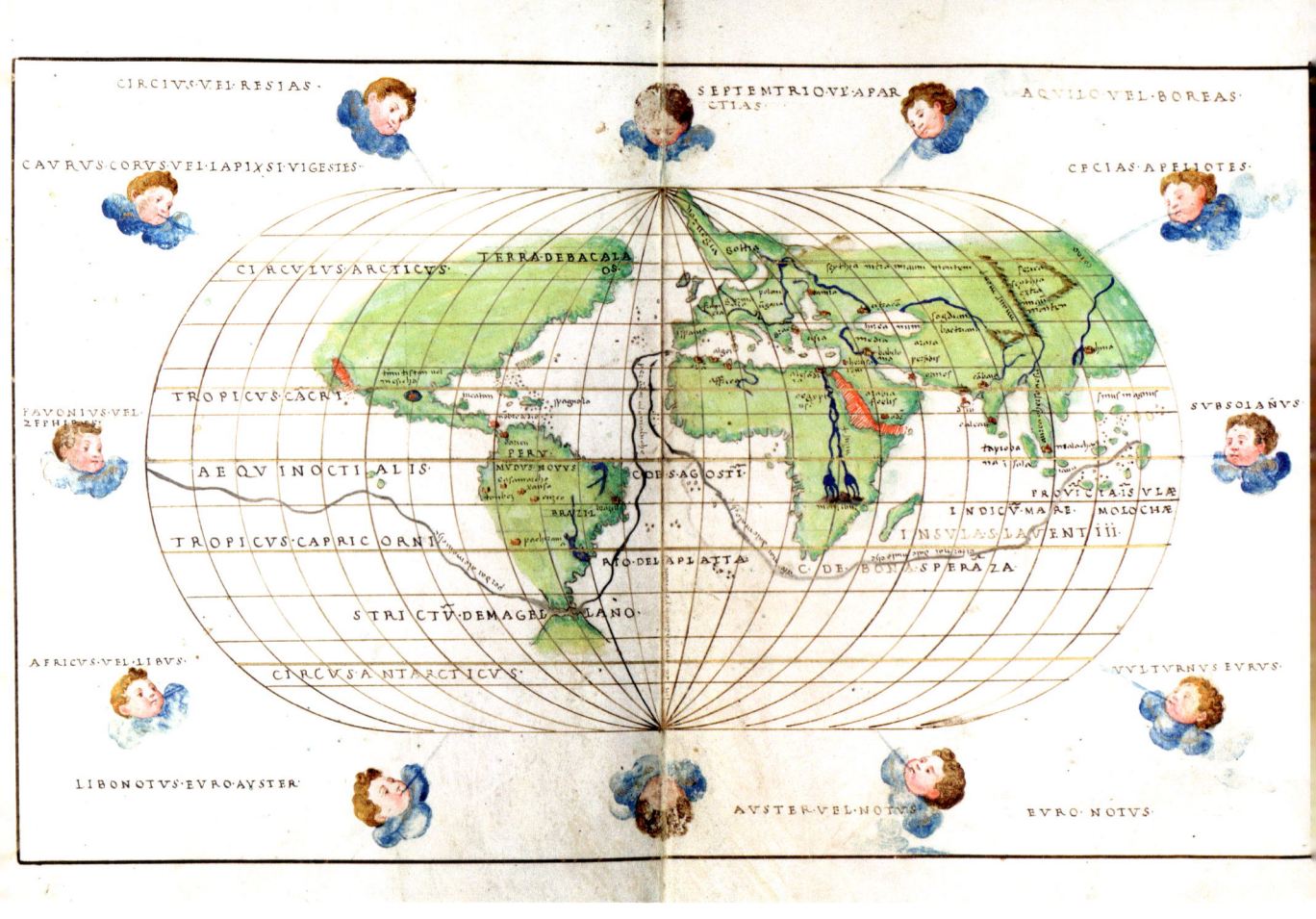

Six years after Spanish *conquistadores* first set eyes on the Pacific Ocean in 1513, five ships commanded by Ferdinand Magellan set out from Spain to discover a west-about route to the Spice Islands, the East Indies. They did so around the southern tip of South America, feeling their way through what would later be named the Strait of Magellan (figs 48, 53). It took his ships just over a month to navigate its twists and turns before they emerged into an unusually benign ocean that they called the Pacific – a name that has drawn hollow laughter from navigators from that day to this. The engraving of Magellan by Johannes Stradanus is one of four plates published in 1589 to celebrate his discoveries and those of Christopher Columbus and Amerigo Vespucci (fig. 49). It is a melange of the observed and the absurd, showing Magellan being guided through the strait by Apollo, the god of light and truth, with Patagonian giants on one side (they would prove a fanciful but tenacious presence in Pacific-voyage narratives up to John Byron's circumnavigation in the 1760s). On the other side are the fires of Tierra del Fuego, which Magellan believed were natural volcanic phenomena, but were actually bonfires built by its inhabitants. Magellan has a compass by his side and is transferring measurements onto an armillary sphere with a pair of dividers. The navigational equipment shown here would have been of limited use in finding his position, but serves, rather, to indicate the inevitable triumph of Western reason over the unknown.

In a seemingly endless crossing of an ocean much larger than had been imagined, Magellan faced ever-growing privations as the food and water ran out and as scurvy, the scourge of Pacific navigation, took a grip. He sighted only two islands, one probably in the Tuamotus and the other one of the Line Islands, but both uninhabited, before finally reaching Guam and then the Philippines, where he became embroiled in a political dispute and was killed in a short and bloody conflict on the beach at Mactan (off Cebu).[3] It was an inauspicious beginning, but the expedition continued after his death and managed to complete the first circumnavigation of the world under his lieutenant, Juan Sebastián de Elcano. Within four decades the

Philippines had been invaded and colonized, and would become the mainstay of Spain's Pacific empire with the establishment of a rich and regular North Pacific trade between Manila and Acapulco, Mexico. The trade was made possible by the discovery that taking advantage of the prevailing winds and currents in the North Pacific, rather than fighting against them, created a circular route for crossing the Pacific: westwards to Manila in a southern latitude and returning to America in a wide sweep to the north. As had previously occurred in the Atlantic, Spain was learning to work with the natural forces of the ocean, although – as J.C. Beaglehole pointed out many years ago – the 'settled highway over the Pacific' was severely limited 'and no man knew what lay south or even north of the regular course'.[4] The limitations were also affected by ship design, for square-rigged, round-hulled and high-sided ocean-going ships of the sixteenth and seventeenth centuries could not sail close to the wind – a severe constraint when trying to cross the Pacific from east to west in high latitudes (fig. 50).

Álvaro de Mendaña, Pedro Fernandes de Quiros and Luís Vaz de Torres followed Magellan to search for the 'Southern Continent' – which the ancients believed had to exist, in order to balance the land masses in the northern hemisphere – and the fabled gold mines of King Solomon, in the second half of the sixteenth century (fig. 52). Their voyages seldom crossed the Pacific more than 15 or 20° south of the Equator. Accurate navigation over such huge distances with very basic instruments was challenging. The ships followed lines of latitude, which were relatively easy to establish from measuring the height of the Sun over the horizon, but a ship's longitude – the distance east or west of a known point – had to be estimated by measuring its speed, compass course, actual direction and force of the current and the strength and direction of the wind. Finding longitude was, at best, informed guesswork, and errors of many hundreds of miles were common. It was a perennial problem, particularly on long voyages, where inaccuracies could accumulate over months at sea without a reliable 'fix'. Philip Carteret famously erred by 200 nautical miles in the position he gave remote Pitcairn Island: Cook then used Carteret's coordinates to find it in 1773, but failed – hardly surprisingly, for his search was the Pacific equivalent of trying to find London while sailing around Paris. Mendaña discovered what he optimistically named the Isles of Solomon on his first voyage, but they proved elusive when searched for on his second. They remained unsighted by Europeans until Carteret's circumnavigation of the 1760s, although he did not identify the islands as the Solomons. Despite their inability to plot positions with any great accuracy, early European voyagers encountered many island groups, including the Marquesas, Society Islands, Tuamotus, Santa Cruz Islands and the Tongan archipelago. These seemingly isolated groups, however, with their small populations, were neither the Southern Continent nor the trading opportunities the mariners sought, and the ships sailed on, often leaving behind a horrific legacy of first encounters.

PERUVIA
Tropicus Capricorni
CHI
LE.
NÆ
PA
R
S.
MARE
PACIFI
CVM
FRETUM MAGA LANICUM
Estrecho de Fernando Magallanes
TERRA DEL FUGO.
MAGELLANICÆ PARS.
El Serro de Potosi
Copaiapo
Coquimbo
Herradura
Chile
R. de Chile
Giuru matas
Talabora
Rio dos Quilocas
Mepenes
Osteco S. Miguel de Tumman
Camechin ganes
Cordaba
Rio de Mecoretas
S. Spirito
Xanaes
Rio de Luxan
S. Ana
Ningatas
Peti.
C.S. Maria
I. Luchengo
I. de Acencam
Hispanicæ leucæ 17½ uni gradui competentia.
Miliaria Italica 70 singulis gradibus respondentia.
Miliaria Germanica, quorum 15 uni gradui respondent
Joannes à Doetechum fecit.

Anno 1585. 6. Ianuarij die, Illust. D. Thomas Candilhe Anglus, fretum hoc Magellanicum attigit & in præsentem formam delineavit, & octauo dicti mensis repperit insulas, quarum unam Franciscus Draco Insulam Bartolomæi, alteram à multitudine piscium quos vocant Penguinos, Insulam Penguinorum, nuncupavit, ex quibus piscibus dictus Candilhe 3. vasa in usum suarū navium sale condivit, suntq́ue grati saporis. Die 9 ad pagū dictum Philippipolim appulit, quem Candilh alio nomine Pagum famelicum nuncupavit, quod ex aliquot cētenis Hispanis quos Petrus Sermientus eo deduxerat vix 22. reliqui essent cæteris fame & inedia confecti.

Hæc pars Peruviana, regiones Chicam & Chilē complectitur, & Regionem Patagonum, cujus incolæ statura reliquos totius orbis populos superant, ex quo Gigantum nomen sortiti sunt, quippe ut plurimum 9. & ad summum 10. pedes longi, facies varijs coloribus ex herbis diversorum colorum expressis pingunt, alioqui colore sunt punicei, Hispani asserunt se in Peruana aut Mexicana nullos Nigritas invenisse, præterquam in quibusdam pagis prope Quarequam.

Fig. 52
Post-Magellan Dutch map:
South America and Magellanica
Joannes A. Doetechum, c.1590
National Maritime Museum
(G279:4/40)

THE ADVANCEMENT OF SCIENCE AND THE INCREASE OF KNOWLEDGE

Fig. 53
Spanish chart of the Straits of Magellan and Le Maire
William Hack, 1685
National Maritime Museum
(P/33(147))

The Dutch – who had been steadily eroding Portugal's eastern monopoly – despatched Willem Schouten and Isaac Le Maire in search of the Southern Continent in 1615. They found a new South American gateway into the South Pacific (later called the Strait of Le Maire), but then followed the familiar low-latitude route across it, although they did sail rather further south than usual and touched on a number of islands previously unseen by Europeans. Abel Tasman was the first to break the east–west pattern in two voyages of 1642 and 1644, entering the Pacific from the Dutch East India Company's base in Batavia (now Jakarta) on Java. Anticipating Cook's search plan for the Southern Continent by more than a hundred years, Tasman sailed deep into the Southern Ocean before heading east and north, reaching Tasmania, New Zealand, Tonga and some outlying Fijian islands before returning to Batavia, having completed what was technically the first circumnavigation of Australia. Although he never saw the mainland on this voyage, Tasman had demonstrated that Terra Australis – or New Holland, as it had been named by the Dutch – could not be the great Southern Continent. Tasman's second voyage in 1644 would enter the Gulf of Carpentaria and chart the northern and north-western Australian coasts before returning again to Batavia. These were important expeditions, even though the English navigator Matthew Flinders would much later dismiss Tasman's charts of the Gulf as 'little better than fairyland', complaining that 'although conjecture had assigned its early examination to Tasman, yet geographers knew not what credit ought to be attached to the form it had assumed in charts'.[5] The Dutch eventually came to the same conclusion as the Spanish: there was no commercial or imperial advantage in continuing to explore the island Pacific. Exploring simply for the purpose of *knowing* would have been an alien concept at the time, and their voyages effectively ceased. The final Dutch foray was that of Jacob Roggeveen, who left Holland in 1721 and, having sighted Rapa Nui on Easter Day, accordingly 'named' it Easter Island. He then sailed on across the Pacific, skirting the Tuamotus and Samoa, but passed too far north of Tahiti to be its discoverer.

But what of England? After the predatory Pacific voyages of Drake and Cavendish at the end of the sixteenth century, which firmly established the rich pickings to be had and demonstrated Spain's inability to police the huge ocean adequately, the English played little formal part in its exploration until John Narborough's naïvely conceived and clumsily executed voyage of 1669–71. This voyage had the 'design' of making a 'Discovery both of the Seas and Coasts of that part of the World, and if possible to lay the foundations of a Trade there'.[6] Spain was rightly suspicious of this ambition, despite English protestations, and the voyage ended in embarrassing failure. It did, however, bring the Pacific once more to England's attention, and more specifically to the notice of its buccaneers and privateers. William Dampier, a member of this 'pack of merry boys', is now often regarded as having introduced a different agenda to Pacific exploration. The title page of his influential and successful *New Voyage Round the World* (1697) promised to go beyond a simple sailor's narrative and closely observe the lands visited, giving accounts of 'Their Soil, Rivers, Harbours, Plants, Fruits, Animals, and Inhabitants; Their Customs, Religion, Government, Trade &c.'[7]

As a result of the book's success, Dampier was offered command of England's first government-sponsored voyage of exploration to North Australia and New Guinea. He was a better naturalist and observer than he was a leader, and left a brief and unimpressed description of Australia's flora and fauna (extremely brief on the latter, as the nearest he got to an animal was a footprint, which 'seemed to be the tread of a beast as big as a great mastiff dog'), and a largely unsympathetic description of Australian Aborigines that coloured Cook's and Banks's later accounts of the east coast. A man of some curiosity, Dampier was also a man of his times: his detached observation of Aborigines quickly degenerated into a colonial master-and-slave relationship when he and his crew equipped them with empty barrels and 'brought these our new servants' to a water hole, indicating that he expected them to carry water to the ship. When they declined, Dampier wryly remarks that he and his crew were 'forced to carry our water ourselves'.[8] He certainly reflected a growing interest in natural history that would flower in the voyages of Cook and his followers, but as Glyn Williams has pithily concluded, 'the voyages of Dampier and his privateer-successors in the Pacific such as Rogers and Shelvocke produced little in the way of geographical discovery'.[9]

What navigational and geographical information the buccaneers did bring back was largely stolen. Navigators' journals (variously called *rutters*, *pilots*, *waggoners* or *derroteros*) were pearls beyond price, containing descriptions of coastlines, tides, islands and harbours that could be more reliable and detailed than published charts. The information these journals contained was not circulated by the Spanish authorities, which were anxious to restrict knowledge of the Pacific as much as possible. The beautifully ornate maps and globes of the Pacific produced in the seventeenth and eighteenth centuries carried little worthwhile information for navigators, but were never intended to go any nearer the sea than the libraries of wealthy individuals and armchair geographers. The journals, however, were a different matter. The pages shown in figs 48, 53 are from a Spanish *derrotero* describing the American-Pacific coastline from Acapulco to Albemarle Island. The journal was captured from the Spanish ship *Rosario* by Dampier's old captain, the buccaneer Bartholomew Sharpe, who later described how 'The Spaniads cryed when I gott the book (farewell South sea now).'[10] Multiple copies of it were created by the chart-maker William Hack, and the copy shown here was presented to James II in 1685.[11]

By the early 1700s, after two centuries of exploration, commerce and plunder by Spanish, Dutch, French and English navigators, geographical knowledge of the oceanic Pacific was still haphazard. The long Spanish-American coast from Mexico to Tierra del Fuego, which was the main artery of Spanish intra-Pacific colonial trade and thus the cruising ground of hopeful buccaneers and privateers, was familiar enough, if not mapped in any great detail. Drake had claimed part of the coast beyond present-day California as New Albion – although exactly where is still a matter for debate. Spain had despatched some exploratory voyages northwards from Central America in the 1500s, but the North-West Passage to link the Western world to the riches of the East – the big goal of most north-bound voyages – had still not revealed itself. These voyages may have reached the latitude of Vancouver Island or thereabouts, although they were not followed up. An intriguing footnote was a letter purporting to have been written by a Spanish admiral, Bartholomew da Fonte, in 1640, and published in an English periodical in 1708, which described his discovery of an inland sea in the north-west of America that was connected to the Atlantic. Interest in the hoax was revived in the mid-eighteenth century and would play its part in formulating the aims of Cook's third voyage in search of a north-west passage.[12]

A part of New Zealand's west coast had been traced by Abel Tasman, his map leaving tantalizingly open the possibility that it could prove to be a promontory of the elusive Southern Continent. Dutch ships had tracked the lines of the north, south and west coasts of Australia. Many of the equatorial Pacific island groups had been encountered, generally fleetingly, but vast tracts outside the usual oceanic routes remained unseen by Europeans. The islands and lands that had been sighted, and even landed upon, were imperfectly plotted and their geographical relationship to each other was unclear. One British geographer complained in the 1750s that their positions were 'scarce of any use', for navigators would 'seldom mention the Longitude or Distances; and rarely the Latitude with any accuracy'.[13] A more balanced view was held by Cook, who in 1770 wrote with criticism born of experience that errors were largely the fault of the 'Compilers and Publishers who publish to the world the rude sketches of the Navigator as accurate surveys without telling what authority for so doing'. The navigators themselves, though, had to share the blame, for he had 'known them lay down the line of a Coast they never have seen and put down soundings where they never have sounded and . . . are so fond of their performances as to [have] pass'd the whole off as sterling under the Title of a *Survey Plan* &c'.[14] Geographical fact could still veer into fiction and vice versa, and the uncharted vastness of the Pacific was in some respects still a region where the implausible and the believable could sit side by side, a situation that writers like Daniel Defoe and Jonathan Swift,

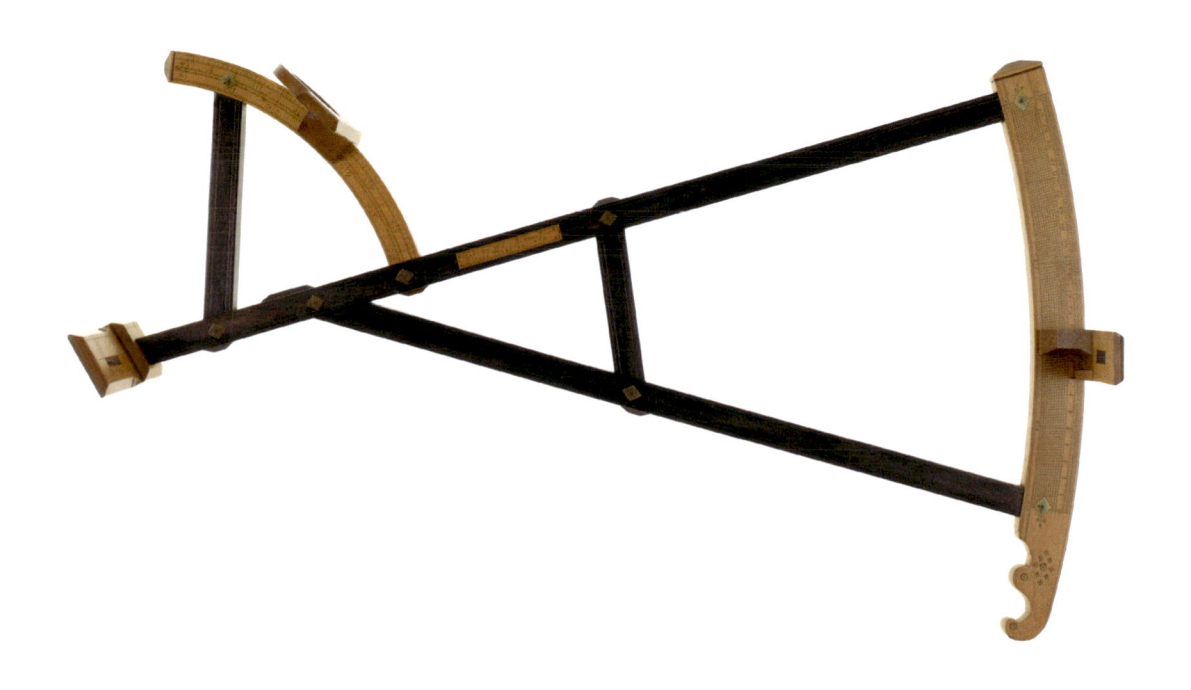

Fig. 54 (above)
Backstaff
Will Garner, 1734
National Maritime Museum,
Caird Fund (NAV0041)

Fig. 55 (left)
Octant
Benjamin Cole, 1761
National Maritime Museum
(NAV1346)

Owned by Lieutenant Philip Carteret, commander of the sloop *Swallow* during its circumnavigation of 1766–69 when Pitcairn Island was discovered.

CAPTAIN COOK AND THE PACIFIC

and financial schemes like the South Sea Bubble, were able to exploit.

From the middle years of the eighteenth century a number of factors – political, philosophical, scientific and mechanical – started to come together. They would, in time, change the Pacific from an ocean of which Europe knew virtually nothing into one of the most thoroughly and systematically explored regions in the world. Its exploration would be shaped by, but also come to stand for, new ways of understanding the world based on knowledge acquired through rational thought, travel, observation and experience, rather than hoary myths and avarice.[15] It was exploration in which commerce and learning appeared to go hand-in-hand, with Europeans encountering, recording and learning from cultures and societies very different from their own. That the lives of indigenous peoples could also be improved by the introduction of European goods, technology and ideas was a well-intentioned – albeit patronizing – corollary assumption, of which both immediate and later negative effects were almost entirely unforeseen. Early English involvement in the Pacific, from Drake to Dampier, had been that of buccaneers and speculators with, at best, semi-official sanction. George Anson's anti-Spanish Royal Naval circumnavigation of the 1740s made it a state affair, and one that brought to the fore all too evidently the inadequacies of both contemporary navigational technology and cartography, especially of the Pacific. It was Anson's subsequent influence as First Lord of the Admiralty during the Seven Years War that saw early moves to remedy both areas of technical endeavour, and, from the mid-1760s until the start of the French Revolutionary War in 1793, Britain played a leading role in Pacific exploration, correcting previous inaccuracies, filling blanks and making itself a standard-bearer for Enlightenment science, scientific method as applied to discovery and the heightened national and international prestige it could bring. An accurate rather than speculative map of the Pacific that could prove or discount the two great geographical questions of the day – the Southern Continent and the North-West Passage – was the great cartographic prize of eighteenth-century science. Drawing that map would introduce the figure of James Cook to the world, enshrining him as one of the great heroes of Enlightenment exploration.

Producing an accurate map of the Pacific depended first of all on the availability of scientific instruments capable of giving a reliable measurement of longitude. This did not begin to happen until the eighteenth century. In 1714, the Board of Longitude was formed by the British government, offering a prize of £20,000 for the development of a workable and accurate means of finding longitude at sea. This was by no means an easy task, and many ideas were presented, deliberated upon and rejected or taken further, as the case might be. The Board soon began to have an effect, however. The first double-reflecting octant, built by John Hadley and designed to measure the altitude of the Sun, the Moon and stars at sea, was trialled in 1733 and was shown to have considerable advantages over the traditional but by now severely limited cross-staff and backstaff: '[the octant] was compact and therefore practical on the unsteady deck of a vessel, and less sensitive to the influence of wind'.[16] Since it was relatively expensive to produce, compared to the backstaff (fig. 54), the octant (fig. 55) did not come to be widely used at sea until the 1750s.[17] The sextant, a further development, received its first sea-trial in 1758. This offered the opportunity to test the lunar-distance method for finding longitude at sea. The measurement of lunar distances had actually been used to establish longitude for well over a century, but up to the mid-1700s it was impractical at sea, partly owing to the motions of the ship, but equally because the necessary calculations required levels of mathematical literacy that were beyond most naval officers, or indeed most people. A set of tables that provided these calculations for observers was developed – based on Tobias Mayer of Göttingen's previous mathematical work – by the astronomer Nevil Maskelyne, who was appointed Astronomer Royal at the Royal Greenwich Observatory in 1765.[18] The tables reduced the average time needed to make the calculation from four hours to one. First published as the *Nautical Almanac* in 1767, the tables for 1768 and 1769 – the only ones that were yet ready – were issued to James Cook for an extended trial on his first voyage. He reported that the method worked well at sea, with an error of no more than half a degree, 'which', wrote Cook, 'is a degree of accuracy more than sufficient for all Nautical purposes'.[19]

All three of Cook's voyages were equipped with the latest scientific instruments, which went far beyond the specific needs of navigation and cartography. These included marine timekeepers (although for the second and third voyages only), logs, sextants (some ten of them), hand telescopes, astronomical telescopes, clocks for land use, astronomical quadrants, transit instruments, portable observatories, artificial horizons, azimuth compasses, dipping needles, variation compasses, barometers (fig. 56), magnets, thermometers, water bottles (for taking deep-water samples), wind and tide gauges, globes and specialist books. In addition, Cook requested for his land surveys a theodolite, a plane table, a brass scale, a double concave glass, a glass for tracing plans, a pair of large dividers, a parallel ruler and a pair of proportional compasses.[20] The investment in this range of scientific instruments was huge. Even with the full support and influence of the Royal Society and the Admiralty, simply managing to obtain them at all was a triumph, for while London was home to some of the finest instrument-makers in the world, the instruments had to be so accurate that they could only be made by their very best craftsmen. Cook was keenly aware of the debt that his voyages owed to the trade:

For the improvements and accuracy with which they make their Instruments, for without good Instruments the Tables [the Nautical Almanac] would loose [sic] part of their use: we cannot have a greater proof of the accuracy of different Instruments than the near agreement of the above observations, taken with four different Sextants and which were made by different workmen, mine by Mr Bird, Mr Wales and Mr Clerkes by Ramsden and Gilberts and Mr Smiths, who observed with the same instrument by Mr Nairn.[21]

Coasts were primarily mapped using 'running surveys', a method of laying down a coastline quickly and tolerably accurately that combined the new and the old: the precise measurement of angles of identifiable parts of the coast taken from a distance offshore, using either an azimuth compass or a sextant turned on its side (which Cook came to prefer for its greater accuracy), and the traditional but less precise dead-reckoning, which plotted the ship's changing position by estimating its speed and course through the water. Where possible, the ship's position would be taken daily by the lunar-distance method on the first voyage and by chronometer as well on the second and third, providing regular checks of the dead-reckoning. Cook did not invent the running survey, but he brought it to new levels of accuracy and rigour. It was work that demanded high levels of concentration over long periods of time. Matthew Flinders, who circumnavigated and charted Australia between 1801 and 1803, and who had been trained by William Bligh, sailing master on Cook's third voyage, described how work would begin at dawn and keep him on deck until dusk, at which point he would steer away from the coast, retire below and transcribe his results, returning to the end of the previous day's survey in the morning. Cook's general habit was also to stop surveying at night, for obvious reasons, but he would continue his course, returning to the survey further down the coast at daylight. The resulting gaps were normally filled in by eye later on. Likely anchorages and promising bays and sounds were mapped in greater detail as the opportunity arose. Using these methods on the first voyage, Cook and his astronomer Charles Green charted 5000 miles of coastline, the largest parts of which were New Zealand and the east coast of Australia.

When Maskelyne's lunar tables ran out at the end of 1769 during the circumnavigation of New Zealand, Cook and Green had to prepare a new set themselves, a tedious affair. Even so, Cook was justifiably proud of the accuracy of large parts of his chart of New Zealand, believing that it would not 'be found to differ materially from the figure I have given it and that the coast affords few or no harbours but what are either taken notice of in this Journal or in some measure point[ed] out in the Chart'.[22]

On Cook's second voyage, he and the two astronomers William Wales and William Bayly – the former on *Resolution* and the latter on *Adventure* – were able to give extended trials to the recently invented marine timekeepers, the 'chronometers' (a term dating only from 1794) that had been made as copies of John Harrison's prizewinning H4 (fig. 57). Larcum Kendall's K1 chronometer was found to be accurate, particularly on Cook's second voyage, the others less so. To Cook's barely concealed dismay, the observations taken on his second voyage revealed that on his first 'the whole of New Zealand is laid down too far East', as the modern overlay of his published chart shows (fig. 58). Cook appears to have been initially resistant to this conclusion, and one can detect a little irritation at what he saw as the pedantry of Wales and Bayly, for he could not 'think the error so great as these two Astronomers have made it, but supposing it is it will not much effect either Geography or Navigation but for the benifit of both I thought proper to mention it though few I believe will look upon it as a capital error'.[23]

When Cook reflected on the many achievements of his second voyage – arguably the most complete of the three – he showed a justifiable pride that 'by twice visiting the Pacific Tropical Sea, I had not only settled the situation of some old discoveries but made there many new ones and left, I conceive, very little more to be done even in that part'.[24] What he does not mention here (although he pays full tribute elsewhere) is that many of his discoveries benefited greatly from information given to him by indigenous peoples, a form of knowledge-transfer entirely in keeping with an Enlightenment ethos. Islanders would sail on all three of his voyages, but the outstanding figure was the navigator and priest Tupaia, who joined *Endeavour* in Tahiti at the invitation of Joseph Banks, despite Cook's initial concern that the likelihood of the Admiralty agreeing to a voyage to take him back to his home was remote. Cook realized very early on that the Polynesian islanders had 'an extensive knowledge of the Islands situated in these seas, Tupaia as well as several others hath given us an account of upwards of seventy'.[25] The islanders navigated through observation of the Sun and stars, wind, clouds, state of the sea and its currents, and the presence of seabirds, in ways that would have been entirely familiar to European seamen. When discussing Polynesian methods of navigation, J.R. Forster (the naturalist on the second voyage) remarked that '*Tupaia* was so well skilled in this, that wherever they came with the ship during the navigation of nearly a year, previous to the arrival of the *Endeavour* at Batavia [where Tupaia would die from fever], he could always point out the direction in which Taheitee was situated.'[26] Tupaia's chart, drawn for Cook and Banks on the voyage from the Society Islands to New Zealand, is centred on

Fig. 56 (opposite)
Marine barometer
Nairne & Blunt, *c.*1774
National Maritime Museum
(NAV0805)

Fig. 57 (left)
Marine timekeeper, H4
John Harrison, 1759
National Maritime Museum
(ZAA0037)

Harrison's longitude-prize-winning marine timekeeper, H4, was not taken on any of Cook's voyages. It was a condition of the prize that a winning mechanical time-keeper could be copied.

THE ADVANCEMENT OF SCIENCE AND THE INCREASE OF KNOWLEDGE

CHART of NEW-ZEALAND,

explored in 1769 and 1770.

by

Lieut: J. COOK, Commander

of

His Majesty's Bark

ENDEAVOUR.

Engrav'd by I. Bayly.

EXPLANATION

The pricked lines shews the Ships Track, and the figures annexed the depth of Water in fathoms.

The unshaded part of the Coast has not been explored.

⚓ *Places where the Ship Anchor'd.*

Rocks above Water.
Rocks under Water.

Var 9° Shews the Compass East Variation in Degrees and Minutes.

In Cook's Strait, the Flood Tide seems strong in from the Southward, and on the days of the New and Full Moon is High-water about 11 o'Clock.

Published as the Act directs 1st Jan'y 1772.

Longitude West from the Meridian of Greenwich.

As laid down by Abel Tasman	1642	
From a chart by W. Wales	1788	
From latest British Admiralty charts	To date	

Overlay prepared by the Hydrographic Department, Ministry of Defence.

1st July 1968

Raiatea, his home island, but uses a system that does not lend itself to European interpretations. Forster annotated a copy that was published in his *Observations Made on a Voyage Round the World*, trying to reconcile Tupaia's chart with Cook's, but ultimately failing, as the two cartographic systems appear to be mutually incompatible. Although some plausible interpretations have been published since then, Tupaia's chart has continued to puzzle scholars to this day, but still serves as a reminder that Europeans were not the first navigators to enter the Pacific Ocean, which had been settled from the west many hundreds – if not thousands – of years before Magellan.[27]

Charles Withers has described Cook's ships as 'mobile "knowledge spaces" – floating instruments of science and geographical expansion', which indeed they were.[28] These vessels of the Enlightenment were also floating classrooms. Cook's 'young gentlemen' received excellent training in seamanship, as one might expect from serving under one of the outstanding seamen of their time. His young officers also received additional and quite specific instruction suited to the voyages' aims: they were taught to draw coastal profiles by the professional artists on board Cook's ships; trained by the astronomers in taking observations, making calculations and drawing maps and surveys; and shown how to observe and describe people, land, flora and fauna and objects by the naturalists. John Elliott, one of the midshipmen in *Resolution*, recalled many years later how he and the other 'young gentlemen' were 'employed in Capt. Cook's Cabbin either copying drawings for him, or drawing for ourselves, under the eye of Mr Hodges. I kept a regular Chart of the Ship's Track, with remarks on it, and Myself, [Henry] Roberts, and [Alexander] Hood formed a very good set for observing the Lunar Observations, as any in the Ship (and very correct).'[29]

What we see here – in the group tests of navigational instruments, the repeated astronomical observations, the intensive training of junior officers and the attempt to bring Tupaia's chart into a European cartographic system, and much more besides – is the arrival of consistency, continuity and order in the new map of the Pacific, qualities that had been so lacking in Europe's earlier disconnected forays. Cook built a cadre of capable young officers who understood the purpose of the voyages, were trained in his methods, confident in their own expertise and familiar with the array of scientific instruments.

Fig. 58
Chart of New Zealand explored in 1767 and 1770 by Lieut. J. Cook, Commander of His Majesty's Bark Endeavour
William Bayly, 1 January 1772
National Maritime Museum (PAI4013)

Cook's chart of New Zealand, with a modern overlay in red showing the true position. The overlay in orange represents Abel Tasman's chart of part of the coast in 1642.

It was a uniquely qualified corps, proud of its own identity: James King referred to his group as 'we discoverers', while an outsider hailed them as 'you men of Captain Cook' to Lieutenant James Burney.[30] When they came to command their own voyages of exploration, as so many would (see chapter 7), they would load their ships with the antiscorbutics recommended by Cook, introduce similar programmes of shipboard hygiene and test their own instruments and observations against Cook's, correcting them as necessary. They had been trained to understand that errors were inevitable, for instruments could be damaged, timekeepers could stop or change their rates, the officers responsible for them could forget to wind them up, and on one occasion the key to a chronometer's box went missing; so they altered Cook's measurements professionally, and with little apology. After George Vancouver made some 'trifling additions' to Cook's chart of Dusky Sound in New Zealand, his surgeon/naturalist Archibald Menzies recorded that they 'drank a cheerful glass to the memory of Cpt. Cook' and went on their way. Vancouver was equally happy to acknowledge that Cook had been right, and himself wrong, in calculating the latitude of Dusky Sound.[31] Accuracy was the overriding principle, but openness and honesty were not far behind.

Some ten years or more after Cook's death a public dispute was carried out in print between George Dixon, who had served as the armourer in *Discovery*, and John Meares, who had no connection with Cook. Both were involved in the burgeoning fur trade on the north-west coast of America. Meares produced a chart of the coast that acknowledged Cook's prior work, but implied that all the subsequent charting had been done by himself, whereas in fact he had copied surveys made by a number of ships already on the coast. Dixon, on the other hand, produced a chart showing which parts he had charted himself and which had been laid down by others, whom he scrupulously named. The 'Dixon-Meares Controversy' is now an obscure footnote in history, but it illustrates that Cook's men understood the principle of building a chart, using and crediting the work of others, and saw themselves to a degree as standard-bearers for a new geographical ethos, which recognized that the improvement of knowledge required due credit to both present and past contributors, even when correcting them.

A significant part of Cook's enshrinement as an arrow of the Enlightenment was the reputation he built for humanitarian exploration, specifically as the conqueror of scurvy, which could decimate a ship's crew as effectively as a sea-fight. His voyages experimented with innumerable cures: his assistant surgeon on *Endeavour*, William Perry, lists 'Sour Krout [sic], Mustard, Vinegar, Wheat, Inspissated Orange and Lemon Juices, Saloup, Portable Soup, Sugar, Melasses [sic], Vegetables (at all time they could possibly be got) . . . Cold Bathing was encouraged and enforced by Example: the allowance of Salt beef & pork was

Fig. 59 (above)
A View of Point Venus and Matavai Bay, looking east
William Hodges, August 1773
National Maritime Museum
(BHC1937)

Fig. 60 (opposite)
View of the Islands of Otaha [Taaha] *and Bola Bola* [Bora Bora] *with Part of the Island of Ulietea* [Raiatea]
William Hodges, 1773
National Maritime Museum
(BHC2376)

abridged from nearly the beginning of the voyage.' There were so many treatments, Perry thought, that it was 'impossible for me to say what was most conducive to our preservation from Scurvy'.[32] Most of these cures shared a single quality: they had no antiscorbutic properties at all. Cook nevertheless achieved extraordinary results against scurvy, not losing a single man to it on his second voyage and receiving the Royal Society's Copley Medal in recognition of the feat. Scurvy was present on all of his voyages, however, and certainly contributed to the death of Tupaia. Cook's success was largely the result of two things: the pattern of his voyaging, in which regular stops gave access to fresh food; and the enforcement of a regime of shipboard health and hygiene, which saw bedding and clothes aired regularly and the crew getting a greater amount of rest than was normal in ships of the time. New research is now helping us understand the full complexity of the disease and its control in the eighteenth century, but Cook's men believed in the antiscorbutics issued to them, even while the evidence of their own eyes showed that the cures were not working.[33] Bligh, Vancouver, Portlock, Flinders and the other commanders after Cook would record their own regular, even obsessive, issue of remedies such as malt wort and sauerkraut in their journals and official voyage narratives, in order to demonstrate that they were following the great man and had taken every possible action to control the disease.

As Withers has observed, 'the facts of exploration, travel and mapping added to the sum of Enlightenment knowledge only when (and if) they made a final move – into print'.[34] The official publications stemming from Cook's three voyages were the primary formal means by which new knowledge of the Pacific was consumed by the public, scientists and – an often forgotten audience – naval officers, especially those who were embarking on their own voyages and needed the latest information on the coasts and islands. Bligh took Hawkesworth's *Voyages Undertaken . . .* on his first breadfruit voyage, although Fletcher Christian held onto it after the mutiny and famously used it to track down Carteret's lost Pitcairn Island. These publications can appear in unexpected places: Bernard Smith first drew attention to the influence of Cook and Douglas's account of the second voyage on Samuel Coleridge's *Rime of the Ancient Mariner*, and Rod Edmond has noted that engravings of the death of Cook turn up in Charles Dickens's *Bleak House* and *Great Expectations*, the latter prefiguring the death of Pip's benefactor, Magwitch.[35] The official literary outputs are considered elsewhere in this volume, but they were in competition with – and sometimes beaten to – publication by unofficial accounts. Cook's were overwhelmingly 'textual' voyages, a fact that has played a part in making them a rich field for scholarly enquiry today. 'James Burney,' writes Anthony Payne, 'who sailed on Cook's second and third voyages and who was later to write *A Chronological History of the Discoveries in the South Sea or Pacific Ocean*, records that on the third voyage a weekly newspaper was prepared on each ship, such was "the literary ambition and disposition to authorship" amongst many of the crews.'[36] A number of the officers and men kept journals with an eye to publication, although they did so with another eye over their shoulders, since unauthorized publication was against regulations and all journals had to be handed in at the end of a voyage. As Payne has suggested, the existence of the many proscribed and official journals reveals an unusually high level of literacy on Cook's ships across the ranks. They also show a

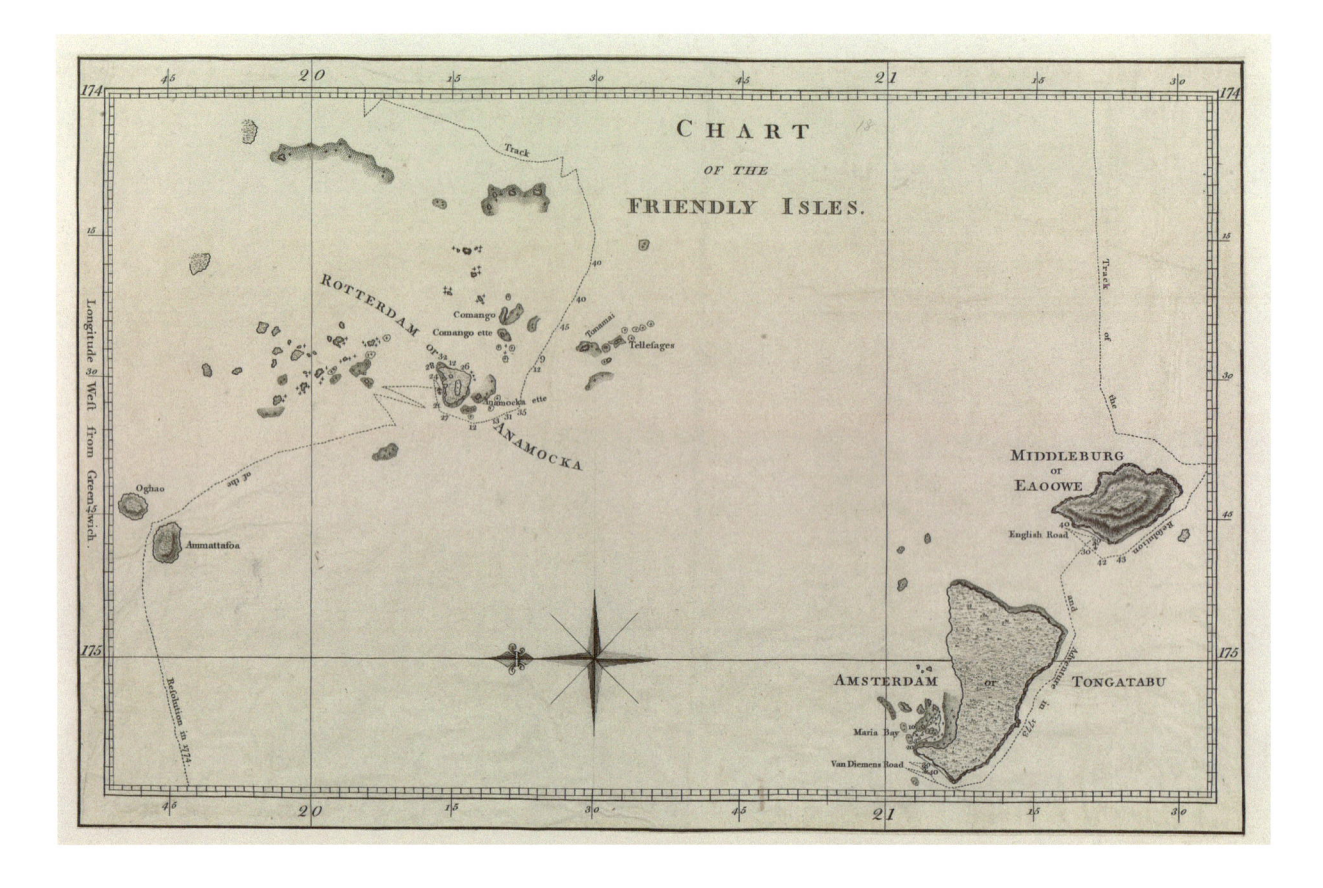

Fig. 61 (above)
Chart of the Friendly Islands, from James Cook and John Douglas's *A Voyage to the South Pole and Around the World* (1777)
William Hodges, 1777
National Maritime Museum (PAI2074)

Fig. 62 (opposite)
Chart of the N W Coast of America and N E Coast of Asia explored in the years 1778 & 1779 from James Cook, John King and John Douglas, *A Voyage to the Pacific Ocean* (1784)
T. Harmer (engraver), after John Webber
National Maritime Museum (PAI4213)

Cook had neither the time nor the weather to chart the complicated north-west coast of America. That job would take his ex-midshipman, George Vancouver, three survey seasons in the 1790s.

keen awareness of the profit that could be made from publication, and serve as a reminder that the public had access to the voyages through unofficial as well as official means.

The high-quality official narratives were lavishly illustrated with maps and engravings, and fetched correspondingly high prices. The third-voyage account was edited after Cook's death by John Douglas and James King, with financial support from the Admiralty, and was sold for £4 14s. 6d., a price 'beyond the reach of the average Englishman', says Lynne Withey, who also points out that, despite this, the first edition sold out in three days. No fewer than four accounts of the last voyage were published before the official book appeared: one in 1781, published anonymously by John Rickman, lieutenant in *Discovery*, which was also translated into German by J.R. Forster and published later the same year; another by William Ellis, surgeon's mate in *Discovery*; a third by Heinrich Zimmermann, an able seaman in *Discovery*; and a fourth by the American marine John Ledyard, also in *Discovery*, whose book *A Journal of Captain Cook's Last Voyage to the Pacific Ocean, and in Quest of a North-West Passage, between Asia and America, Performed in the years 1776, 1777, 1778, and 1779* was published in America in 1783.[37] The fact that all four illicit accounts stemmed from *Discovery* rather than *Resolution* may suggest that the ordinance against journal-writing was less rigorously applied in the junior ship: its first lieutenant was the highly literate and unconventional officer James Burney. These unauthorized volumes were cheaper and contained fewer and lower-quality illustrations and maps. However, they made Cook's voyages accessible to less wealthy segments of society, and could hold information that did not find its way into the official narrative. The engraving of the death of Cook in Rickman's *Journal of Captain Cook's Last Voyage* (1784) shows the explorer ignominiously sprawled face-down on the rocks, in an altogether more believable end than that represented by Webber and Zoffany.

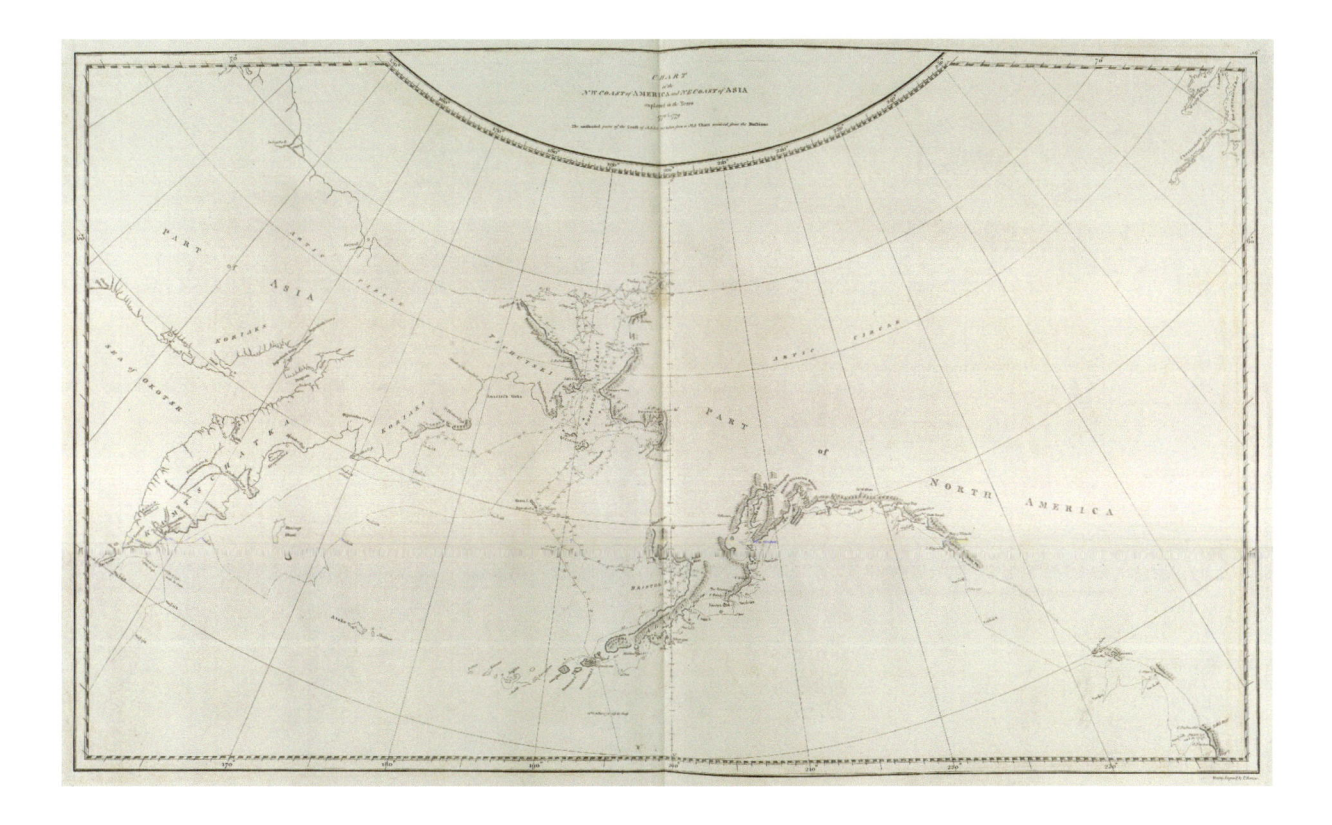

The scientific world also had its informal structures, which helped to gather and disseminate the voyagers' knowledge of the Pacific. The Royal Society Dining Club, founded in the seventeenth century, held a regular series of talks for its members on Thursday evenings, repairing later to the Mitre on Fleet Street and, from 1780, to the Crown and Anchor in the Strand. Each member was allowed to bring a guest, although Sir Joseph Banks, the President from 1778 to his death in 1820, could invite more. Among the many famous names invited were a number of veterans of Pacific exploration: James Cook, Omai, Daniel Solander, Tobias Furneaux, Charles Clerke, James King, William Bligh, George Vancouver, Edward Riou, Matthew Flinders, William Broughton and George Bass.[38] It is a reasonable assumption that they were invited by Banks.

As Glyn Williams has recently shown in *The Death of Captain Cook: A Hero Made and Unmade*, Cook emerged as a new type of maritime hero, one who doggedly pushed Europe's geographical knowledge of the Pacific to its practical limits, defeating freezing conditions, ignorance, treacherous coasts and disease rather than destroying Britain's enemies at sea.[39] Henry Roberts's map of the world, which was drawn for Cook, Douglas and King's *A Voyage to the Pacific Ocean* (1784), demonstrates just what had been achieved geographically on Cook's three voyages. In a little over ten years the existence of a Southern Continent in temperate climes had been scotched: the map shows Cook's systematic and punishing approach to the geographical myth, circumnavigating the world in three gigantic legs at high latitudes – the highest that it would have been humanly possible to reach at that time, but over-wintering in the tropics. Resolving the existence of a North-West Passage finally proved impossible after the ships were nearly trapped in a solid wall of ice, but the map nevertheless shows the stubborn

Fig. 63 (above)
*Midshipman James Ward,
c.1759–1806*
Attributed to John Webber,
1776–80
National Maritime Museum
(BHC3077)

Fig. 64 (opposite)
'A chart of the southern hemisphere; shewing the tracks of some of the most distinguished navigators, by Captain James Cook, of His Majesty's Navy.'
from James Cook and John Douglas's *A Voyage Towards the South Pole, and Round the World.*
Captain James Cook, 1777
National Maritime Museum
(PBC4700)

Fig. 65 (overleaf)
A General Chart Exhibiting the Discoveries made by Captn James Cook in this and his two preceding Voyages, with the Tracks of the Ships under his Command
W. Palmer, after
Lt Henry Roberts, 1784
National Maritime Museum
(G201:1/5)

resolve to return and seek new ways through the frozen sea – the last without Cook, of course. It depicts the Hawai'ian islands, first seen by midshipman James Ward (fig. 63) in 1778, which provided Cook with a better and more convenient winter base than Tahiti for his explorations of the North Pacific and Arctic Sea: the intertwined tracks of *Resolution* and *Discovery* appear to form a hub or knot around the islands where he would die.

Roberts's *General Chart* also picks out the routes through the islands of the 'Tropical Seas' that formed a marginal part of Cook's orders, but which would capture the imagination of Europe. It shows how New Zealand became a nodal point, the tracks of his three voyages criss-crossing around his base in Queen Charlotte Sound. Finally, it shows the survey of Australia's east coast and *Endeavour*'s route back to England through the Torres Strait and the East Indies, a route partly mirrored by *Resolution*'s and *Discovery*'s return on the last voyage.

There were only three substantial grey areas. It was not clear whether Van Diemen's Land (now Tasmania) was connected to mainland New Holland, a question that would later be answered by George Bass and Matthew Flinders at the end of the eighteenth century. Cook's track up the north-west coast of America – a spectacularly challenging coastline – left long gaps that it would take George Vancouver three surveying seasons to chart accurately between 1791 and 1795 (fig. 62). The Sea of Japan – again a region that formed no part of Cook's orders – was sailed past after his death, but not entered: it would later form one of the primary objectives for the voyages of the French explorer Jean-François de la Pérouse and of William Broughton.

Although the voyages are inextricably entwined with Cook's contemporary – and lasting – reputation as an explorer of genius and humanity (even if his record, in terms of punitive violence, was less than perfect in some circumstances), he did not draw this map on his own. As this chapter has demonstrated, he would not have been able to measure the extent of the Pacific, or survey its coasts and islands, without people like John Harrison and Jesse Ramsden, who were masters in their own field of making scientific instruments, just as Cook was in their practical use. Although Cook seemed to have an uncanny knack of knowing where islands should be, his survey of Polynesia would not have been possible without the help of Tupaia, Hitihiti and other unnamed islanders. Cook was also well served by his officers and crews, many of whom would accompany him on two or even three of his voyages; and a few sailed again with his immediate successors. They were in many ways typical products of Britain's eighteenth-century Navy, which was approaching the peak of its power and global influence, but Cook made them realize what they could achieve, which exceeded the usual demands made on naval officers. Finally, he owed a great debt to Joseph Banks, J.R. and George Forster and his other scientific gentlemen, and to the accompanying artists who generated the rich visual record of the expeditions. They may not always have been ideal companions for voyages of such length, but they undoubtedly taught Cook much and formed a trusted bridge with the scientific communities, and the wider audience, of both Britain and Europe.

A CHART OF THE SOUTHERN HEMISPHERE;

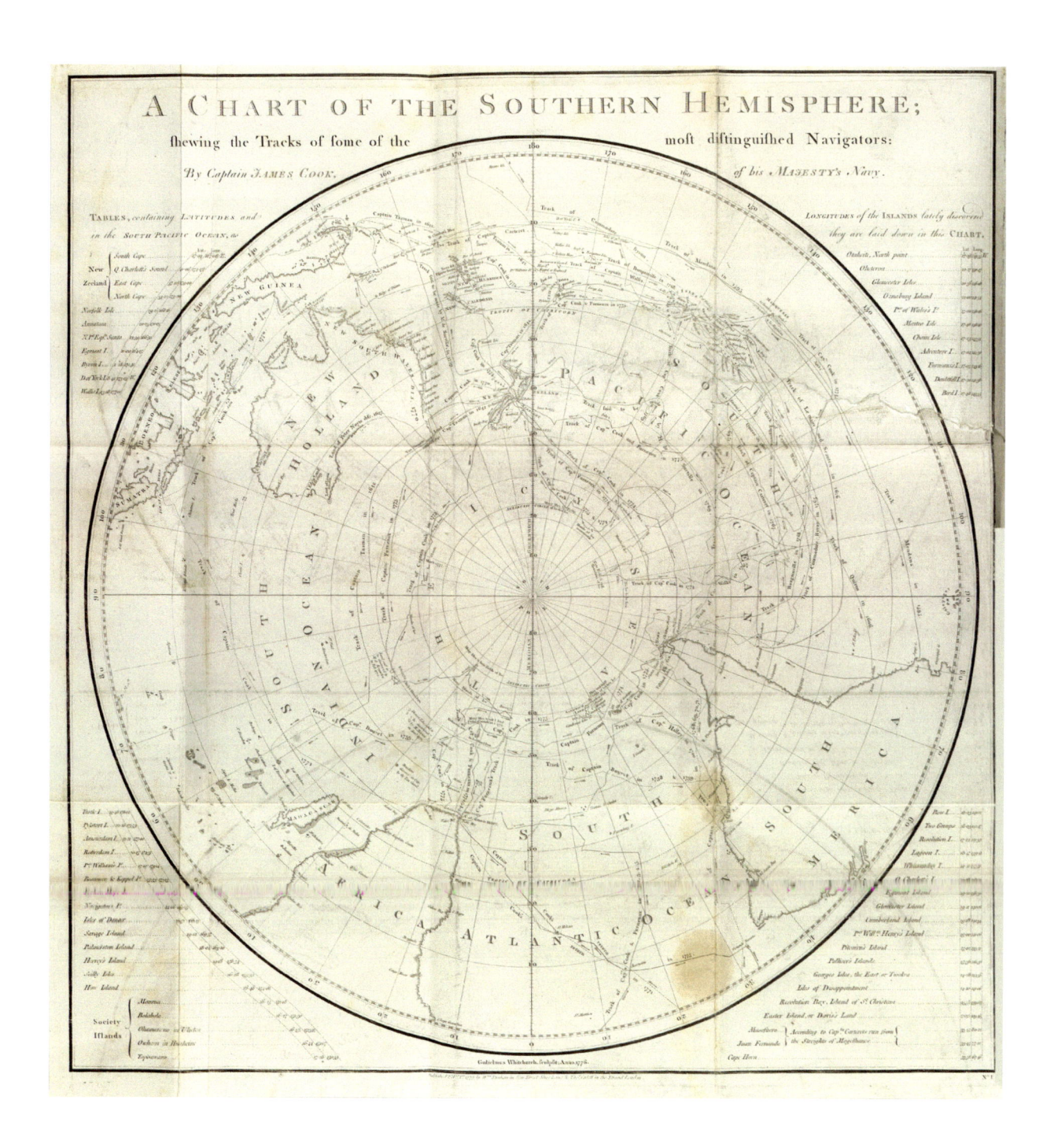

shewing the Tracks of some of the most distinguished Navigators:

By Captain JAMES COOK, of his MAJESTY's Navy.

CAPTAIN COOK AND THE PACIFIC

THE ADVANCEMENT OF SCIENCE AND THE INCREASE OF KNOWLEDGE

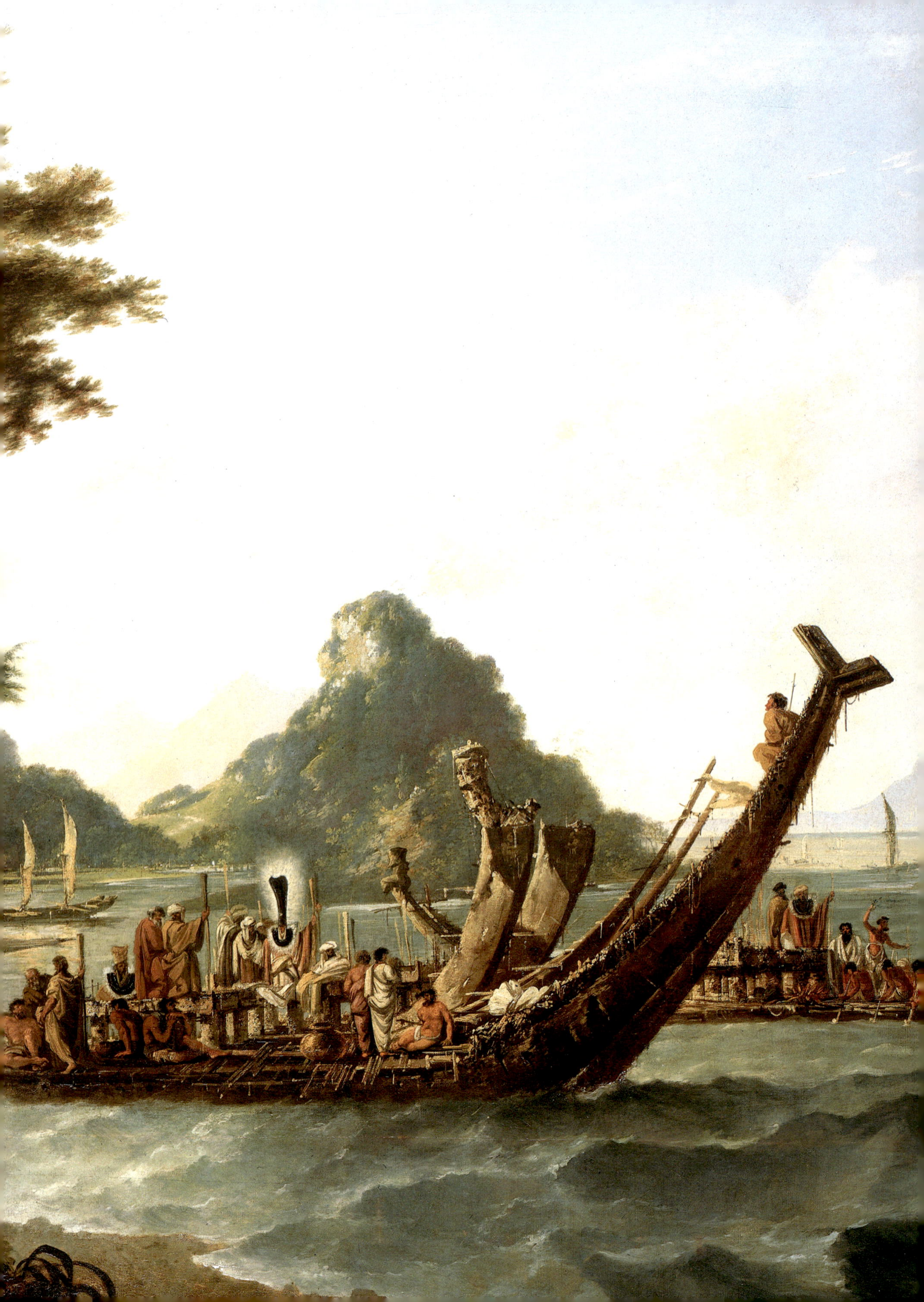

Chapter 4
Cook's Pacific
Exploration and Encounter
John McAleer

The official impetus for Cook's voyages was, to a large extent, the acquisition of scientific knowledge. But one of their most significant results was the new knowledge and information they gathered about the people of the Pacific. The accounts of contemporary travellers and the collections of Western museums are full of incidents and objects that demonstrate the enduring European fascination with this oceanic world, its people and its environment. Or, rather, its peoples and environments, because the Pacific was – and is – a geographically vast and varied place, encompassing many societies and cultures within a huge expanse of ocean. To take one example: the prints derived from Cook's third voyage demonstrate this extraordinary range, as well as the new vistas on the Pacific that the expedition presented to European viewers. Cook had travelled extensively in the South Pacific (the 'South Seas') before, of course, with Tahiti, Tasmania and New Zealand known principally through the work of William Hodges on, and after, the second voyage. The third voyage expanded that oceanic world to include views by John Webber of places such as the little-known Friendly Islands (today's Tonga), 14 prints of coastal Siberia, ten of Alaska, 13 of the Sandwich Islands (Hawai'i), where Cook died, and six from Nootka Sound on Vancouver Island.[1] Simply leafing through the official voyage account brought readers into contact with people and places across an entire oceanic hemisphere and from pole to pole. Through such images – as well as artefacts and even Pacific people themselves – Cook's voyages brought back more information than any voyages before them about an ocean that was little known. We will later explore the ways in which this affected European, and particularly British, society, but this chapter attempts to give a sense of some of the ways in which Cook and his crew encountered, engaged with and presented the Pacific world.

THE PACIFIC WORLD

The Pacific Ocean is so vast and variegated that no single voyage, or even group of voyages (let alone the resultant publications), could possibly hope to capture or experience it in all its forms. Within its bounds are Micronesia, Melanesia and Polynesia, which make up Oceania – European designations that the people of the region have themselves adopted.[2] More languages are spoken in the Pacific (over 1000) than in the Americas or Europe, and nowhere have people made themselves more at home in the sea than in the Pacific Islands.[3] It is worth pausing briefly to consider the scale, complexity and variety of this ocean.

Although earlier European travellers called it the 'South Sea' – *mar del sur* – the Pacific reaches from the Arctic to the Antarctic. It is fringed by long continental shorelines and encompasses some 25,000 islands.[4] Unlike the Europeans who first came into their midst, Pacific navigators regarded the sea as connecting its islands rather than separating them, and the boundaries between sea and shore were as fluid as the ocean itself.[5] In the words of Epeli Hau'ofa, a Tongan writer, this was a 'sea of islands' whose inhabitants comprised an ocean-based community of seafarers.[6] They all relied on the ocean for food and raw materials. They travelled, migrated and fought on its waters, creating their own distinctive maritime spaces – what Damon Salesa calls 'native seas' – that often ranged over millions of square miles.[7] These people saw in the Pacific a vital, life-affirming spiritual force. Little wonder, then, that islanders across its huge, watery expanse considered themselves to be *kakai mei tahi* or 'people from the sea'.[8]

Furthermore, the Pacific has always been characterized by the movement of people. Pre-modern trade circuits can be found across Melanesia, western Polynesia and coastal South-East Asia.[9] Two-thousand-year-old bronze objects from Asia have been found in the Manus archipelago in Papua New Guinea, long inhabited by a highly mobile maritime people.[10] Other archaeological evidence confirms the long history of

Fig. 66
The War-Boats of the Island of Otaheite and the Society Isles, with a View of Part of the Harbour of Ohaneneno, in the Island of Ulietea, one of the Society Islands (detail), William Hodges, 1776
National Maritime Museum (BHC2374)

86 CAPTAIN COOK AND THE PACIFIC

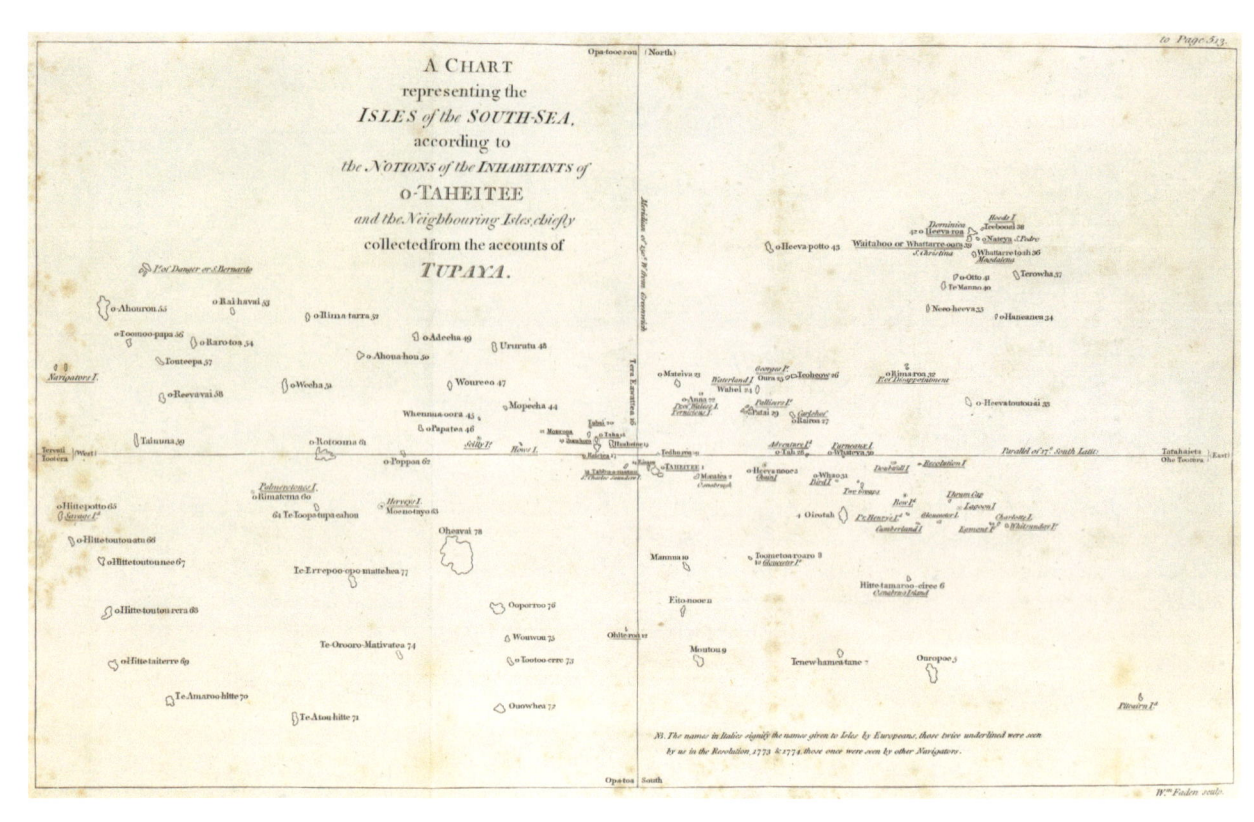

economic and social interactions across vast distances: the discovery of basalt adzes in Samoa and the southern Cook Islands suggests that contact has existed between these places for millennia.[11] Samoan mats and Fijian canoes were in use as far away as Tonga, and Polynesian migrants found their way to South America, where bones of their chickens have been identified.[12] In this sense, the expeditions of Cook and his European contemporaries came rather late and were relatively inconsequential affairs. The greatest navigations of the Pacific were those of the Polynesians, not the Europeans who bumbled their way through the ocean, missing almost all of its islands except Guam until the last third of the eighteenth century. In Cook's approving words, the maritime prowess of the Polynesians made them 'the most extensive nation upon earth'.[13] His admiration was further demonstrated in his estimation of Tupaia, the navigator and *arioi* (priest) originally from the island of Raiatea in the Society Islands, of which Tahiti is the largest. Cook found him to be 'a very intelligent person and to know more of the Geography of the islands situated in these seas, their produce and religious laws and customs of their inhabitants then [sic] any one we had met'.[14] Tupaia travelled with Cook to both New Zealand and Australia, acting as the expedition's interpreter, and the latter's faith in him was borne out in the map that resulted from their encounter and collaboration (fig. 67). The even greater significance of Tupaia as an all-too-brief 'voyage artist' in the Western mode has also finally been recognized. For it is now known that his was the hand, long identified only as 'The Artist of the Chief Mourner', to whom Banks – or perhaps Sydney Parkinson – made available watercolours, paper and probably guidance, from which he produced just a few images, his drawing of the figure of a Tahitian 'chief mourner' in full ceremonial dress being the most striking.

Long before European incursions, advanced local maritime technologies, particularly of vessel design and navigation, made it possible to traverse the world's most expansive ocean and find small and difficult island targets.[15] The major 'canoes' of indigenous design and construction were indeed striking examples of their technology. The *drua* in Fiji, the *kalia* in Tonga and the *'alia* in Samoa were large, asymmetric double-hulled vessels up to 100 feet in length, capable of safely carrying more than 250 people on shorter voyages, and 100 people with several tons of goods, over 1000 miles of ocean. Built by stitching together large planks of wood, without any need for iron, these were hardly canoes at all, but large and powerful sailing vessels. Fleets of upwards of 100 were reported to be capable of transporting thousands of people.[16]

As it became clear that Pacific Islanders had migrated and moved across the ocean for millennia, interest in their maritime culture increased among travelling Europeans. John Webber's fine watercolour *A View in Ulietea* [Raiatea] (fig. 68), exhibited at

Fig. 67 (opposite, above)
'A Chart representing the isles of the South Sea, according to the notions of the inhabitants of O'Taheitee and the neighbouring isles, chiefly collected from the accounts of Tupaya' in Johann Reinhold Forster, *Observations made during a Voyage round the World* (1778)
National Maritime Museum (PBH6670)

Fig. 68 (opposite, below)
A View in Ulietea [Raiatea]
John Webber, 1787
National Maritime Museum. Presented by Captain A.W.F. Fuller through The Art Fund (PAJ2966)

Fig. 69 (above)
A Native of Otaheite, in the Dress of his Country, plate III from Sydney Parkinson's *Journal of a Voyage to the South Seas*
Sydney Parkinson, c.1773
National Maritime Museum (PBC4680)

the Royal Academy in 1787, further demonstrates the developing European interest in the kinds of vessels and systems of navigation that enabled islanders to undertake such extensive voyages in and around the islands. The *pahi*, fitted with outriggers and capable of travelling considerable distances under difficult conditions, incorporated a thatched hut for the protection of people and provisions on long journeys. It was further safeguarded by votive figures carved on the heads of its stem and sternpost.[17] The combination of the practical and the spiritual, captured in the drawing, was a focal point for European fascination with the new maritime world that was opening to their view. To explore these cultural encounters, it is first worth looking at the personal interactions between Pacific Islanders and their European visitors, including how the former were represented by the voyage artists. From there, one can widen the scope to examples of the ways in which the cultural practices, material culture and maritime technology of the Pacific world were represented by European travellers in their accounts and images of the voyages, as well as in the objects they collected and brought back to Europe.

PERSONAL ENCOUNTERS

On one level, European encounters with the Pacific and its peoples were deeply personal: European sailors came into direct and close contact with Pacific Islanders, on an individual basis. Many of these meetings were recorded as simple, matter-of-fact interactions. In some instances, however, they led to wider musings on peoples and to detailed descriptions of their appearance and cultural practices. In turn, these were often closely linked to judgements about the islanders' perceived 'state of civilization'. One such incident was recorded in both words and image by Sydney Parkinson, when a group of 'locals' visited the *Endeavour* as it anchored at Tahiti on the first voyage: 'There were some people of distinction in double canoes; their cloaths [sic], carriage, and behaviour evinced their superiority. I never beheld statelier men, having a pleasant countenance, large black eyes, black hair, and white teeth.'[18] Parkinson's visual record of their appearance conforms to European 'ethnographic conventions' by concentrating on the dress, adornments and gestures of the islanders, to the exclusion of the wider pictorial context or landscape settings (fig. 69).

Another such meeting, which resulted in more extended philosophizing about the nature and condition of the islanders, took place on the second voyage and was recounted by George Forster. On Tuesday 5 October 1773, 'the captain's friend Attahha or Attagha [often called 'Otago'] came on board in one of the first canoes, and breakfasted with us'.[19] Forster not only recorded the friendly interaction but, like Parkinson, gave an insight into the details of Attahha's clothes: 'He was drest in mats, one of which, on account of the coolness of the morning, he has drawn over his shoulders.'[20] But a keen eye for observation, as well as his intellectual and academic predilections, then encouraged him to go beyond the merely empirical and make broader judgements about the ship's visitor: 'He resembled all other uncivilised people in the circumstance that his attention could not be fixed to one object for any space of time, and it was difficult to prevail on him to sit still, whilst Mr Hodges drew his portrait' (fig. 70).[21] In this incident, Forster alluded to the central role played by visual culture in making and taking such records back to Europe. The work of Hodges and his colleagues was one of the principal means by which information about the Pacific and its inhabitants was fixed, and subsequently disseminated (in the prints derived from their work):

> An excellent print, executed by Mr Sherwin, has been made from his drawing, which expresses the countenance of the chief, and the mild character of the whole nation, better than any description ... [It] represents Attahha in the action of thanksgiving, laying a nail on his head, which he had received as a present.[22]

Fig. 70 (opposite))
Otago (native from the Pacific)
John Keyse Sherwin, after
William Hodges, 16 July 1776
National Maritime Museum
(PAI2076)

The figure here was
more properly called Attahha.

Fig. 71 (right)
Landing of Captain Cook at Middleburg, Friendly Islands
After William Hodges, after 1774
National Maritime Museum.
Presented by Captain A.W.F. Fuller
through The Art Fund (BHC1906)

Fig. 72 (below)
Medal commemorating Captain James Cook's second voyage
W. Barnett, 1772
National Maritime Museum
(MEC1385)

Forster described the same Tongan thanksgiving gesture elsewhere in his journal, explaining how the recipient would hold the item momentarily over his head while pronouncing the word *fagafetai*. The image also draws attention to Attahha's missing little finger. In a journal entry for 7 October 1773, Cook remarked: 'Most of the people of these isles wanted either one or both of their little fingers . . . We could not learn the cause of this mutilation with any degree of certainty but judged it to be on account of the death of their parents.'[23] Subsequent research has revealed that this was not a sign of mourning, as Forster assumed, but represented a sacrifice to a god for the recovery of a sick relation who was superior in rank.[24] The potential confusion caused by this misunderstanding underscores the complex cultural encounters experienced by European sailors travelling with Cook.

While Hodges presents a somewhat decontextualized image of a single islander in the portrait of Attahha ('Otago'), his work often went beyond the representation of individuals, to give a richer view of the encounters between Europeans and Pacific societies. His painting of the *Landing of Captain Cook at Middleburg, Friendly Islands* (fig. 71), for example, captures something of the possibilities and uncertainties surrounding the initial contact between people of different cultures. Cook is represented going ashore, apparently led by one of the islanders, who holds a branch aloft to symbolize peaceful intent. Both the European travellers and Pacific Islanders are depicted as inquisitive and welcoming. Just as with the individual portraits, this work is indicative of the complex and multifaceted nature of encounter and exchange during the voyages, although with the caveat that no such painting is a 'photographic record', but represents the incident at a further conceptual remove, in this case through the lens of Hodges's thorough training in European classicism.

COOK'S PACIFIC   89

Of course, locals often had different views of these strange people arriving in their midst. In 1769, for example, a young boy named Horeta Te Taniwha watched Cook's *Endeavour* sail into a harbour in New Zealand: 'We lived at Whitianga and a vessel came there, and when our old men saw the ship they said it was an auta, a god, and the people on board were tupua, strange beings or "goblins".'[25] Elsewhere, Maori people referred to European travellers as 'shallow-rooted shellfish', moving with the tide and lacking a sense of place. Some terms for outsiders used by Pacific Islanders evoked the varying degrees of alarm sparked by the foreigners' alien appearance and sudden arrival. Samoans and Tongans called newcomers *papālagi* or 'sky-bursters', to account for the ships that appeared on the horizon as if they had broken through the sky.[26] When the ships of the third expedition touched on the Siberian coast in August 1778, the British sailors found the initial hostility of the people was due to the fact that they were mistaken for Russians – a reminder of the numerous Pacific–European encounters in the period.[27]

One of the most commonplace and prosaic ways in which Europeans and Pacific Islanders met and learnt about each other was through the bartering that went on between them. Many of the Pacific items surviving in Western museums today were acquired through this process. Some even explicitly refer to it or were created to facilitate it, such as the medal of which the National Maritime Museum holds an example.

Beyond this, gift-giving was an important part of many Pacific cultures and became a central feature of Cook's encounter and engagement with islanders. The barter system and the complex rituals of gift-giving and receiving should not be seen as implying that islanders were unaware of the riches that surrounded them. They frequently drove hard bargains and were both active and tenacious in negotiating the terms of exchange with European interlopers. When Cook tried to acquire basic supplies on the north-west coast of America, for instance, he was exasperated to discover that the locals 'had such high notions of everything the country produced'. They regarded basic natural supplies as 'being their exclusive property . . . the very wood and water we took on board they first wanted us to pay for'.[28]

Barter exchanges demonstrate the possibilities and opportunities of cross-cultural encounters in the Pacific, but many such interactions were undoubtedly imbalanced in terms of power relations. Cook, for example, was prone to taking hostages, such as Poedua (fig. 82), in order to enhance his bargaining position.[29] The sexual exploits of European sailors were notorious, even in their own day, and more extreme forms of power imbalance are recorded in the voyage accounts. For instance, the image of 'Otegoongoon, Son of a New Zealand Chief' by Parkinson, with his face 'curiously tataow'd' (fig. 78), undoubtedly gives us an insight into the cultural practices of Pacific people, but the text accompanying the image records the violence that often attended interactions: 'Otegoowgoow, son to one of their chiefs, was wounded in the thigh' during one of his encounters with Cook's men.[30]

Fig. 73 (left)
Tereoboo, King of Owyhee, bringing Presents to Captain Cook
Benjamin Thomas Pouncy (engraver), after John Webber, c.1788
National Maritime Museum (PAJ1499)

Fig. 74 (opposite)
A young woman of Otaheite, bringing a present
Francesco Bartolozzi (engraver), after John Webber, c.1785
National Maritime Museum (PAF6440)

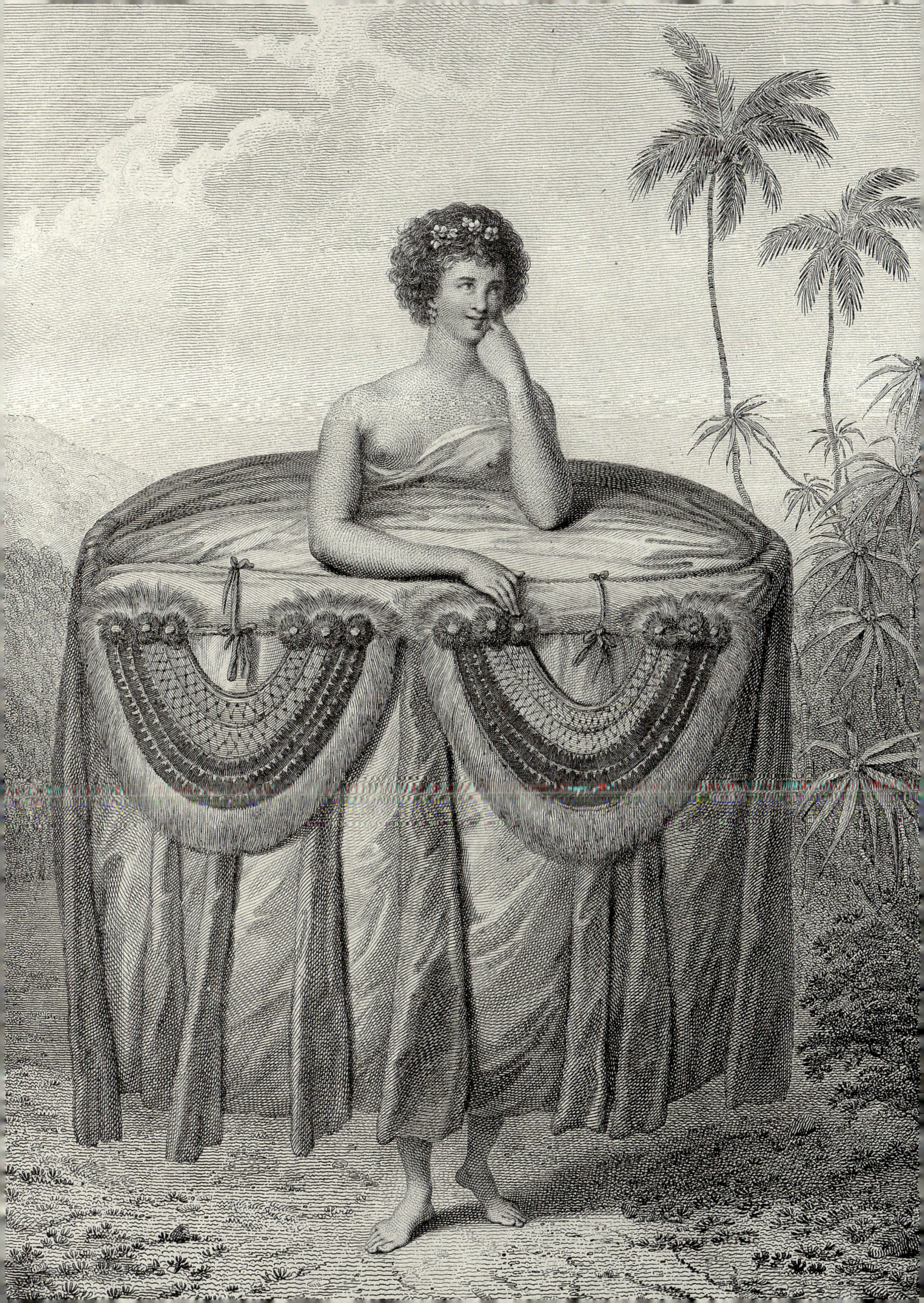

Fig. 75
Tahiti, Bearing South-East
William Hodges, 1775
National Maritime Museum
(BHC1935)

COOK'S PACIFIC

It is also important to recognize that European voyages of exploration produced not only European–Pacific encounters. Among their other effects, Cook's voyages also brought islanders from very different Pacific regions into close contact with each other. For men like Omai or Mahine who joined the *Resolution* during its cruise of 1774, these were also voyages of exploration, as they took the opportunity to observe and engage with many islands, places and peoples well beyond the usual range of their voyaging. Omai visited places as diverse as Rapa Nui, the Marquesas, New Caledonia and New Zealand. Undoubtedly people from those places garnered – through conversation with him, for example – some knowledge of Tahiti. More broadly, social and cultural affinities were discovered or rediscovered, comparisons made and knowledge reshaped. As Nicholas Thomas points out, the kind of knowledge acquired through this process was similar in scope to the comparative understanding of manners and customs acquired by people like Joseph Banks, the Forsters and even Cook himself.[31] In this way, the voyages encouraged and facilitated cultural encounter on a vast scale, often leading on all sides to broader reflections on, and reassessments of, the place of particular Pacific Island communities in the 'grand chain' of civilization.

VISUAL ENCOUNTERS

One of the most important ways in which such encounters were mediated for travellers and the wider British public alike was through the various portraits of Pacific Islanders produced by the voyage artists. If we are to judge from the official account of the second voyage, in which there are 18 full-page prints of them, these portraits were seen as a crucial element in the record.[32] We will consider the voyage artists' representation of land- and seascapes in chapter 5, but their images of people do several significant and related things. As well as providing a sequential record of inter-cultural encounter as it occurred on individual voyages, they give us an insight into the kind of philosophical and cultural assumptions that Europeans invested in them, both in their creation and in the way the European audience subsequently conceived Pacific society through them. In other words, they tell us as much about the European travellers, their hopes and fears for what they expected to find in the Pacific, and their own views about human societies, as they do about the reality of what they discovered there.[33] Here it is important to remember that such images could be portraits of named individuals, but could also be produced according to a kind of 'ethnographic convention' whereby the artist (and engraver) concentrated on the dress, accoutrements, facial expressions and gestures of the people being depicted rather than placing them in any broader spatial context.[34] In this sense, then, these types of images are less portraiture and more ethnographic depiction. In other words, they do not convey a sense of the unique, individual personality, but rather attempt to capture generic information about the physical 'type' being depicted, its characteristic dress, appearance and personal accoutrements.

Sydney Parkinson produced one of the most famous and enduring images of the entire *Endeavour* voyage when he drew the head of a chief from New Zealand, with his 'face curiously tataowd, or mark'd, according to their Manner' (fig. 77).[35] Parkinson's written account is meticulous in its attention to detail, and finds visual expression in the related print that circulated widely in Europe when the voyage returned, providing a powerful and lasting visual archetype for the Maori. One afternoon, when *Endeavour* was becalmed, six Maori canoes 'filled with people' approached it. Some of the Maori in them were 'armed with bludgeons made of wood, and of the bone of a large animal'. European interest in the maritime and martial technology of the Pacific was, as we will see, a strong theme, but in this account Parkinson soon concentrated on the people who approached the vessel:

> *They were a spare thin people, and had garments wrapt about them made of a silky flax, wove in the same manner as the cotton hammocks of Brazil, each corner being ornamented with a piece of dog-skin. Most of them had their hair tied upon the crown of their heads in a knot, and by the knot stuck a comb of wood or bone. In and about their ears some of them had white feathers, with pieces of birds skins, whose feathers were soft as down; but others had the teeth of their parents, or a bit of green stone worked very smooth. These stone ornaments were of various shapes. They also wore a kind of shoulder-knot, made of the skin of the neck of a large sea-fowl, with the feathers on, split in two length-ways. Their faces were tataowed, or marked either all over, or on one side, in a very curious manner; some of them in fine spiral directions like a volute, being indented in the skin very different from the rest.*[36]

The practice of tattooing, detailed so powerfully both in Parkinson's forensic description and in his famous related print, was one that would have a deep impact on European culture in subsequent centuries. In terms of displaying their own encounter with the Pacific, it offered European sailors one of the most distinctive and permanent ways to advertise their travels. The notorious mutiny on the *Bounty* has, among its many other legacies, left us one of the most complete records of Polynesian tattooing among eighteenth-century British seafarers. William

Bligh made a careful record of the appearance of the mutineers who expelled him from his ship in April 1789. Most were tattooed: Fletcher Christian, the master's mate and leader of the mutiny, had marks on his chest and buttocks; midshipman Peter Heywood was heavily tattooed with, among other things, the three-legged crest and Latin motto of the Isle of Man; the able-bodied seaman John Millward had a feathered Tahitian gorget on his chest; and the ship's youngest crew member, 15-year-old Thomas Ellison, bore on his right arm his name and the date he first saw Tahiti, 25 October 1788. It was only at the end of the century that tattooing lost its exotic associations and became a widespread emblem of the ordinary seafarer's trade.[37]

On Cook's second voyage, William Hodges paid equally close attention to human and personal details – unfamiliar clothing or adornment and weapons in particular – when he portrayed a *Family in Dusky Bay, New Zealand* (fig. 76), and something similar occurs in his depictions of a man and woman from Easter Island and New Caledonia, respectively (figs 80, 81). The red-chalk drawings (now in the National Library of Australia, Canberra) on which the prints of a man and woman were based were executed during a nine-day stay in September 1774. In some ways these are generic images, aimed at recording a representative type of islander, yet they are also deeply sympathetic depictions. The image of the woman, with her pierced, elongated earlobe and three parallel tattooed strips running below her lower lip, demonstrates Hodges's skill in capturing and conveying the individuality of these unnamed sitters.[38]

John Webber, the Anglo-Swiss artist on the third voyage, graphically documented the expedition's progress, recording the landscapes, inhabitants, costumes and dwellings encountered.

Fig. 76 (above)
Family in Dusky Bay, New Zealand
Lerperniere (engraver), after William Hodges, 1777
National Maritime Museum (PAH3197)

Figs 77 and 78 (overleaf)
Parkinson's *Journal of a Voyage to the South Seas*
Sydney Parkinson, c.1773
National Maritime Museum (PBC4680)

Plate XVI, *The Head of a Chief of New-Zealand, the face curiously tataowd, or mark'd, according to their Manner*, and plate XXI, *Head of Otegoongoon, Son of a New Zealand Chief, the face curiously tataow'd*

Plate XVI

S. Parkinson del. T. Chambers Sc.

The Head of a Chief of New-Zealand, the face curiously tataow'd, or mark'd, according to their Manner.

Plate XXI.

S. Parkinson del.   T. Chambers Sculp.
Head of Otegoongoon, Son of a New Zealand Chief, the face curiously lataow'd.

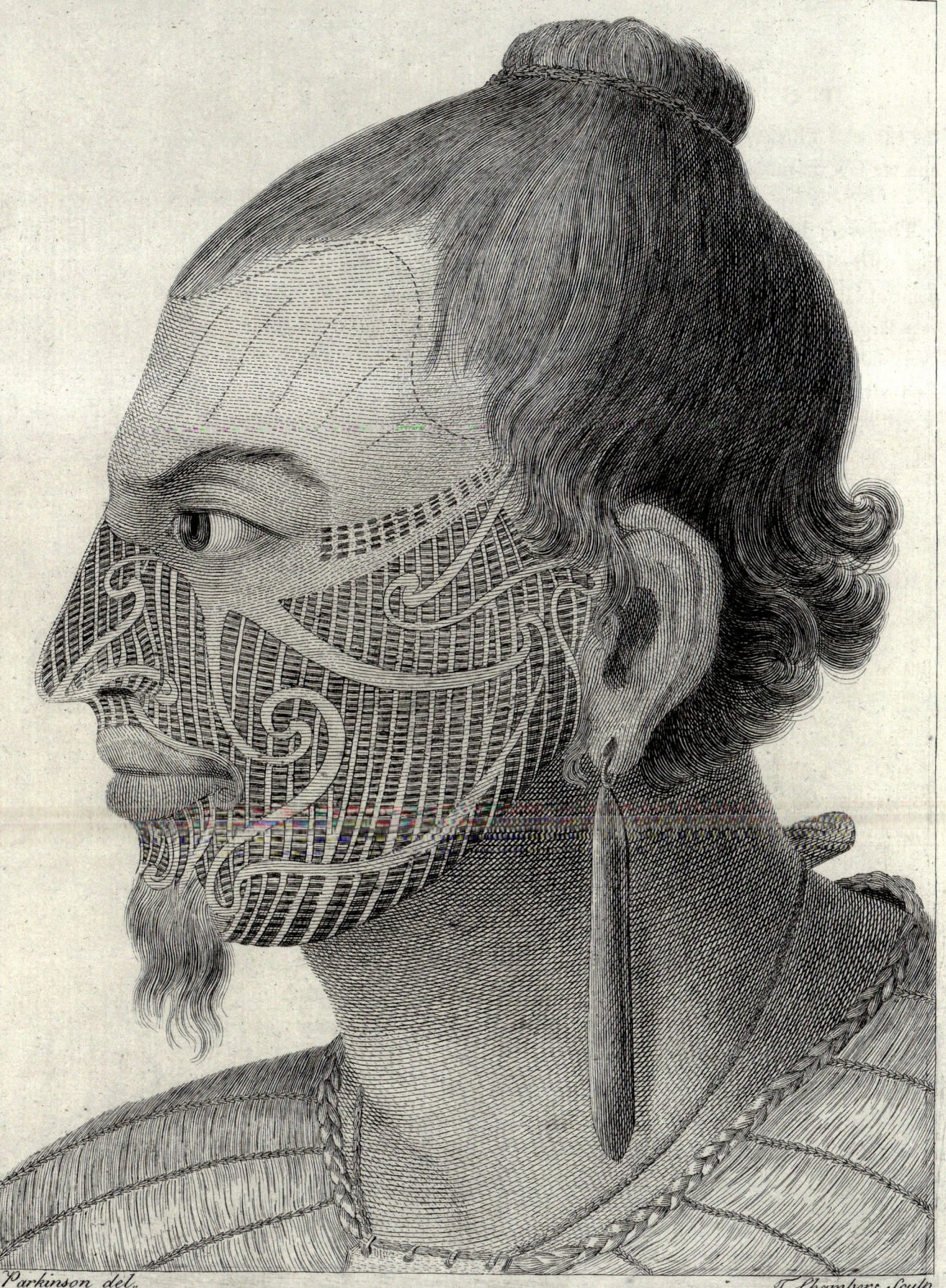

A MAN of the SANDWICH ISLANDS, with his HELMET.

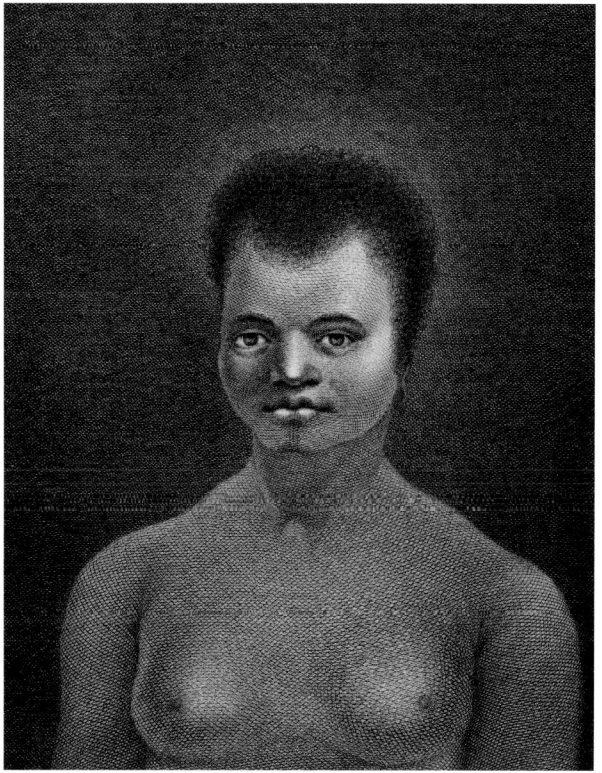

His portrait of Poedua, the 19-year-old daughter of Orio, chief of the Haamanino district of Raiatea (Ulietea) in the Society Islands, is a complex work that incorporates classical references but is also very reflective of the kinds of encounters that took place on the voyages (fig. 82). The 'princess' is shown with her head slightly inclined, looking out of the picture to meet the gaze of the viewer. She wears a white drape of *tapa* cloth beneath her bare breasts and her long black hair cascades over her shoulders. Cape-jasmine blossom is positioned in her hair above her ears. Her right arm falls by her side and she holds a fly-whisk in her right hand; her left arm rests across her hips. Her arms and hands are covered with small tattoos. She is shown against an imaginary tropical background of sky and distant mountains, with a plantain tree positioned on the left.

The portrait was made possible by an incident that occurred after Cook moored his ships at Raiatea on 3 November 1777.

On 24 November, two sailors deserted from the *Discovery*. To bring about their return, Cook enticed on board Orio's son and daughter, Ta-eura and Poedua, and the latter's husband, Moetua, all of whom he planned to hold hostage in Captain Clerke's great cabin until the two deserters were returned. This was a tactic he had used before with relative success, but was also indicative of the tensions and travails that beset the voyages, as well as the increasingly irascible and erratic state of Cook's mind during the third voyage. Nevertheless, the incident gave the voyage artist, John Webber, an opportunity to make a study – now lost – of the young noblewoman, renowned for her beauty and grace as a dancer, and in fact in the early stages of pregnancy at the time.

The portrait of Poedua, one of the earliest images of a Polynesian woman produced by a European painter for a Western audience, reflects the conventional demands of the Royal Academy, where the prime version was exhibited in 1785 –

Fig. 79 (opposite)
A man of the Sandwich Islands with his helmet
John Keyse Sherwin (engraver), after John Webber, c.1784
National Maritime Museum (PAF6446)

Fig. 80 (above, left)
Man of Easter Island
Francesco Bartolozzi (engraver), after William Hodges, 1777
National Maritime Museum (PAI2083)

Fig. 81 (above, right)
Woman of New Caledonia
John Hall (engraver), after William Hodges, 1776
National Maritime Museum (PAI2115)

COOK'S PACIFIC

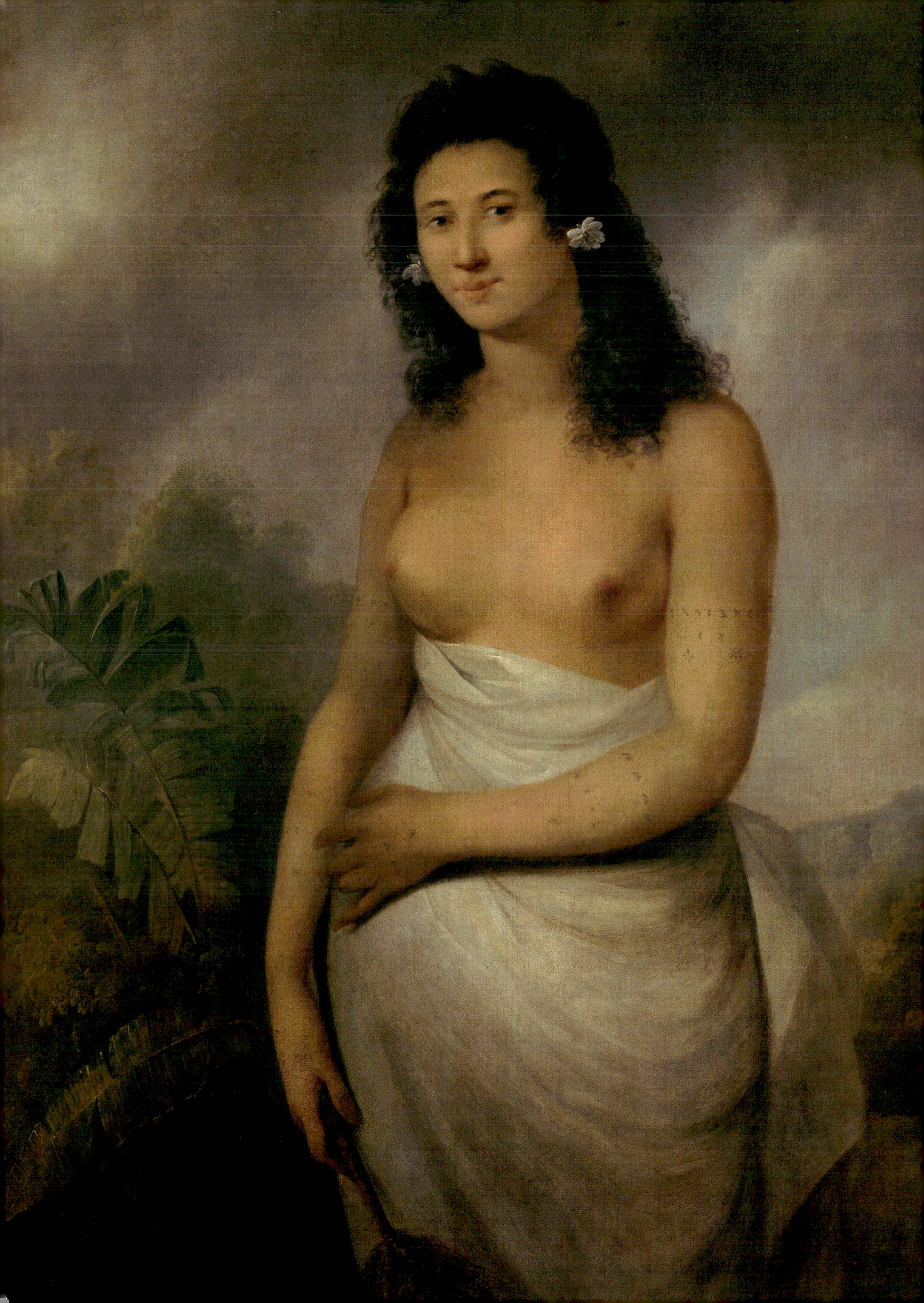

with two other versions known, all painted after the voyage in London. Webber adapted the pose from a classical sculpture, the so-called 'Venus Pudica' (or 'modest Venus'), which would have been well known to viewers in Britain. The enigmatic smile playing on Poedua's face and the lush vegetation and sultry sky contribute to the painting's mood of sensuous eroticism. They also serve to create an idealized version of beauty. In doing this, of course, Webber implies that the Pacific itself is a kind of classical idyll, and his image was undoubtedly influenced by portraits of another Pacific Islander, Omai (see fig. 153), who had returned to Britain in the *Adventure* from Cook's second voyage. Sir Joshua Reynolds exhibited his famous full-length portrait of Omai at the Royal Academy in May 1776, while another by Nathaniel Dance found widespread circulation in print form (fig. 83). Depicted wearing loose robes, the exotic visitor is marked out by the swags of classically inspired drapery as a Rousseau-esque 'noble savage', a concept that developed in the late seventeenth and eighteenth centuries.[39]

PACIFIC PRACTICES

Encountering Pacific people meant encountering their very different habits and ways of life. The recording of new customs and cultural practices was, as we have seen, a key part of the published voyage accounts. In the same way that personal encounters and visual images provide insights into European expectations, prejudices and preoccupations, so the way in which cultural practices were described and depicted reveals the attitudes prevalent during this period of intense interaction and learning. Often the accounts are extensive, even exhaustive, and they frequently cover events and scenes in great detail. For example, Sydney Parkinson described a *morai* on the Society Island of 'Yoolee-Etea' (or 'Ulietea', now Raiatea) (fig. 84). Inside this burial place, the party found 'a whatee, or altar, with a roasted hog, and fish upon it, designed as an offering to the Ethooa, or god'; Parkinson goes on to give a detailed description of the scene before him.[40] The human element was evident again, as he provided an image of the *heiva* or priest who attended the ceremonies in the *morai*, and further, painstakingly detailed information on his appearance (fig. 85). This account described him as:

> cloathed [sic] *in a feather garment, ornamented with round pieces of mother-of-pearl, and a very high cap on his head, made of cane, or bamboo, the front of which is feather-work; the edges beset with quills stripped of the plumage. He has also a sort of breastplate, of a semicircular shape, made of a kind of wicker-work, on which they weave their plaited twine*

Fig. 82 (opposite)
Poedua, the Daughter of Orio
(b. c.1758–d. before 1788)
John Webber, c.1784
National Maritime Museum
(BHC2957)

Fig. 83 (above)
Full-length portrait of a native, possibly Omai
Francesco Bartolozzi (engraver), after Sir Nathaniel Dance, 1774
National Maritime Museum
(PAI2071)

COOK'S PACIFIC   101

in a variety of figures: over this they put feathers of a green pigeon in rows; and between the rows is a semicircular row of Shark's teeth. The edge of the breast-plate is fringed with fine white dog's hair.[41]

A similar interest in the religious customs and beliefs of the people encountered on the second voyage is manifest in some of the prints made after William Hodges's original sketches. In one, a priest-like figure wearing striking regalia is shown with a foreground of luxuriant vegetation and a lagoon (fig. 86). Despite the exotic nature of these ceremonies, artists like Hodges often included more commonplace detail to anchor the depictions in reality for their viewers. In the image of 'Afia-too-ca, a Burying Place in the Isle of Amsterdam' (today's Tongatapu), he incorporates a man carrying a pole laden with bananas and coconuts, in an attempt to normalize or even domesticate the scene (fig. 87).[42]

As well as manifesting a strong interest in the objects and artefacts used by Pacific Islanders, John Webber was also interested in depicting indigenous architecture. He showed the structures of the buildings he encountered, their methods of construction and the materials with which they were made (fig. 88).[43] Even descriptions of the bleakness of the Siberian landscape, which contained 'neither tree nor shrub', still included details of skin tents and clothing (fig. 89).[44] In this image, Webber employed one of the most common strategies

Fig. 84 (above)
A Morai, or Burial Place, in the Island of Yoolee-Etea, plate X from Sydney Parkinson's *Journal of a Voyage to the South Seas*, c.1773
National Maritime Museum (PBC4680)

Fig. 85 (opposite)
An Heiva, or kind of Priest of Yoolee-Etea, & the Neighbouring Islands, plate XI from Sydney Parkinson's *Journal of a Voyage to the South Seas*, c.1773
National Maritime Museum (PBC4680)

102 CAPTAIN COOK AND THE PACIFIC

Plate XI.

S. Parkinson del.   T. Chambers Sculp.

CAPTAIN COOK AND THE PACIFIC

used by artists to reassure European viewers of the accuracy and authenticity of the scene before them: the inclusion of Europeans. These figures serve to guarantee the truthfulness of the image, particularly helpful when depicting an unusual practice or event. Webber's image of a human sacrifice in Tahiti, for example, includes the choric figure of Cook, with identifiable expedition companions, including Webber himself (fig. 90). Their presence in this case serves to verify not only the reliability of representation of such an apparently barbaric act, but also – in their poses – their diplomatically restrained revulsion towards it, as scientific observers acting on behalf of a 'higher' level of civilization.[45]

Details of domestic affairs and rituals were of particular interest. Johann Reinhold Forster, for example, was keen to observe and record as much as he could about the people he came across. This interest in the everyday life and ordinary objects of Pacific Islanders lent itself to the collecting of artefacts of material culture, as well as their depiction by the voyage artists. This trend is evident in some of the items collected by the crews and now in the collection of the NMM. A fly-whisk made from a spear point, with the wooden handle carved into a series of seven notches and coconut fibres bound to the handle, was probably collected from Tonga in 1777 (fig. 91). A breadfruit-pounder with a crossbar handle (fig. 92) was collected on the third voyage and provides evidence for the

Fig. 86 (opposite, above)
A Toupapow with a Corpse on it, Attended by the Chief Mourner in his Habit of Ceremony
William Woollett (engraver),
after William Hodges, 1777
National Maritime Museum
(PAI2073)

Fig. 87 (opposite, below)
Afia-too-ca, a Burying Place in the Isle of Amsterdam
William Byrne (engraver),
after William Hodges, 1777
National Maritime Museum
(PAI2078)

Fig. 88 (above)
The Inside of a Winter Habitation, in Kamtschatka
William Sharp (engraver),
after John Webber, 1779
National Maritime Museum
(PAI4210)

COOK'S PACIFIC 105

preparation of this food: 'They make two or three dishes by beating it with a stone pestil till it makes a paste, mixing water or cocoa nut liquor or both with it and adding ripe plaintains [and] sour paste'.[46] Meanwhile, a wooden feast bowl collected at Nootka Sound, also on the third voyage, is both a work of art and a useful accoutrement (fig. 93). It has a seal's head at one end and its back flipper at the other, and is inlaid, around the rim of the shallow central dish, with shell spots. The bowl probably also fulfilled a practical function, being used to distribute dishes to guests during feast celebrations, and the choice of motif here reflects the importance of the sea in the local diet of a region in which harvesting protein and laying down summer fat against winter cold, and using blubber as a source of oil for other purposes, were all critical. Seals, when they could be killed, were a rich bounty in this way.

Many of the objects brought back from the Pacific were not part of any 'official' collection, but rather the result of individual sailors' specific interests. William Griffin, cooper on the *Resolution* for the third voyage, collected a wooden neck-rest with three legs (fig. 94), as well as a war club and a wooden spear with the point carved into five wooden notches or barbs (fig. 95). Griffin's collection highlights a common interest in the martial aspects of the cultures they encountered, since weapons were by far the most numerous objects among the curiosities and local artefacts acquired by the crew.[47] Sydney Parkinson's account of the first voyage is exhaustive in its detailed description of the kinds of weapons with which Tahitians equipped themselves:

Their weapons are a kind of club, and long wooden lances. They have also bows and arrows. The former are made of a strong elastic wood. The arrows are a small species of reed, or bamboes, pointed with hard wood, or with the sting of the ray-fish, which is a sharp-bearded bone. They also make use of slings, made of the fibres of the bark of some tree, of which, in general, they make their cordage too: some of them, as well as their slings, are neatly plaited. Their hatchets, or rather adzes, which they call towa, are made by tying a hard black stone, of the kind of which they make their paste-beaters, to the end of a wooden handle; and they look very much like a small garden hoe: and the stone part is ground or worn to an edge.[48]

The NMM's collections include a host of paddle-shaped war clubs collected on the second voyage (fig. 100). A hand club, or *kotiate* (fig. 96), was also collected on the second: a loop of rope (now missing), passed through the hole in the straight handle, secured the club to the owner's wrist, and the carving of the grotesque head terminating the handle may have first attracted the attention of European sailors. The war club collected in October 1773 on Cook's second voyage gives an insight into the effects of such exchanges on local societies (fig. 98; see also fig. 99).

Fig. 89 (opposite)
Captain Cook's Meeting with the Chukchi, 1778
Pen, ink and watercolour, by John Webber, c.1778
National Maritime Museum
(PAJ2908)

Cook's brief encounter with the inhabitants, shown dancing for his welcome, of the Kamchatka peninsula of Siberia.

Fig. 90 (above)
A human sacrifice, in a morai, in Otaheite
William Woollett (engraver), after John Webber, c.1784
National Maritime Museum
(PAF6439)

Fig. 91 (left)
Tongan *fue* (fly-whisk) made from reusing an altered spear point
Before 1777
National Maritime Museum
(AAA2833)

COOK'S PACIFIC

Fig. 92 (left)
Tahitian breadfruit-pounder, collected on Captain Cook's third voyage
Before 1777
National Maritime Museum (AAA2832)

Fig. 93 (left, below)
Seal-shaped feast bowl, believed to have been collected on Cook's third voyage at Nootka Sound
Before 1778
National Maritime Museum (AAA2836)

Fig. 94 (below)
Tongan wooden neck-rest with three legs, collected by William Griffin, cooper on the *Resolution*
Before 1777
National Maritime Museum (AAA2851)

Figs 95–100 (opposite; clockwise from left)
Tongan wooden spear and war clubs
Before 1773
National Maritime Museum (AAA2850 [95]; AAA2834 [96]; AAA2830 [97]; AAA2840 [98]; AAA2841 [99]; and AAA2838 [100])

108   CAPTAIN COOK AND THE PACIFIC

COOK'S PACIFIC 109

Its cylindrical handle and four-sided head were made of toa wood and shaped using iron tools. However, the iron tools traded to Tongans during this expedition encouraged the development of their wood-carving: items collected on the third voyage are more elaborately and intricately decorated (fig. 97), indicating the speed and facility with which Pacific Islanders adapted to new tools and techniques.

MARITIME TECHNOLOGY

William Hodges's pen, ink and wash sketch of the *Resolution* in the Marquesas Islands (see fig. 33) is a testament to the interest in the maritime cultures of the Pacific people that was manifest in many of the European travellers who came to the region. A brief four-day respite there allowed Hodges to sketch the coastline and to juxtapose the European vessel with a Marquesan canoe in the background.[49] As with the general interest in Pacific material culture, the focus on ships and their variety symbolized European travellers' awareness of the complexity and sophistication of the societies they were encountering on their voyage (figs 101–4). There is evidence that Polynesians themselves recognized the value of the kinds of visual records that Hodges was able to provide. As the Tahitian fleet prepared to attack Moorea, Tarevatoo, brother of the war leader Tu, suggested to Cook that Hodges draw the scene.[50]

Hodges's *The War-Boats of the Island of Otaheite*, exhibited at the Royal Academy in 1777, is a comment on the status of local indigenous political and military powers (figs 66, 105). The painting suggests that the sea craft of the Pacific Islanders

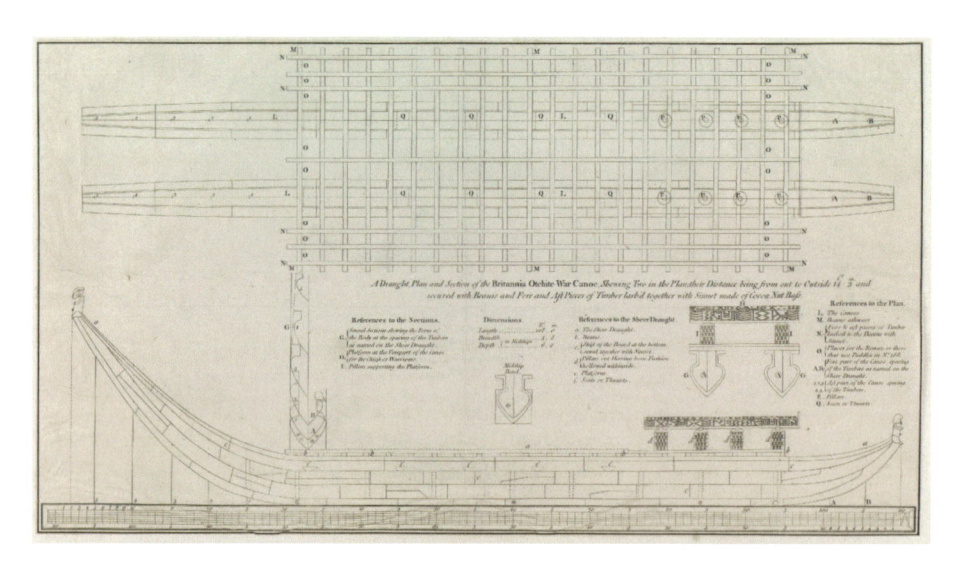

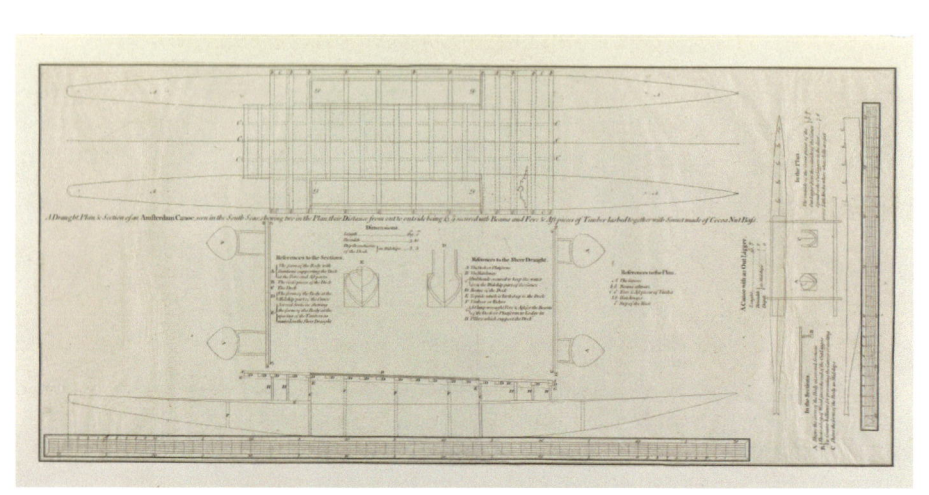

Fig. 101 (left)
A Draught, Plan and Section of the Britannia Otahite War Canoe
W. Palmer (engraver), after William Hodges, 1777
National Maritime Museum (PAI2095)

Fig. 102 (left, below)
A Draught Plan & Section of an Amsterdam Canoe, seen in the South Seas
William Hodges, 1777
National Maritime Museum (PAI2079)

Fig. 103 (opposite, above)
A war canoe of New Zealand
Sydney Parkinson, 1770
National Maritime Museum (PAJ2165)

Fig. 104 (opposite, below)
Boats of the Friendly Isles
William Watts (engraver), after William Hodges, 1777
National Maritime Museum (PAI2102)

COOK'S PACIFIC 111

112 CAPTAIN COOK AND THE PACIFIC

can be used to compare societies and their relative technological development. Specifically, in relation to the scene depicted in this image, Cook calculated that there were 160 large double canoes and 170 smaller double canoes, carrying 7760 men in total. George Forster candidly noted the effect of this display of maritime prowess on the visiting European witnesses. 'All our former ideas of power and affluence of this island were so greatly surpassed by this magnificent scene, that we were perfectly left in admiration.'[51]

Tahiti, Bearing South-East offers a similar meditation on the state of maritime technology and the society that produced it (fig. 75). The painting may have been envisaged as a printmaker's 'model' for the published voyage account, and was probably completed from the large coastal profile drawing (now in the British Library) made on the spot during a second visit to Tahiti between 22 April and 14 May 1774. It shows a view of Matavai Bay and the island from the north-west, with Mount Orofena in the distance, together with Point Venus and One-Tree Hill.[52] The scene is diffused with the light from the rising sun on the left of the painting. Various Tahitian boats can be seen in the foreground: a small outrigger sailing canoe on the far left, the coastal craft in the centre with two figures on board, and the war canoe on the far right with its dominant stern. Hodges was known to 'make drawings of every thing curious', which was encouraged by his own and Cook's orders.[53] The related painting attempts to compare the boat-types, with the island used as an appropriate backdrop. Specifics of construction, decoration and sailing method, as well as types of local dress, can all be gleaned from his image. It offers separate studies of the individual craft, as well as depicting the atmospheric effects over the island on the same canvas.

MATERIAL CULTURE

In addition to accounts and images, people in Europe gained critical insights into Pacific Island culture through the collecting and display of objects. As we will see, examples of tangible, three-dimensional artefacts were highly sought-after in Europe and caused quite a stir when they were exhibited. They offered evidence of the exotic, as well as of expanding boundaries, both politically and intellectually. In Britain, objects and material culture also played a crucial role in educating, and illustrating encounters for, the public. Many of these items were reproduced in prints to accompany the voyage accounts. There was often a desire to illustrate individual Polynesian artefacts that lacked European equivalents, such as fly-whisks, stone adzes, taro-mashers and *tapa* (cloth)-beaters. These were usually presented without a local visual context, in the tradition of European natural history illustration.[54] They offered an ethnographic, almost scientific manual of items as diverse as spears, bows and arrows, head-dresses and necklaces and specimens of cloth (figs 107–09).

A room of such 'curiosities' was described in John Feltham's *Picture of London for 1806*. This proto-guidebook to the capital focused on the variety of objects that could be seen there, and his account of the Sandwich Room in the British Museum – whose name commemorated the patron of Cook's voyages – is indicative. Visitors would find themselves in a space:

> *ornamented on the sides with flaxen mantles from Nootka, or King George's Sound, and New Zealand, made by the people to whom the use of a loom is totally unknown; above which are the war-clubs, adzes, and paddles, of New Caledonia, Otaheite and the Friendly Islands . . . Here also are several beautiful specimens of matting from the Sandwich Islands, which, in strength, firmness, and beauty, excel the similar productions of the world; daggers, in shape like to that which afterwards put a period to Captain Cook's existence; cava bowls; feathered and other necklaces; cordage; adzes; chissels* [sic]*; hand-weapons; fishing-hooks and lines, helmets, with wicker linings; feathered cloaks; drums; models of canoes; idols; and innumerable other rarities.*[55]

The very fact of the presence in London of such specimens, which 'excel the similar productions of the world', and their public display in so prestigious a context, was highlighted as crucial in conveying a sense of the people and places represented, and by implication the wider importance of their 'discovery' over simply their novelty. For Feltham, as for many others, the material culture gave 'a clearer conception of the people who make and use them, than can ever be obtained from descriptions'.[56] Before returning to consider the ways in which such objects and displays influenced European perceptions of the Pacific, we will turn to the sea- and landscapes of the region and explore the ways in which European artists represented these for audiences at home.

Fig. 105 (opposite, above)
The War-Boats of the Island of Otaheite and the Society Isles, with a View of Part of the Harbour of Ohaneneno, in the Island of Ulietea, one of the Society Islands
William Hodges, 1776
National Maritime Museum (BHC2374)

Fig. 106 (opposite, below)
View of Murray's Islands with the natives offering to barter, October 1802
William Westall, *c.*1805
National Maritime Museum (ZBA7942)

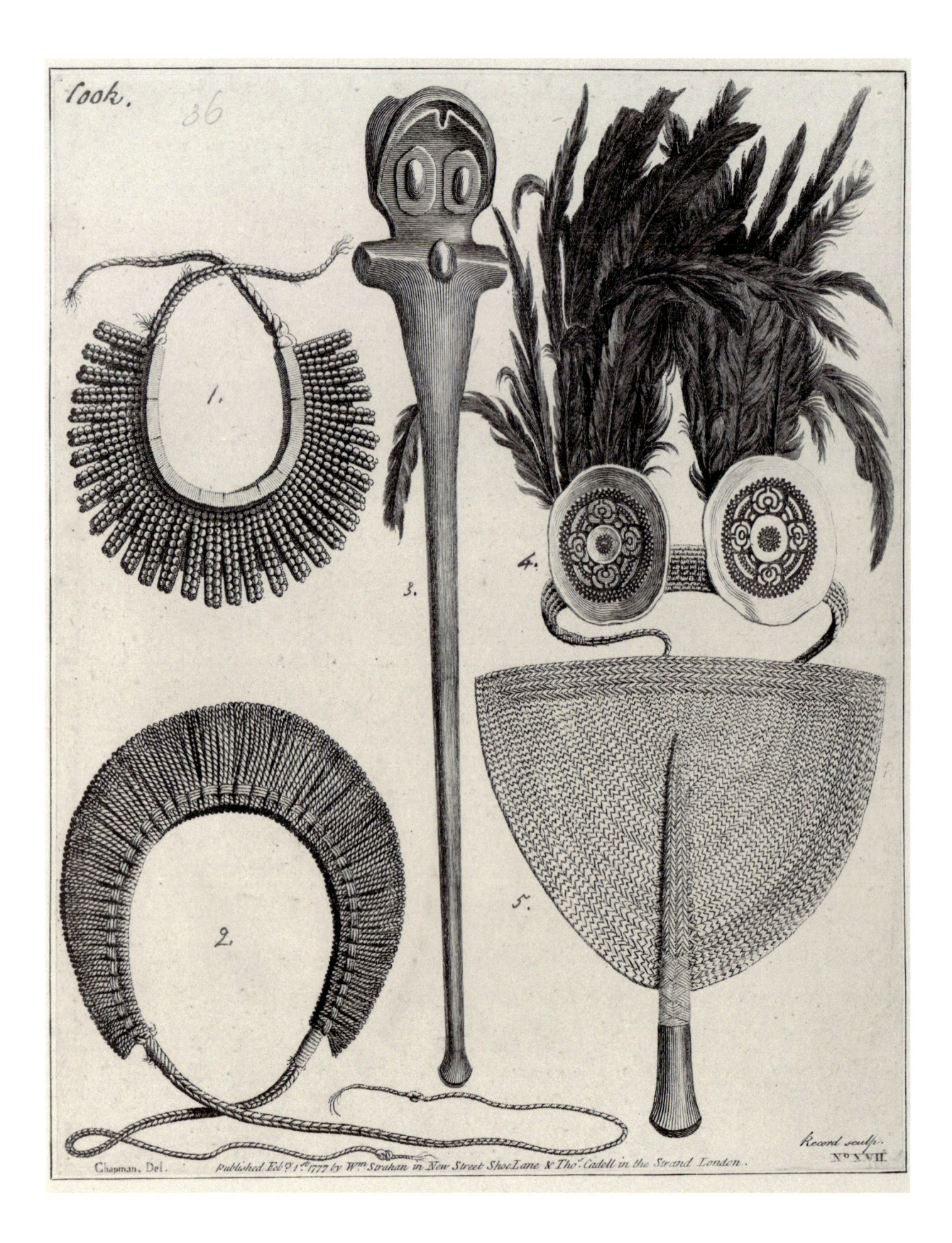

114 CAPTAIN COOK AND THE PACIFIC

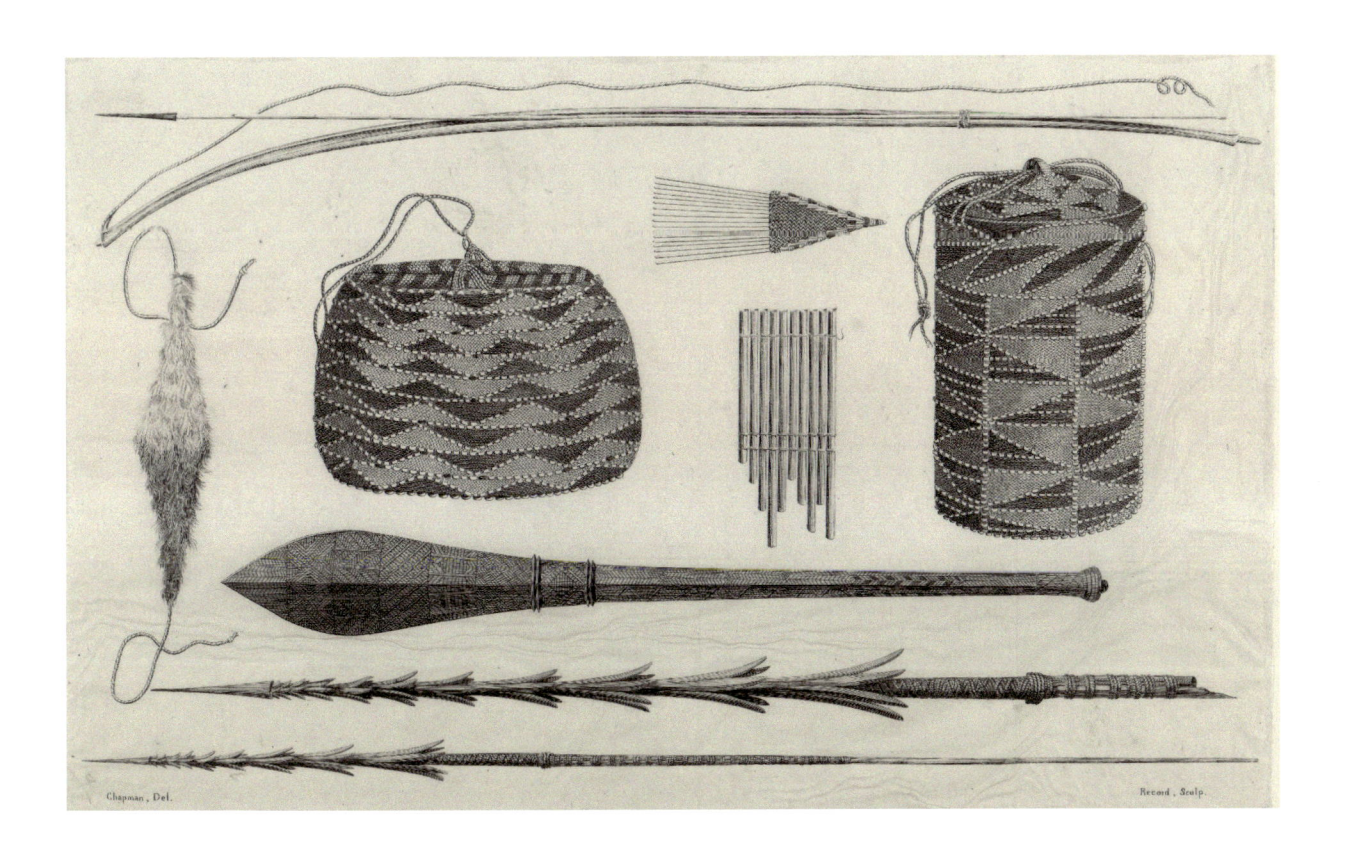

Fig. 107 (opposite)
*Native head dress,
necklace and other items*
John Chapman,
after William Hodges, 1777
National Maritime Museum
(PAI2092)

Fig. 108 (above)
*Various articles belonging to
South Pacific natives, including
spears and bow and arrow*
John Chapman, after
William Hodges, 1777
National Maritime Museum
(PAI2080)

Fig. 109 (right)
Specimen of tapa *or bark cloth
from Polynesia*
1834
National Maritime Museum
(ZBA5494)

COOK'S PACIFIC 115

Chapter 5
Visualizing the Pacific
Art, Landscape and Exploration

John McAleer

One of the principal selling points of Cook's voyages, partly in conceiving them but much more fully realized in what they delivered, was that they 'visualized' the Pacific. The map-making impulse discussed in chapter 3 was not the only way they did so, for the work of the artists who accompanied them is perhaps one of their most enduring legacies. John Hawkesworth assured readers that the account of the *Endeavour* expedition was 'illustrated and adorned with a great number of cuts, from which every class of readers, whether their object is knowledge or pleasure, will find equal advantage, as they consist not only of maps and charts, drawn with great skill and attention, but of views and Figures, designed and executed by the best artists in this country' (by which he primarily meant those who converted the voyage sketches to prints).[1] These processes of seeing, recording and preserving for posterity the landscapes, seascapes and people of the Pacific in pictorial form are, of course, closely related to the scientific impulses underlying the expeditions. But just as the voyages were undertaken as the result of a variety of factors, so their visual results have bequeathed a rich and complex legacy of acute observation and artistic innovation, and these images have exercised a lasting influence on European ideas of the region.

When John MacPherson introduced William Hodges to Warren Hastings, the British Governor General in India, MacPherson described the artist's mission as being 'to bring . . . home upon paper . . . the most curious appearances of nature and art'.[2] This motivation could be applied equally well to Hodges's time on Cook's second voyage to the South Pacific (figs 111–13). Hodges was not unique: John Webber, a 'draughtsman and landskip painter', was engaged on the third voyage 'for the express purpose of supplying the unavoidable imperfections of written accounts, by enabling us to preserve, and to bring home, such drawings of the most memorable scenes of our transaction, as could be executed by a professed and skilled artist'.[3] In this respect, Cook's voyages were not new. William Dampier's voyage to New Holland at the turn of the eighteenth century included 'in the ship . . . a person skilled in drawing'.[4] In the preface to the publication that resulted from this voyage,

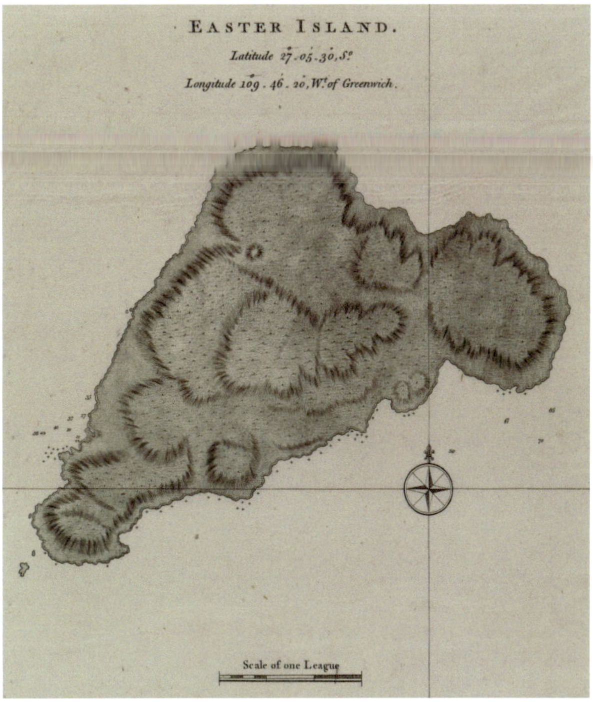

Fig. 110 (opposite)
A Cascade in the Tuauru Valley, Tahiti (detail)
William Hodges, *c*.1775
National Maritime Museum
(BHC2373)

Fig. 111 (right)
Easter Island
Unknown engraver,
after William Hodges,
1797
National Maritime Museum
(PAI2082)

117

Fig. 112 (above)
View in the Island of Tanna
[Tana]
William Woollett (engraver),
after William Hodges, 1777
National Maritime Museum (PAJ1509)

Fig. 113 (left)
View in the Island of New Caledonia
William Byrne (engraver),
after William Hodges, 1777
National Maritime Museum (PAJ1510)

Fig. 114 (opposite)
View of Port Lincoln, South Coast of Australia
(study for illustration of Flinders's voyage, 1801–03)
William Westall
National Maritime Museum. Presented by Captain A.W.F. Fuller through The Art Fund (PAG9778)

A study for an illustration of Port Lincoln, surveyed by Matthew Flinders in February 1801, during his circumnavigation of Australia.

118   CAPTAIN COOK AND THE PACIFIC

Dampier outlined the value of the images produced by his artistic travelling companion:

> *I have by this means been enabled, for the greater satisfaction of the curious reader, to present him with exact cuts and Figures of several of the principal and most remarkable of those birds, beasts, fishes and plants which are described in the following narrative.*[5]

George Anson's highly successful account of his anti-Spanish circumnavigation of 1740–44 in the *Centurion* was similarly interleaved with a plethora of prints depicting 'views of land, soundings, draughts of roads and ports, charts, and other materials for the improvement of geography and navigation', though these derived from the surveys and sketches of naval personnel, not from 'artists'.[6] In total, Anson's *Voyage Round the World* (1748) comprised 42 prints and gave a new visibility to the Pacific that would only be surpassed several decades later by the fruits of Cook's voyages. What is more, publishers recognized the independent attraction of illustrations of the scenes witnessed by these intrepid travellers – separate sets could be purchased for seven shillings.[7] For Anson and many of his contemporaries, the value of the images revolved around 'the great accuracy they were drawn with' and the guarantee of authenticity and truthfulness thus bestowed:

> *For they were not copied from the work of others, or composed at home from imperfect accounts, given by incurious and unskilled observers, as hath been frequently the case in these matters; but the greatest part of them were drawn on the spot with the utmost exactness.*[8]

The artists who accompanied Cook took these pictorial achievements to new heights, combining accuracy and truthfulness with a vibrant and dynamic aesthetic response to the Pacific.

While these artists played an important role in recording Pacific people and societies, as already discussed, this chapter focuses on their engagement with the landscapes they encountered, the newness and novelty with which they managed to

VISUALIZING THE PACIFIC   119

imbue them and the artistic and aesthetic trends that influenced how they were presented. There was often a fruitful dialogue with the scientific experts on board. John Webber collaborated with Cook and the surgeon's mate (and *de facto* naturalist) William Anderson: they consulted his drawings to remind themselves of what they had seen and he read their accounts to remind himself of details for his finished work.[9] There were difficulties, too: artists struggled with cramped conditions, poor lighting and lack of materials. These spartan constraints were only magnified in relation to producing fine art, but, remarkably, a group of dedicated and talented individuals succeeded in recording Cook's voyages in this way for posterity. They often made hastily conceived sketches, jotting down the bare minimum with rapid pencil-and-wash impressions – lines of the landscape, the tones and effects of light and shade – and these formed the basis for more elaborate compositions, some worked up on the voyage and others later in London for publication.

Applying what John Crowley has called a 'topographic imperative' to their work, these artists served to underscore in their images the variety of landscapes in the Pacific – not some barren or watery wasteland, but a rich and varied fretwork of islands, beaches and seascapes (fig. 114).[10] But just as charting, mapping or describing a place helps to translate and interpret it, so too do the pictorial representations. The images produced by artists like Sydney Parkinson, William Hodges and John Webber are not neutral, but reflect the concerns and preoccupations of the people making the record and, often, of the people for whom it was made.

THE FIRST VOYAGE

If the scientific focus of the voyage was partially a result of Sir Joseph Banks's presence on board the *Endeavour* and his interest in botany, so too was its artistic legacy. In addition to 20 tons of luggage and scientific equipment, Banks brought a retinue of artists and draughtsmen to record the sights and scenes they encountered.[11] The presence of Alexander Buchan and Sydney Parkinson suggests that, from the beginning, Banks had in mind an illustrated publication of the scientific results of the expedition.[12] Both artists died during the voyage but, before his death at Batavia, Parkinson produced more than 800 drawings of plants and animals (now in the collection of the Natural History Museum, London). These images became instrumental and influential in conveying the achievements of the voyage to audiences in Britain.

Sydney Parkinson was born in Edinburgh around 1745 and showed an early talent for drawing natural history subjects. After moving to London, his flower paintings were exhibited by the Free Society of Artists in 1765–66. A fellow Scottish Quaker,

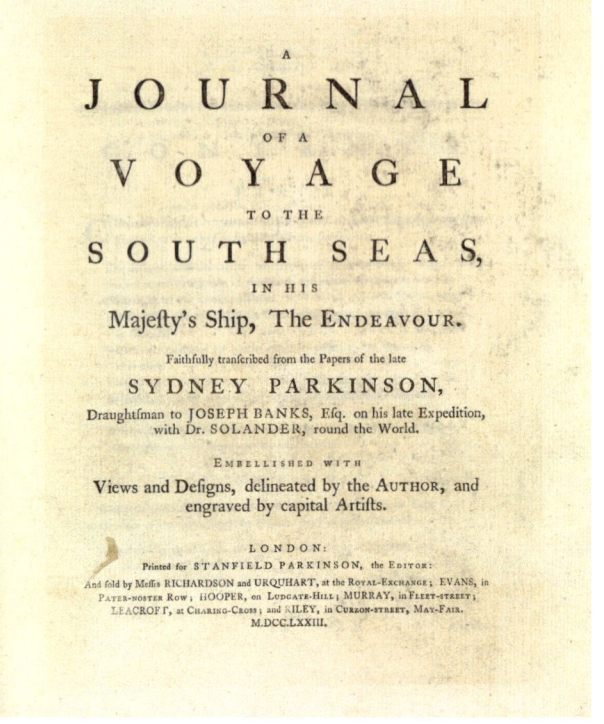

James Lee, helped to bring Parkinson into contact with Sir Joseph Banks, who commissioned the young artist to depict some of the creatures collected on his 1766 expedition to Labrador and Newfoundland. Evidently Parkinson's work impressed, as he was recruited to Banks's entourage for the *Endeavour* voyage in July 1768. Initially tasked with providing just the botanical illustrations, he was eventually given – by the untimely death of the landscape draughtsman Alexander Buchan, on arrival at Tahiti – the responsibility for landscape, too. Displaying great diligence and flair for his wider brief, Parkinson had made at least 1300 drawings, many more than Banks had expected, by the time the ship called at Batavia for repairs on the voyage home. Among Parkinson's corpus of sketches and drawings were the first in Western history of Australian landscapes and of their inhabitants at first hand, as well as vocabularies that he compiled of the languages spoken in Tahiti and New Holland.

Parkinson's reputation is less exalted than that of William Hodges or John Webber, perhaps because he did not survive

Figs 115, 116 (opposite)
Title page and frontispiece of Sydney Parkinson's *Journal of a Voyage to the South Seas*
After Sydney Parkinson, c.1773
National Maritime Museum
(PBC4680)

Fig. 117 (above)
Venus Fort, Erected by the Endeavour's People to secure themselves during the Observation of the Transit of Venus, at Otaheite, plate IV from Sydney Parkinson's *Journal of a Voyage to the South Seas*
After Sydney Parkinson, c.1773
National Maritime Museum
(PBC4680)

VISUALIZING THE PACIFIC   121

Fig. 118
View of the North Side of the Entrance into Poverty Bay, & Morai Island, in New Zealand and *View of another Side of the Entrance into the said Bay*, plate XIV from Sydney Parkinson's *Journal of a Voyage to the South Seas*
After Sydney Parkinson, c.1773
National Maritime Museum (PBC4680)

the expedition, but also because he was, quite strictly, a 'flora and fauna' man, trained neither as a portraitist nor as a landscape artist. Unfortunately, his artistic legacy was also somewhat overshadowed by the protracted and bitter legal dispute between Banks and Parkinson's brother, Stanfield, over the rights to the dead artist's effects, including the landscape drawings he had done for himself rather than Banks, and by the fact that his personal illustrated journal, much admired on board, simply disappeared with his death and was not seen again, then or since. Stanfield eventually succeeded in publishing *Journal of a Voyage to the South Seas* (1773) based on his brother's other notes and sketches (figs 115, 116). This account, partial though it is, offers an insight into Parkinson's work and the aesthetic and artistic perspectives he brought to the voyage.

In some ways, the landscapes in *Journal of a Voyage* are typical of European views of the extra-European world at the time. For example, the image of the 'Venus Fort' (fig. 117) shows great similarities to coastal profiles being produced of forts in West Africa. Concerned with defence, security and enclosure, there is little sense of the wider Pacific world lying beyond the fort or, in the case of this image, beyond the frame of the print. The accompanying description confirms this interpretation, noting how the 'temporary fort for our accommodation on shore' to observe the transit of Venus had 'a fosse, with palisadoes [sic], next the river; guns and swivels mounted on the ramparts; and within, we had an observatory, an oven, forge, and pens for our sheep . . . Centinels were also appointed as usual in garrisons, and military discipline observed'.[13]

Parkinson's coastal profiles also conformed to the needs of maritime exploration by providing visual information about the topography and morphology of the land. In two views of

122 CAPTAIN COOK AND THE PACIFIC

Fig. 119 (left)
View of a curious Arched Rock, having a River running under it, in Tolago Bay, on the East Coast of New Zealand, plate XX from Sydney Parkinson's *Journal of a Voyage to the South Seas*
After Sydney Parkinson, *c*.1773
National Maritime Museum (PBC4680)

Fig. 120 (below)
View of the great Peak, & the adjacent Country, on the West Coast of New Zealand, plate XXII from Sydney Parkinson's *Journal of a Voyage to the South Seas*
After Sydney Parkinson, *c*.1773
National Maritime Museum (PBC4680)

Poverty Bay in New Zealand, for example (fig. 118), his treatment conformed to Anson's advice that 'exact views of land are the surest guide to a seaman, on a coast where he has never been before'.[14] Anson wanted the Admiralty to ensure that every ship carried a person capable of 'drawing such coasts, and planning such harbours, as the ship should touch at, and in making such other observations of all kinds', a diktat that he was in the perfect position to enforce when he joined the Admiralty Board and later became First Lord in 1751.[15] Parkinson's proficiency as an artist only enhanced the record value of these profiles and, unlike previous voyage accounts and their accompanying images, his work could claim authority by virtue of his professional artistic presence – albeit not primarily as a view-painter – in front of the scenes being presented to the European viewer.

Parkinson's engagement with the landscapes of the Pacific went beyond the workaday exigencies of a naval draughtsman.

VISUALIZING THE PACIFIC   123

His images show signs of a developing interest in solving the problem faced by all European artists there: that of dealing with the burst of tropical colour, light and atmosphere that greeted them in the South Seas. However, unlike problems such as a lack of materials or unsuitable working conditions, solutions were at hand. These involved working with the kinds of aesthetic tropes and ideals of beauty being developed in Europe at the time. Parkinson's interest in the Picturesque, for example, can be identified in the way he rendered dramatic features like caves, cliffs and waterfalls.[16] He was particularly impressed by the 'wild and romantic' scenery in New Zealand, which he recorded in words and accompanying images.[17] His *View of a curious Arched Rock, having a River running under it, in Tolago Bay, on the East Coast of New Zealand* (fig. 119) represented 'a very uncommon view, peculiarly striking to a curious spectator'.[18] In another image he remarked on a *hippa*, or place of retreat, which was 'situated on a very high rock, hollow underneath, forming a most grand natural arch, one side of which was connected with the land; the other rose out of the sea. Underneath this arch a small vessel might have sailed.'[19] And in another scene, 'rocks and mountains, whose tops were covered with snow, rose in view one above another from the water's edge: and those near the shore were cloathed with wood, as well as some of the valleys between the hills, whose summits reached the clouds' (fig. 120).[20] This sensitivity to European artistic conventions took Parkinson's work beyond the mere coastal profile and placed it at the heart of a developing, if still immature, approach to painting and visual representation in the tropics.

THE SECOND VOYAGE

Banks had planned to join Cook's second voyage together with three topographical artists and the respected society portrait painter Johan Zoffany. However, following a row with the Admiralty over the fitting out of the *Resolution*, Banks withdrew, taking his entire entourage with him.[21] Fortunately for the voyage's artistic legacy, the void was filled by William Hodges. Born in 1744, Hodges was trained in the classical landscape style through his apprenticeship with Richard Wilson. The intervention of Lord Palmerston obtained him the post of draughtsman on the Pacific expedition. From 1772 to 1775 Hodges made sketches, drawings and some vivid oil studies of the islands and indigenous people there and on his return; for some years thereafter, he was gainfully employed, turning what he had gathered into more permanent records. The Admiralty employed him for a further two years after the voyage to make large finished oil paintings, several of which he exhibited at the Royal Academy, including *The War-Boats of the Island of Otaheite* (see figs 66, 105) in 1777. He also supervised the engraving (by William Woollett and others) of plates from his work as illustrations for the official account of the expedition, published in 1777. Inspired by this encounter with the wider world, Hodges travelled to India in 1779, where he came under the patronage of Warren Hastings. He remained in India for some six years, recording scenes of interest and architectural landmarks, from which he also derived a major publication and an even more successful exhibition at the Academy than he had had for his Pacific work.

On the second Cook voyage, the artist was part of the official travelling party. As a result, Hodges had his own instructions: to make 'drawings and paintings of such places in the countries you may touch at in the course of the said voyage as may be proper to give a more perfect idea thereof than can be formed from written descriptions only'.[22] His representations of the Pacific are vivid records of British exploration and present a

Fig. 121 (opposite)
View of part of Owharre [Fare] *Harbour, Island of Huahine*
William Hodges, 1774
National Maritime Museum
(BHC2418)

Fig. 122 (below)
View of Resolution [Vaitahu] *Bay in the Marquesas*
William Hodges, 1774
National Maritime Museum
(BHC2419)

Fig. 123 (left)
A View of the Cape of Good Hope, Taken on the Spot, from on Board the Resolution, *Capt Cooke*
William Hodges, 1772
National Maritime Museum,
Ministry of Defence Art Collection
(BHC1778)

Fig. 124 (opposite)
View of the Province of Oparree [Pare], *Island of Otaheite, with part of the Island of Eimeo* [Moorea]
William Hodges, *c*.1776
National Maritime Museum
(BHC1936)

variety of different insights into its engagement with the people and places of this ocean, influenced by the scientific impulses of his travelling companions, but also by the aesthetic predilections of the time, both in their production and the way in which these encounters were understood by people in Britain.

Hodges grasped the opportunity afforded him, his work tracing and fixing the progress of Cook's particularly peripatetic second voyage as it snaked its way around and across the vast expanses of the Pacific Ocean. Most of the finished oil paintings became the property of the Admiralty and are now among the treasures held by the National Maritime Museum.[23] Hodges's images are so wonderfully various that it is difficult to categorize them under any one heading. Indeed, they fulfil a number of purposes – often simultaneously in the same image. However, they might be considered under three rubrics: scientific records, aesthetic views and encounters.

Cook's main purpose on the second expedition was to solve the conundrum of the so-called 'Southern Continent', about which geographers and philosophers had argued for centuries. Belief in its existence was predicated on the supposed need for a large land mass in the southern hemisphere to 'balance' the land in the northern hemisphere. In addition to answering questions like this, Cook's mission was to gain further knowledge of the central Pacific Islands. In this, Hodges's records were important navigational aids. One of Hodges's tasks on the ship was the routine training of officers to make coastal profiles, of which he drew many. The impact of this work on his own painting is evident in his small oil studies of the islands and coastlines encountered by the expedition, but they also preserve something of Hodges's artistic character (figs 121, 122). Most seem to have been painted on the spot and are strikingly unconventional departures from the artistic tradition of landscape painting. In all of his coastal views, Hodges records the effect of tropical sunlight as well as the precise contours of the shoreline. They reveal his efforts to express the relatively unchanging nature of the landscape, in contrast to the constantly changing nature of the water and the weather. Above all, they show a European artist's attempts to come to terms with the effects of light in the southern hemisphere, combining thinly painted skies and shimmering water with thick impasto, to render the headlands and local craft. In this, Hodges's art might be regarded as the visual expression of Johann Reinhold Forster's written attempts to capture the visual experience of intense tropical sun:

> *The setting sun commonly gilds all the sky and clouds near the horizon with a lively gold-yellow or orange; it is, therefore, by no means extraordinary to see, at sun-setting, a greenish sky or cloud, and it may be observed frequently in Europe. But as the rising and setting sun causes, between and about the tropics, the tincts* [sic] *of the sky and clouds to be infinitely brighter than anywhere else, it happens now and then, that all the appearances of the sky and clouds are more striking and brilliant, and therefore more noticed.*[24]

In Hodges's hands, coastal profiles become artistic renditions. Towering, craggy peaks are silhouetted against a vast sky and calm water, and dappled light effects are produced as the vegetation picks up and reflects the differential fall of the light. Vivid colours and an almost magical interaction between water, sky and land lend vibrancy and freshness to a genre that could be stale and sterile.

Another type of scientific response to the landscape is evident in the view of Table Mountain from Table Bay, which Hodges painted in the first three weeks of November 1772. The expedition sighted Table Mountain on 30 October 1772 on its way to the Pacific, and remained at Cape Town until 23 November to take on bread, wine and other provisions, and to overhaul and caulk the ships.[25] As he would do throughout the voyage, Hodges spent his time making drawings and coastal views.[26] But this painting, showing the southern side of Table Mountain with the tiny, brilliant white buildings of Cape Town at the edge of Table Bay, was his most ambitious piece made on the voyage. By tradition, he painted it from Cook's cabin on *Resolution*, although – since a closely related drawing exists – it is perhaps more likely that he painted it in more spacious shore quarters, given its size and the other normal activity on board. It shows *Adventure* at anchor, with Green Point at the western extremity of the bay. The Dutch fort on the left is lit by the sun and imitates the square solidity of the mountain above. To the right, the tower of the church dominates. In many ways, the image is closely akin to the coastal profiles that he would execute when the expedition reached the Pacific. Cook was happy to be able to have it 'properly packed up with some others and left with Mr Brand, in order to be forwarded to the Admiralty by the first ship that should sail for England'.[27]

(The 'others' probably included a small view of Funchal, Madeira, certainly done on board, out of sight of suspicious local authority.) The painting was displayed at the Free Society of Artists exhibition of 1774, a year before Cook's ships returned home. It acted as a striking, if rather cumbersome postcard informing the Admiralty and others in Britain that the expedition had successfully reached the southernmost tip of Africa; however, it also served another purpose.

The painting goes beyond the naval tradition of the topographically accurate coastal profile of settlements and fortifications. Here Hodges displays a real interest in capturing the atmospheric phenomena and climatic conditions of the Cape. It is a strikingly realistic rendering of the scene, with the shafts of sunlight breaking through the cloudy sky and the vast mass of Table Mountain. The squalls of rain and the gusts of sea-breeze that Hodges evokes in the painting actually occurred when he was preparing this work, as detailed in the ship's meteorological records kept by William Wales from 4 to 14 November 1772.[28] Wales's observation that 'it is not uncommon to see the top of this mountain, alternately, clear and covered with thick clouds five or six times a day' finds visual translation in this work.[29] For the rest of the voyage Hodges demonstrated a keen interest in combining climatic atmosphere with accurate topography.

In a panoramic view showing Tahiti's northern shoreline with neighbouring islands in the distance, Hodges demonstrated a real engagement with the reality of the landscape and climate as it appeared to him (fig. 124). It is one of his most complex paintings, with an unusual composition and a particularly free style. The impressive scale (30 x 48½ in), loose handling of paint and compositional cropping all add up to a challenging of the classical formulae and a striving for naturalism and realism. The placement of the islanders' canoe, cut off by the bottom edge, is very unusual and adds to the sense of vastness. Hodges's emphasis on light, air and colour, combined with the absence of classicizing and literary elements of the larger post-voyage paintings, has led to suggestions that this was painted on the voyage, and as a pair with another same-size panel painting of *A View of the Monuments of Easter Island* [Rapanui] (fig. 132). However, recent evidence suggests that they were probably painted in London around 1776, and that their breadth of handling and brightness are more accountable to being designed as over-door paintings, necessarily seen at a distance.[30] In these coastal profiles and his capturing of 'typical' climatic phenomena, Hodges showed how visual images could also be scientific records. They transcribed, fixed and recorded landscapes and seascapes for posterity, providing a rich visual archive of data for future travellers and anyone else who might be interested.

On some occasions, however, a realistic study of light and atmosphere was insufficient to convey the majesty of the scenes being encountered. In these instances, Hodges's classical training and his familiarity with European aesthetic categories were brought to bear on the landscapes of the Pacific (figs 125, 126). This was the case in New Zealand in May 1773. Late in the after-

Fig. 125 (below)
View in Dusky Bay with a Maori Canoe
William Hodges, *c.*1775
National Maritime Museum
(BHC1907)

Fig. 126 (opposite)
Waterfall in Dusky Bay with a Maori Canoe
William Hodges, *c.*1775
National Maritime Museum
(BHC1908)

noon of the 17th, a few days out from Dusky Sound and en route to rendezvous with the *Adventure* in Queen Charlotte Sound, the *Resolution* encountered four waterspouts off Cape Stephens. The incident elicited alarm from ordinary sailors but, characteristically, it was recorded in meticulous detail by George Forster:

> On a sudden a whitish spot appears on the sea in that quarter, and a column arose out of it, looking like a glass tube; another seemed to come out of the clouds to meet this, and they made a coalition, forming what is commonly called a water-spout... Our situation during all this time was very dangerous and alarming; a phenomenon which carried so much terrific majesty in it, and which connected as it were the sea with the clouds, made our oldest mariners uneasy and at a loss how to behave... We prepared indeed for the worst, by cluing up topsails; but it was the general opinion that our masts and yards must have gone to wreck if we have been drawn into the vortex.[31]

A View of Cape Stephens in Cook's Straits with Waterspout is Hodges's response to the 'terrific majesty' of the incident (fig. 127).[32] In one respect, it is an attempt to offer an objective, scientifically accurate rendition of the scene. Indications of the specific location are offered by a focus on the distinctive, rugged coastline, overgrown with vegetation and inhabited by indigenous animal life, such as seals (which climb to safety on the far right) and exotic waterfowl. But the painting goes well beyond this to convey the sublime elements in the scene: obscurity, darkness and immensity. All of these are conveyed with dramatic contrasts of tone, scale and form. In a further indication of the sublimity of the scene, the ravening sky is pierced by flashes of lightning that have struck the Maori *pa*, or fortified settlement, which is precariously balanced on the promontory to the right.

The strange group observing the scene from the rocky foreground promontory further complicates the scene. The woman with European features faces away from the viewer; naked to the waist, she holds a baby. The man gestures towards

Fig. 127
A View of Cape Stephens in Cook's Straits with Waterspout
William Hodges, 1776
National Maritime Museum (BHC1906)

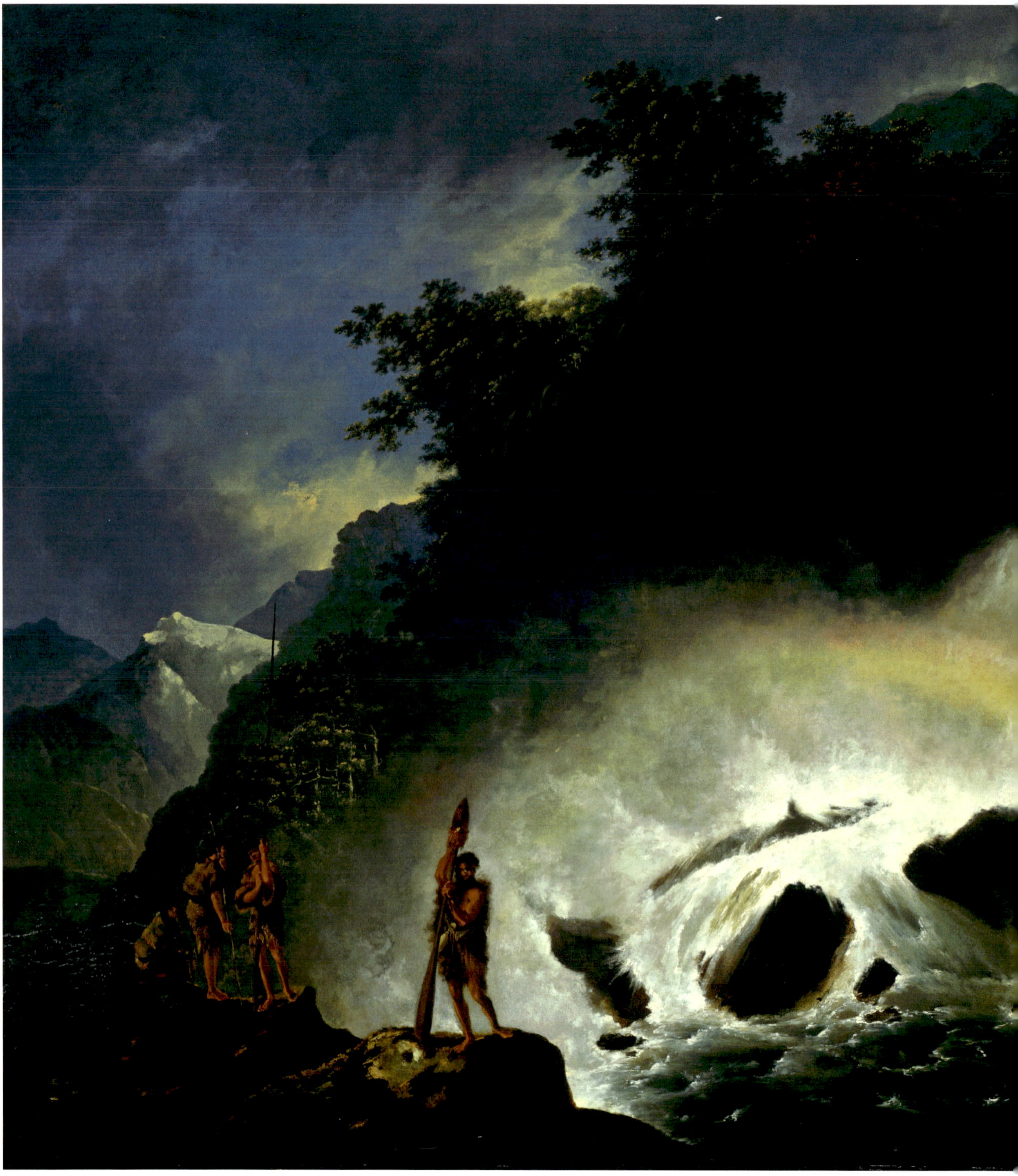

the water-spout in the left foreground. They signify both the specific – as a particular Maori family – and the universality of mankind. In creating such an image, Hodges followed his classical training and even includes specific visual references from the work of his teacher Richard Wilson, as well as from artists such as Joseph Wright of Derby and Gaspard Dughet. The biblical overtones of the male observer's gesture, pose and costume may not be entirely fanciful or coincidental. In adverting to the deliverance of the ship, the group reinforces the Deluge-like quality of the episode.[33]

Something similar occurs in *Dusky Bay* [Cascade Cove] (fig. 128). Hodges developed this large painting after his return to England in 1775, combining the drama and beauty of the New Zealand landscape with a sensitive portrayal of a Maori family that Cook's company encountered in the bay. Here Hodges elided a number of different events during the ship's stay there. The rainbow and waterfall were discovered when some of Cook's party sailed to an inlet from Dusky Bay in April 1773. This sight, perceived as 'one of Nature's most romantic Scenes', prompted them to name it 'Cascade Cove'. They climbed high to look down on a rainbow produced by the rays of the noon sun refracted in the cascade's spray. The cascade was loud and reverberating, and birdsong completed the beauty of the wild and romantic spot. Johann Reinhold Forster takes up the story:

> *We observed, in the several inlets and arms forming this spacious bay, sometimes cascades rushing rapidly down, and falling from vast heights before they met with another rock. Some of these cascades with their neighbouring scenery, require the pencil and genius of a* SALVATOR ROSA *to do them justice: however the ingenious artist, who went with us on this expedition has great merit, in having executed some of these romantic landscapes in a masterly manner.*[34]

Fig. 128
Dusky Bay [Cascade Cove]
William Hodges, 1775
National Maritime Museum
(BHC2371)

VISUALIZING THE PACIFIC   133

Fig. 129 (above)
A Waterfall in Tahiti
William Hodges, c.1775
National Maritime Museum
(BHC2372)

Fig. 130 (opposite)
A Cascade in the Tuauru Valley, Tahiti
William Hodges, c.1775
National Maritime Museum
(BHC2373)

The inherent sublimity of nature is also apparent in two views of waterfalls painted by Hodges (figs 129, 130). These works – which may themselves be studies for a larger pair – were developed from small sketches made quickly on the spot in May 1774, when the *Resolution* visited Tahiti for a second time. In *A Cascade in the Tuauru Valley, Tahiti* (figs 110, 130), a river cascades over the basalt rock in the centre foreground. George Forster called it 'dark, wild and romantic', an impression Hodges managed to preserve.[35] The dark foreground contrasts strongly with the distant mountain partly obscured by mist and, unusually for Hodges, he has not used a depiction of figures to mediate the scene. Hodges's appreciation of the wildness of the idyllic setting is evident from the delight with which sublime landscape elements have been deployed to elevate and transform the picture.

The work of voyage artists was particularly interesting to the public for the sorts of insights it gave into indigenous cultures. Like Parkinson before him, Hodges depicted the European travellers' encounters with Pacific Island people and their ways of life. He showed a variety of daily activities as well as events of greater cultural and social significance, as a way of visualizing these people and places for European audiences. He conveyed something of the technical sophistication of their maritime craft, their interest in and facility with a variety of weapons, and hinted at the complex cultural and social structures that defined their societies.

For example, Hodges's images of *Tahiti Revisited* (a relatively modern title; fig. 131) and *A View of the Monuments of Easter Island* [Rapanui] (fig. 132) were complex comments about the state and development of civilization to be found on Tahiti and Easter Island respectively, as well as about European knowledge of them. The first, *Tahiti Revisited*, depicts the initial anchorage at Tahiti on the voyage and reveals the beauty and peace of Vaitepiha Bay. This is underscored by the inclusion of several female figures who bathe in the river in the foreground. Their presence transforms the landscape into a sensual paradise with erotic charms, while the composition lends an aura of classicism to the work.[36] The statue, or *tii*, that presides prominently over the figures in the right foreground as they prepare to bathe further reinforces the notion of the exotic and erotic.[37] Hodges has attempted to introduce moral purpose and dignity to the landscape by presenting an image of Tahitian society untouched by European contact, since the image includes no hint of Cook's party. However, the inclusion of the shrouded corpse on the far right implies that even in such a tropical idyll, death is present. Hodges used this personal interpretation of Tahiti to hint at the island's heady temptations.

In the image of Easter Island, the fallen rocks in the foreground are indicative of the dark undertones suffusing the painting. Beneath the statues in the foreground, Hodges has depicted a bone, a skull and a surveyor's instrument, possibly a quadrant, used to measure the upright statues. Hodges was sent ashore at Easter Island from 14 to 17 March 1774, to record and observe. He made two walking tours of the island. None of his Easter Island drawings are now known, but later engravings suggest that some were made. The whole party was struck by the beauty and strangeness of the island and its monuments. On the second of these walking tours, along the south-eastern coastline, the party encountered the stone sculptures. According to Johann Reinhold Forster, they 'marched to an elevated place and stopped a little in order to refresh, or to give Mr Hodges an opportunity of drawing some stone-pillars at a distance'.[38]

Forster believed, incorrectly, that the stones marked a burial place, which might account for the prominent skull and bones. In this dramatic landscape, several inhabitants of the island are visible in the middle distance. The unusual figures introduce an air of mystery and the unknown to the interpretation of the painting. Hodges suggests a continuous human link from the ancient to the modern, combined with the sublime terror of the unknown.

Fig. 131 (above)
Tahiti Revisited
William Hodges, 1775
National Maritime Museum (BHC2396)

The title now used for this picture is a fairly recent one: an inscription on the back (now covered by a lining canvas) identifies it as a view of Vaitepiha Bay.

Fig. 132 (overleaf)
A View of the Monuments of Easter Island [Rapanui]
William Hodges, c.1776
National Maritime Museum (BHC1795)

THIRD VOYAGE

John Webber, a London-born Anglo-Swiss artist, was attached to the third voyage. Like Hodges before him, Webber had some training in the classical fine-art tradition before he embarked with Cook. In 1767, after being raised largely by his aunt in Bern, he was apprenticed to Johann Ludwig Aberli, a leading Swiss topographical artist, who trained him in landscape watercolour drawing and may have introduced him to oils. In 1770, Webber went to study painting at the Académie Royale in Paris. Five years later, he returned to London and enrolled as a student in the Royal Academy Schools. He also painted portraits and mythological subjects, submitting works for the Royal Academy exhibition in 1776. These so impressed Daniel Solander, botanist on Cook's first Pacific voyage, that Solander sought him out and recommended Webber to the Admiralty as draughtsman for Cook's impending third voyage in the *Resolution* and *Discovery*.

Webber was duly appointed at 100 guineas a year on 24 June 1776 and, on 12 July, he sailed from Plymouth in Cook's *Resolution*. His fame rests largely on his fine topographical and ethnographic work from the voyage, planned with publication in mind. Guided by William Anderson, he drew natural history subjects, too. Webber returned in October 1780, having completed more than 200 drawings and some 20 portraits in oils. He was reappointed by the Admiralty at £250 a year to redraw and direct the engraving of 61 plates and several unsigned coastal views for the official account, which appeared in June 1784. Webber also painted other views for the Admiralty, his last payment being in July 1785, and published sets of voyage prints. One of these was reissued in aquatint from about 1808 under the title *Views in the South Seas*, and they continued to sell into the 1820s, proving the enduring appeal of Cook's encounter with the Pacific.

The fascination with Cook, the Pacific and its people never really subsided. In 1785, Webber assisted his friend Philippe-Jacques de Loutherbourg in designing scenery and costumes for the celebrated pantomime *Omai, or, A Trip Round the World* at the Theatre Royal, Covent Garden. He resumed exhibiting at the Royal Academy in 1784, with two views of the China Sea and *A Party from His Majesty's ships* Resolution *&* Discovery *shooting sea-horses* [walruses]*, Latitude 71 North, 1778* (see fig. 146). The following year, Webber exhibited his portrait of the Society Islands beauty Poedua (see fig. 82), together with views of Macau and Krakatoa. All 22 paintings that he exhibited from 1784 to 1787 were subjects from Cook's voyages, with seven more (to 1791) among the 25 pictures he exhibited from 1788 to 1792.

LEGACIES

Parkinson, Hodges and Webber were intimately connected with the voyages, by virtue of their presence on the vessels, but artists did not necessarily need to have travelled with Cook to engage with landscapes of the Pacific. For example, John Cleveley, the Younger, was responsible for a number of scenes that depicted the South Seas, even though he had never set foot on Captain Cook's ships. By virtue of his training under the artist Paul Sandby at Woolwich, Cleveley was engaged to produce engravings from drawings made on Cook's second voyage. Later, through his brother James, who was a carpenter on the *Resolution* on Cook's third voyage, Cleveley had access to some of the artwork produced on it. From this he attempted to capitalize on the ready market for South Seas images by composing a set of four original prints, including one of the death of Cook in Hawai'i. When they were issued in 1787–88, the publishers claimed authenticity for them by saying they were based on drawings by James himself, though these are not known and other elements seem to have gone into them. John presumably had the advantage of James's advice, so they are the kind of composite image produced from more primary ones, voyage accounts, and personal *viva voce* information, and they themselves prompted onward copying. The NMM has oil versions of three of them, once thought to be John Cleveley's original images, but – given that he was primarily (though not solely) a watercolourist – more likely to be good copies from the prints by another hand. The view of *Resolution* and *Discovery* at Morea (fig. 133), for example, shows the island's mountains rising into a clouded sky. The bay stretches into the picture space from the right, leaving part of the foreground to the depiction of the shore. In the central middle ground, *Resolution* and *Discovery* are at anchor, dominating the calm waters. Islanders and Europeans can be seen on boats and among the houses and palm trees, where they are busily engaged in exchanging goods. Here, as in the others of the set (fig. 134), including that of Cook's death, alien encounter is tamed and framed, metaphorically and literally, to suit the tastes of the prosperous middle-order British print market.

The example of Hodges and Webber was not confined to representations of Cook's voyages. Their work had a direct influence on the way that images of subsequent exploration endeavours were represented, exhibited and displayed to the public. William Westall, the official voyage artist on Matthew Flinders's expedition to chart the coast of Australia, produced nine oil paintings based on his experiences. These were exhibited to some acclaim at the Royal Academy in 1810 and 1812, and they show the clear and direct influence of Hodges and Webber.[39] Westall's painting *View of Sir Edward Pellew's Group, Northern Territory, December 1802* (fig. 135), depicts North

VISUALIZING THE PACIFIC

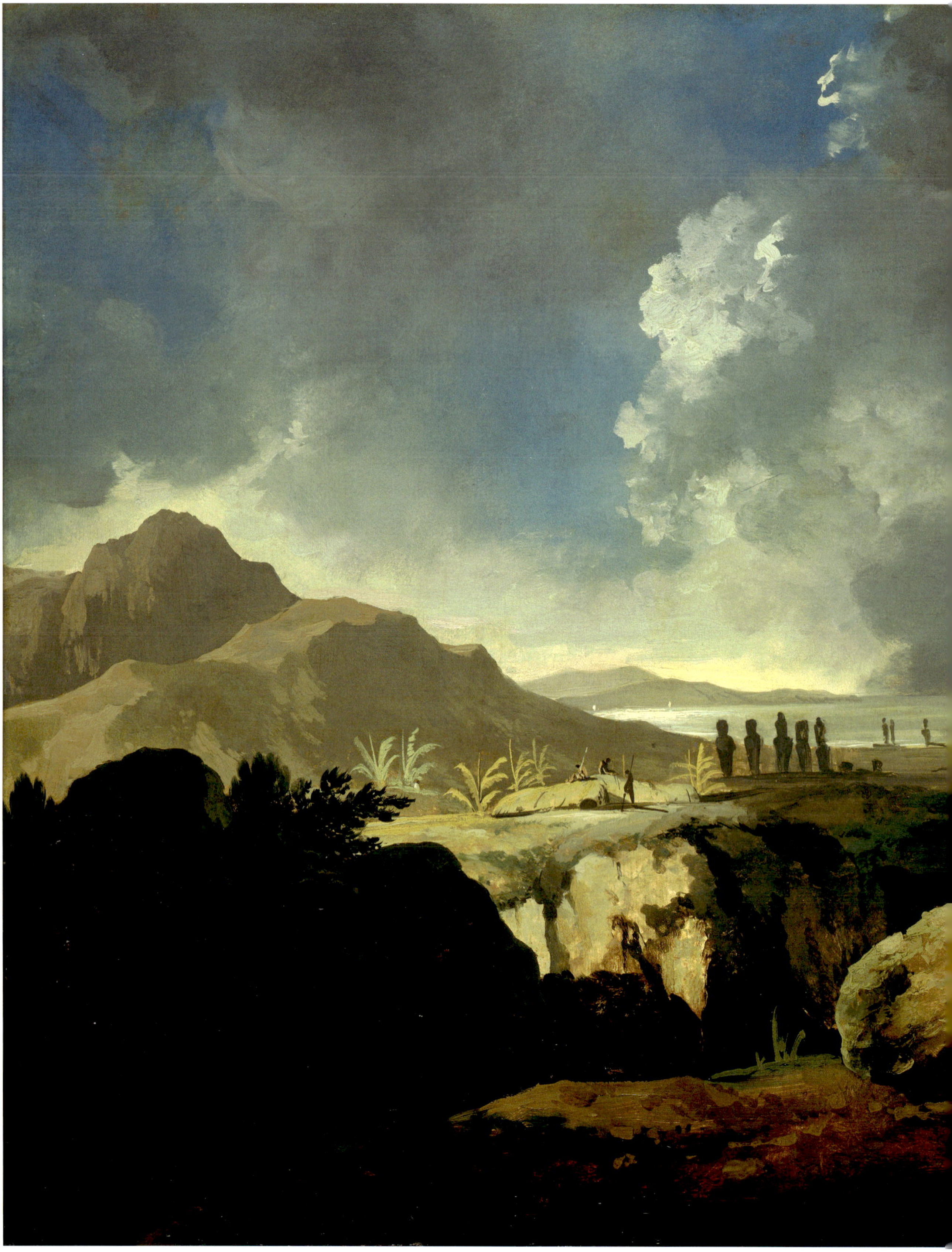

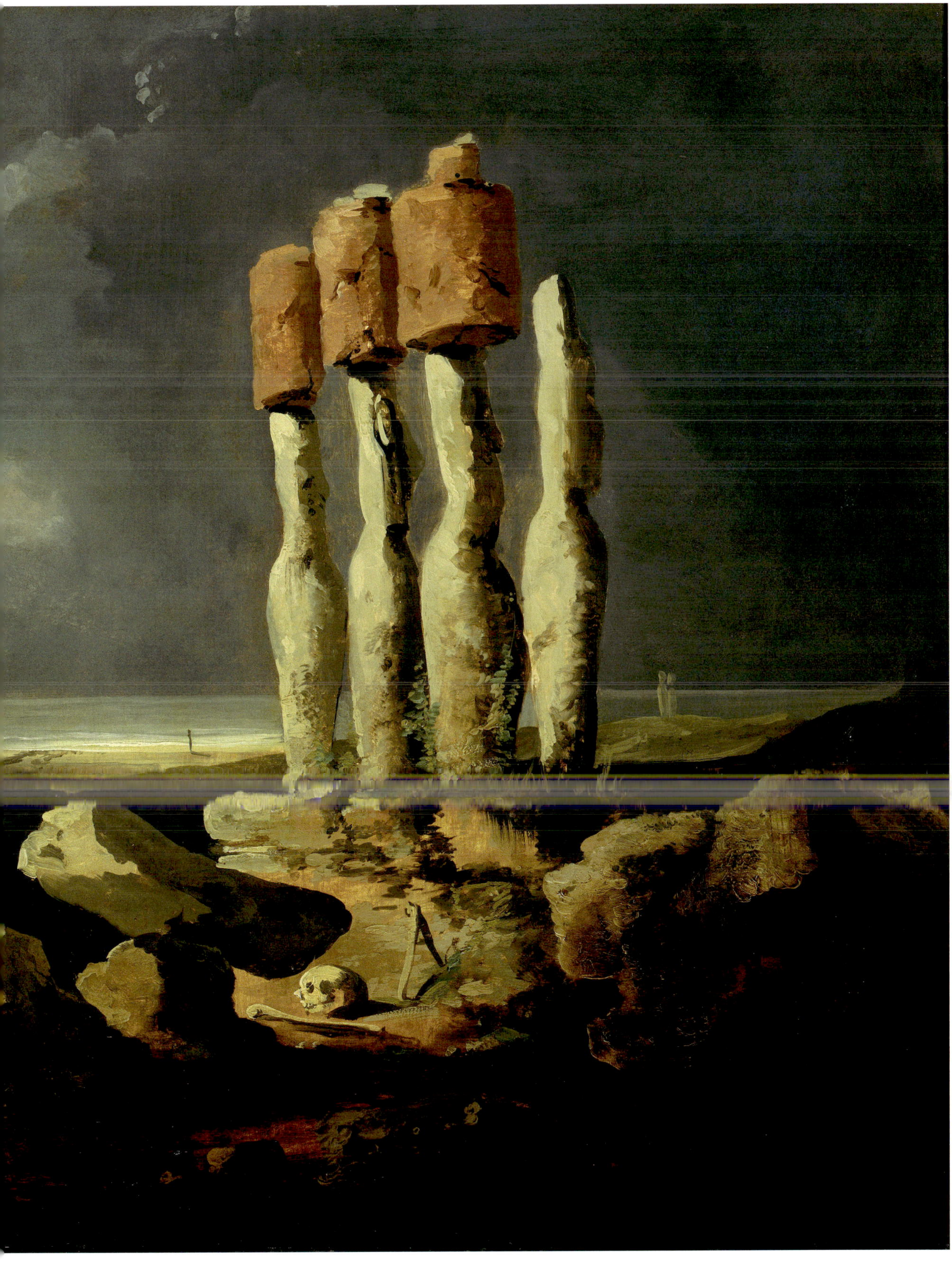

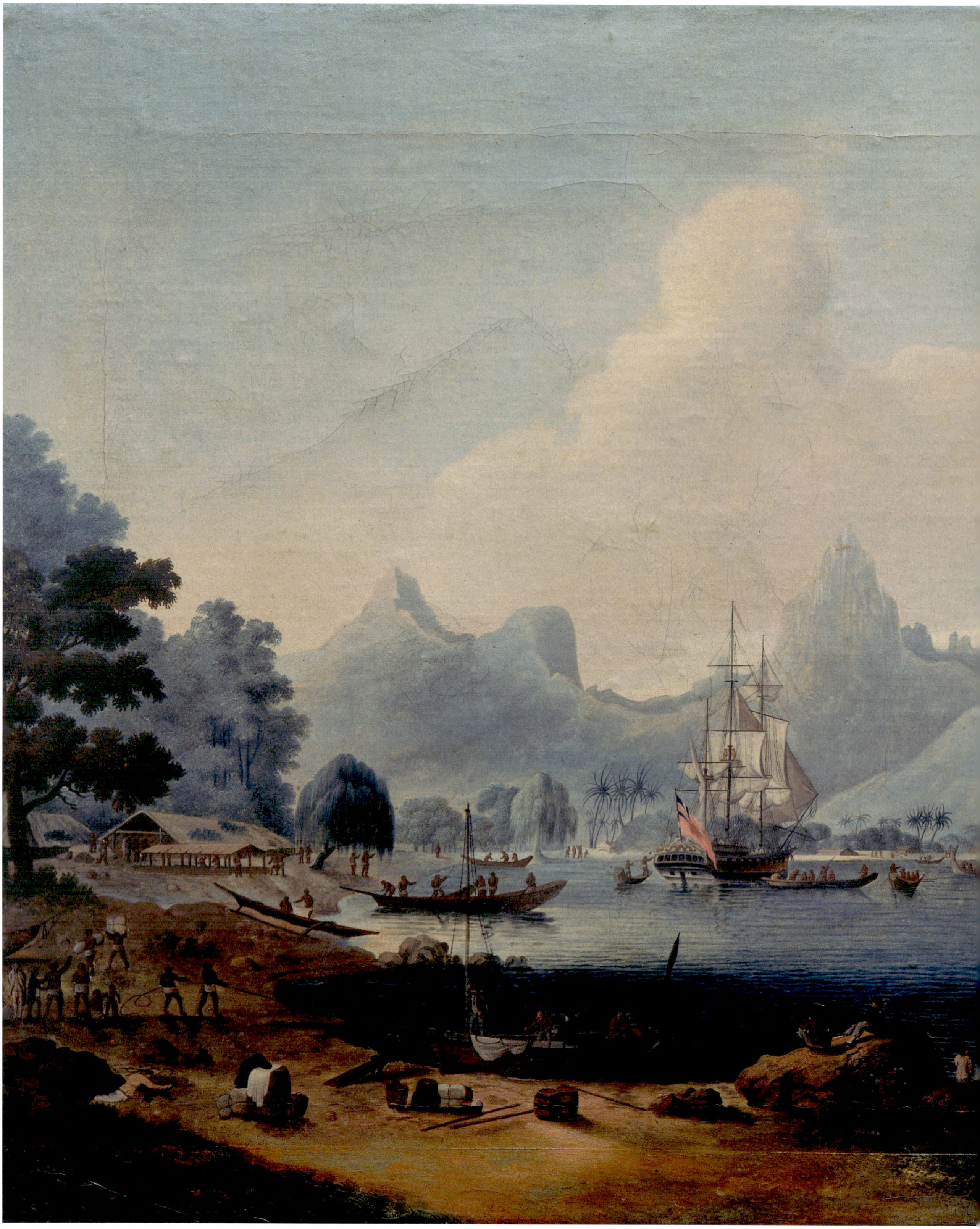

Island, Australia – or, for the Yanyuwa indigenous people, *Murrkunbiji*. The painting offers a near-idealized view of the beach and includes a small shelter under which two stone objects stand. These stones are *kundawira*, powerful memorial stones associated with the death rites of the Yanyuwa. They do not appear in the original drawing, so they represent an active choice on Westall's behalf to include such traces of the indigenous landscape. As they are placed centrally, the viewer's gaze is drawn to them, and their inclusion demonstrates a sensitivity to the cultural presence of indigenous people that was not always displayed by his contemporaries. Westall was also the first person to copy cave paintings accurately on Chasm Island, off the coast of Groote Eylandt in the Gulf of Carpentaria.[40] However, this work is perhaps most striking because of the way in which it conveys a sense of the bleached blue skies of tropical north Australia. It defies the standard image of the Picturesque with its darker foreground receding to lighter background; instead we get a sun-drenched vista typical of this part of Australia.[41]

The impact of these images from the Pacific was felt even further afield, as they influenced the ways in which British travellers elsewhere responded and reacted to the landscapes and people they encountered. As an indication of how influential they were, Bishop Heber linked them to the emblematic island of this period in a letter to his mother from Ceylon:

> *Here I have been more than ever reminded of the prints and descriptions in Cook's 'Voyages'. The whole coast of the island is marked by the same features, a high white surf dashing against the coral rocks . . . Low thatched cottages scattered among the trees, and narrow canoes, each cut of the trunk of a single tree, with an out-rigger to keep it steady, and a sail exactly like that used in Otaheite. The people, too, who differ both in language and appearance from those in Hindostan, are still more like the South Sea islanders, having neither turban nor cap, but their long black hair fastened in a knot.*[42]

Fig. 133
Resolution and *Discovery at Morea*
Style of John Cleveley, the Younger, 1780s
National Maritime Museum. Presented by Captain A.W.F. Fuller through The Art Fund (BHC1896)

VISUALIZING THE PACIFIC

Fig. 134 (above)
Resolution and Discovery in Tahiti
Style of John Cleveley, the Younger, 1770–90
National Maritime Museum. Presented by Captain A. W. F. Fuller through The Art Fund (BHC1939)

Fig. 135 (opposite)
View of Sir Edward Pellew's Group, Northern Territory, December 1802
William Westall, early 19th century
National Maritime Museum (ZBA7944)

The Pellew Islands are in the Torres Strait, between northern Australia and New Guinea.

The wider cultural impact of Cook's voyages, as well as the images produced of them and the publications derived from them, could hardly have been more explicit.

For the artists employed to record the sights and scenes of Cook's travels, the success of these voyages had an enormous influence on the direction of their future careers. The letter of introduction carried by William Hodges to India presented him as 'a celebrated painter who was round the world with Captain Cook', suggesting that the Governor General, Warren Hastings, and anybody else in a position to patronize him would be familiar with the voyages.[43] But the impact of Cook's voyages extended far beyond the careers of the artists who travelled with him. With the help of their work, as well as the information and artefacts gathered on the voyages and the accounts published on their return, the expeditions influenced how people in Europe engaged with the wider world in the eighteenth century.

Not everyone appreciated the visual results of the Cook voyages, and many still questioned whether such images were really art at all; but at the Royal Academy exhibition of 1788, Sir Joshua Reynolds, the Academy's first president, thought that the 'best landskips' on display were those of William Hodges.[44] Reynolds's opinion is important for our understanding of these works of art, but it also informs us about something else: the ways in which the results of Cook's voyages were exhibited and displayed to the wider public. The next chapter explores this theme.

A Striking Likeness of the Late
CAPTAIN JAMES COOK, F.R.S.

His FIRST VOYAGE, *performed in the Years* 1768. 1769. 1770. 1771.
SECOND VOYAGE, 1772. 1773. 1774. 1775.
THIRD VOYAGE, 1776. 1777. 1778. 1779. 1780.

He was Born *at Marton in the North Riding of Yorkshire Nov.r 3d 1728, & Unfortunately Killed by the Savages of the Island, O Whyhee, Feb.y 14th 1779.*

Chapter 6
Exhibiting the Pacific
Collecting, Recording and Display

John McAleer

The voyages of James Cook were part of a burgeoning European curiosity about the world and its people. They were inspired by and contributed to the Enlightenment, with its apparently disinterested quest for knowledge. The expansion of European empires, of which Cook's travels are such a potent symbol, meant that the wider world was the focus of popular public attention as never before. Thomas Bankes, in the preface to his *Universal Geography* (1784), remarked that the 'latest discoveries appear to engross conversation from the politest circles and throughout every class of the Kingdom'.[1] In Britain, the growth and consolidation of a maritime trading empire gave this interest added impetus by bringing people, items and commodities to the country from every corner of the globe. The American colonies, the Caribbean plantations and the East India Company's commercial ventures in Asia provided numerous examples. This phenomenon was so pronounced that it appeared to author and statesman Edmund Burke as if all the regions and people of the world were 'at the same instant under our view'. For those living at the heart of Britain's empire, it seemed that 'the great map of mankind is unroll'd at once';[2] but for Burke and his contemporaries, Cook's voyages of exploration were the most novel and intriguing source of information about their world, offering exciting new perspectives on the natural world and humanity's place in it. These voyages – or, more precisely, their various results, in the form of objects, images and the cultural products they inspired – were crucial conduits for presenting the new possibilities of the Pacific to people in late- eighteenth-century Britain.

Cook and his crew returned to Europe with an extraordinary range of knowledge, experience and physical objects. In doing so, they contributed to an existing culture of knowledge-harvesting and dissemination. The clearest impact of various European voyages to the Pacific in the second half of the eighteenth century can be seen in the additions to the information about the South Pacific available in the *Encyclopaedia Britannica*. This swelled from a meagre seven lines in the first edition (1768–71) to an impressive 40 double-column pages by the third edition of 1788–89.[3] As this example highlights, the vast distances covered by the expeditions, the range and diversity of the people they encountered and the voyages' scientific achievements and artistic breakthroughs could only be made known to a wider public through the exhibition, publication and display of these results. This chapter assesses the various channels through which information about the Pacific permeated British cultural consciousness. Paintings by voyage artists such as William Hodges and John Webber adorned the walls of influential art exhibitions. Lavishly illustrated official travel narratives and cheaper, more popular accounts were eagerly anticipated by the reading public, while audiences were exposed to narratives of exploration in theatres, panoramas and music halls. The presence of indigenous people in Britain also powerfully confirmed that exploration literally brought the world back to Europe. The extraordinary cultural cachet of these voyages, and their percolation into wider society, can be seen in the National Maritime Museum collections.

Fig. 136
A Striking Likeness of the Late Captain James Cook, F.R.S... Frontispiece to Anderson's Large Folio Edition of the Whole of Captn Cook's Voyages &c.
Thornton (engraver), after William Hodges, 1781
National Maritime Museum (PAD2896)

While this image from a secondary edition of Cook's voyages derives from Hodges's portrait of Cook (see fig. 234), that was probably not its direct source.

The popularity of certain objects, such as the hand-held terrestrial globes that made it possible to carry in one's pocket the story of exploration and Britain's maritime engagement with the wider world, demonstrates that Cook's voyages had a much wider influence and impact than their initial scientific objectives might suggest.[4] One particular example, a papier-mâché and plaster globe made by John Cary in 1791 and 'agreeable to the latest Discoveries', depicts the route of Cook's three voyages and carefully records his place of death (fig. 156). It also incorporates the track of Constantine Phipps's voyage towards the North Pole in 1773 and Alexander Mackenzie's explorations in Canada of 1789, thereby setting the exhibition of Pacific exploration in a broader chronological and geographical context. The globe is a physical embodiment of this 'golden age' of exploration, and this object – one that could be displayed in the palm of one's hand – reinforced the idea of exploration as a distinctly British activity and one intimately and inextricably connected to rising imperial ambitions.

TEXTS

Publication was probably the easiest way of conveying new-found knowledge about the Pacific to a broad cross-section of the public. Travel literature was one of the most popular and wide-reaching literary genres of the eighteenth century.[5] In *The Complete English Gentleman*, first published in 1730, Daniel Defoe reflected that, such was the penchant for travel literature:

> [a man] may take a tour of the world in books, he may make a master of the geography of the universe in maps, atlases and measurements of our mathematicians. He may travel by land with the historians, by sea with the navigators. He may go round the globe with Dampier and Rogers, and kno' a thousand times more doing it than all those illiterate sailors.[6]

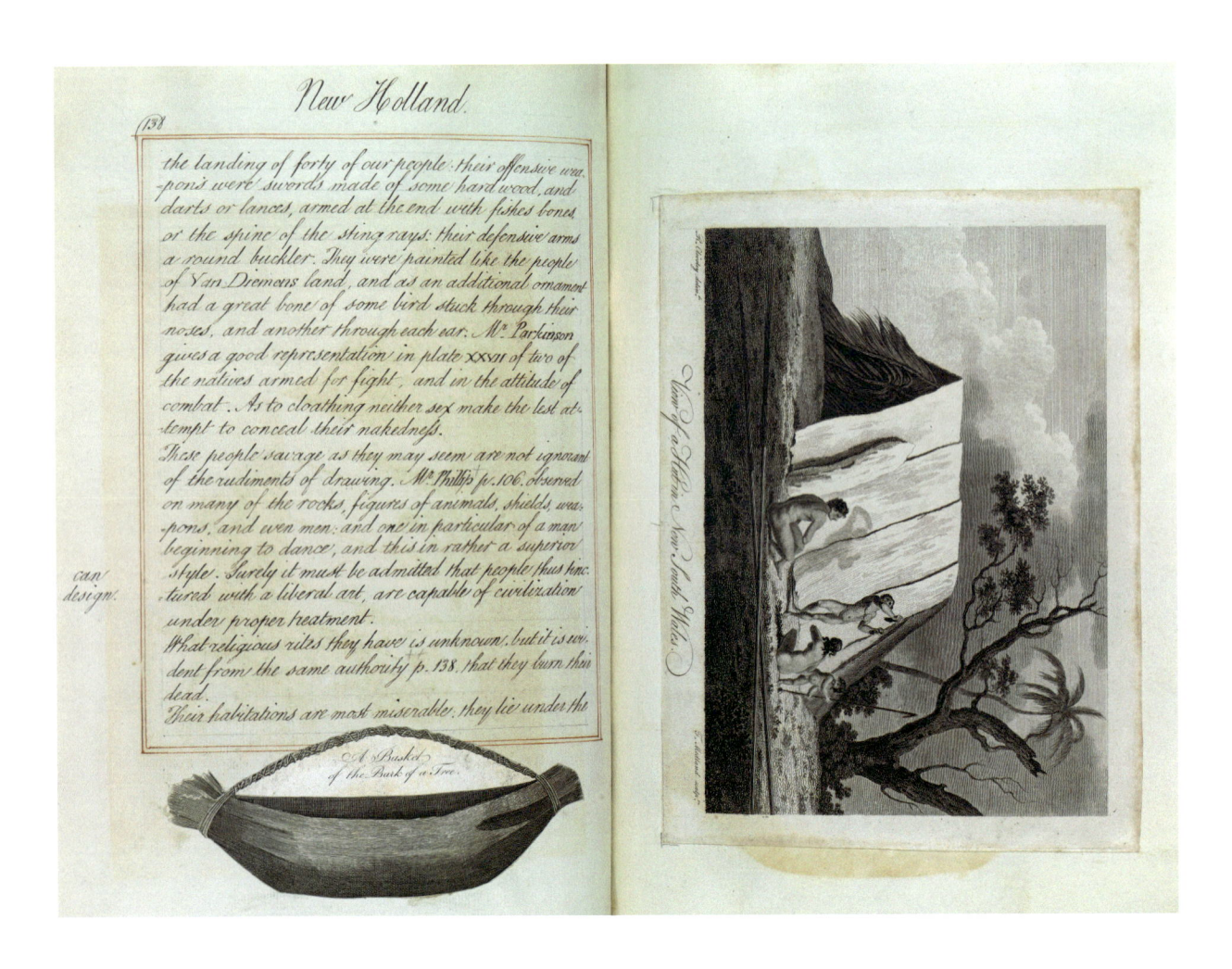

CAPTAIN COOK AND THE PACIFIC

Thomas Pennant, the Welsh naturalist, traveller and antiquarian, offers a real-life example of Defoe's armchair traveller. Pennant was inspired to compile his *Outlines of the Globe* by the wealth of travel writing then in circulation. The 22 manuscript volumes of his *Outlines* were derived from accounts and illustrations drawn from other travel accounts, including those of James Cook (fig. 137).

Travellers recorded and recounted their experiences from the local to the global, offering a personal guarantee of veracity by publishing their accounts and vouching for their truthfulness. This focus on the individual also offered readers a very personal engagement with the circumstances of travel, and gave them an insight into the traveller's motivations and feelings while undertaking the journey. Few readers could hope to follow in the wake of these global travellers, making their accounts all the more popular for a domestic audience. The painting of an unknown gentleman in the NMM collection, holding a copy of a book inscribed 'Cook's voyages', further emphasizes how the publication of travel and exploration literature brought news of the expanding boundaries of Britain's global endeavours to the reading public.[7] The man's eagerness to be depicted holding the volume also demonstrates the extraordinary impact of the voyages on British society.

It was not just audience demand that helped to fuel such an array of popular publications. A key measure of the success of any scientific expedition was whether its results were seriously published, and how influential that account proved to be with the public. The scope and success of Cook's voyages followed a pattern that characterized the eighteenth-century publishing world. Britain undoubtedly led the way when it came to advertising the fruits of her voyages in the Pacific. Although only four English expeditions reached the South Seas in the first 20 years of the eighteenth century (Dampier's twice, Rogers's and Shelvocke's), they inspired six books and several pamphlets and other published pieces. In contrast, there were just three books describing the much more numerous French voyages.[8] William Dampier's *New Voyage around the World* was an instant best-seller, going into five editions in six years; 50 years later the *Voyage* of Admiral George Anson (1748) went through five editions in seven months and had more than 1800 pre-publication subscribers (fig. 138).[9] Even before the 1748 authorized version of Anson's account became a, five unofficial versions of the voyage had already been published.[10] Indeed, a copy was taken on board the *Endeavour*, as both Cook and Banks refer to it in their accounts of the voyage.[11] Perhaps unsurprisingly then, these publications became a model for the narratives of the great explorations later in the century.

Published travel accounts, however partial and incomplete, added to the 'imperial archive' of information being steadily accumulated throughout the eighteenth century.[12] Publication added gravitas to, and guaranteed the authenticity of, exploration narratives that often recounted tall tales of travel in distant lands, with European explorers surrounded by unusual people,

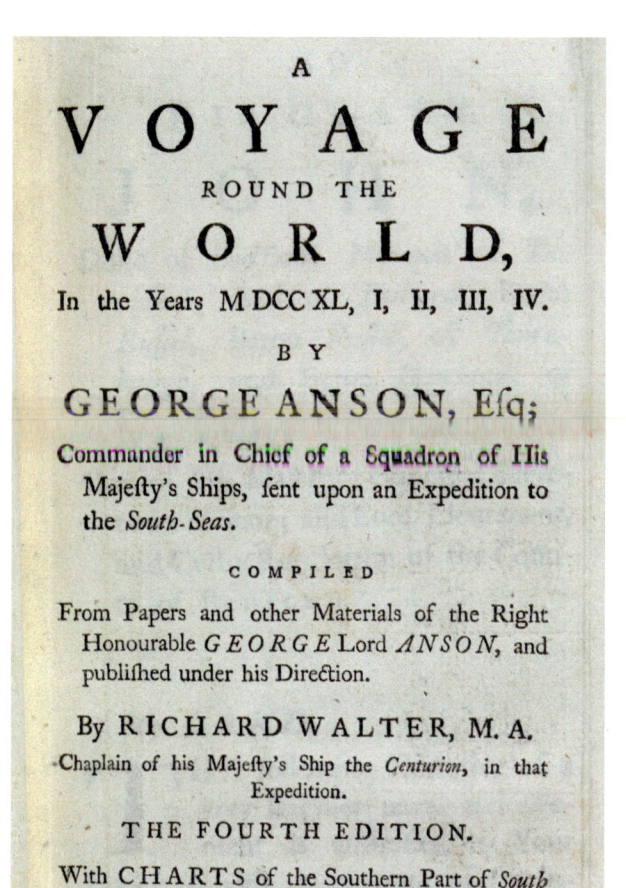

Fig. 137 (opposite)
Pages from Thomas Pennant's
Outlines of the Globe, 1787–92,
including a print by Robert
Cleveley, 'View of a hut in New
South Wales', 1792
National Maritime Museum (P/16/18)

Fig. 138 (left)
Title page from *A Voyage
Round the World in the Years
MDCCXL, I, II, III, IV*
George Anson, 1748
National Maritime Museum
(PBD3287)

unfamiliar landscapes and strange customs.[13] At the end of his first voyage, Cook urged that an account of the voyage should be 'published by authority to fix the prior right of discovery beyond dispute'.[14] Conversely, the absence of a published account could do untold damage to reputations and the perceived veracity of exploration voyages. The importance of publication as a guarantee of accuracy is borne out by the controversy surrounding Vitus Bering's voyage of 1741 (the Second Kamchatka or Great Northern Expedition). Doubts about its significance surfaced partially because the Russian authorities failed to issue any proper account of the voyage until 1758.[15] Similarly, Alessandro Malaspina's considerable achievements in the same region of the north Pacific are still relatively unknown because, for a whole host of complex political reasons, his account lay unpublished for almost 100 years.[16]

Publication and subsequent distribution and consumption were a crucial conduit by which exploration endeavours entered public consciousness. It was often an intensely personal engagement, as readers perused the intimate and apparently unmediated (though, in reality, heavily edited) words of someone who had travelled to the distant corners of the globe. James Cook's expeditions were particularly important in bringing exploration to a wide public audience, and the secondary published versions derived from them offer strong evidence of the interest in exploration at all levels of society. Within three years of Cook's return from his first expedition, European society was transfixed by illustrated accounts of the voyage. Official and unofficial versions proliferated: more than 100 editions and impressions of Cook's voyages were published between 1770 and 1800.[17] Even before they appeared in print, public interest in the publications was high, as Horace Walpole remarked wryly in May 1773: 'at present our ears listen and our eyes are expecting East Indian affairs, and Mr Banks's voyage'. Walpole expected that the publishers, who had paid a huge advance for the account of Cook's first voyage, would 'take due care that we shall read nothing else till they meet with such another pennyworth'.[18]

John Hawkesworth, a self-educated author and editor, was selected to chronicle this expedition (the published volume also contained accounts of the voyages undertaken by Wallis and Byron to the Pacific in the 1760s) (figs 139, 140). On the recommendation of Charles Burney, supported by David Garrick (the great actor, but also a playwright), Hawkesworth was appointed by Lord Sandwich, First Lord of the Admiralty, to compile the official account of the voyages to the South Pacific, based on the journals kept by Cook and others. He was given exclusive access to the voyage journals by the Admiralty and free rein to strike a deal with a publisher. Hawkesworth proved to be as much a sleek businessman as a literary talent. Recognizing the huge public appetite for such accounts, the publishers Strahan and Cadell gave him the enormous sum of £6000 for his manuscript, which was lavishly illustrated with plates from the work of the late Sydney Parkinson, or otherwise concocted (as no artists had accompanied the Byron or Wallis expeditions).[19] Written in the first person, Hawkesworth's narrative of the voyages of Byron, Wallis, Carteret and Cook appeared in 1773 and was rapidly reprinted. Although expensive and aimed at an educated and elite audience, it quickly aroused widespread interest in the Pacific, and its contents reached a wide section of the British public.[20] It became the most popular title in the Bristol Library from 1773 to

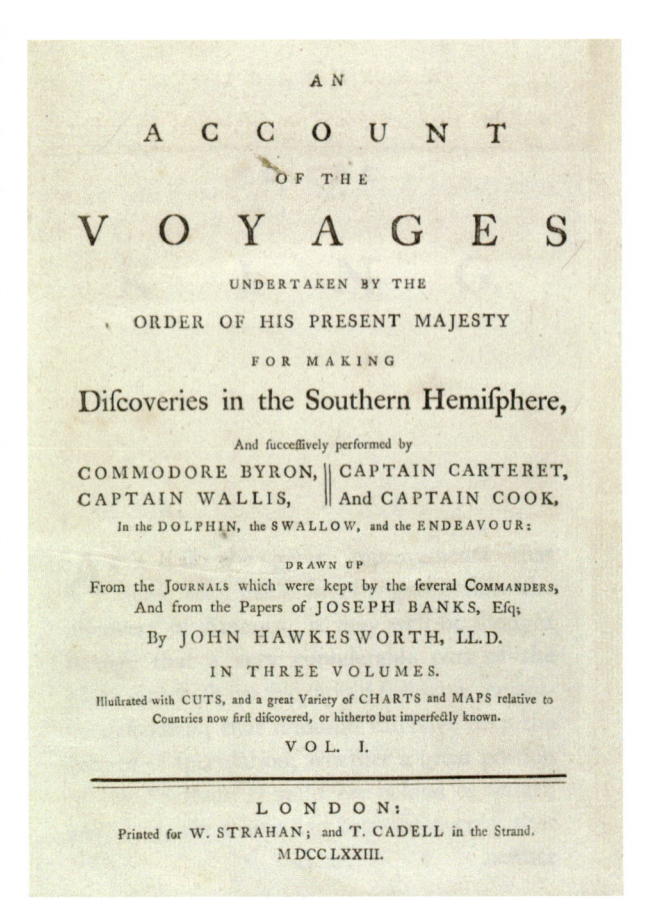

Figs 139, 140
Title page (right) and map (opposite) from *An Account of the Voyages Undertaken by the Order of His Present Majesty for Making Discoveries in the Southern Hemisphere, and Successively Performed by Commodore Byron, Captain Wallis, Captain Carteret, and Captain Cook, in the Dolphin, the Swallow, and the Endeavour*
John Hawkesworth, 1773
National Maritime Museum (PBG1195)

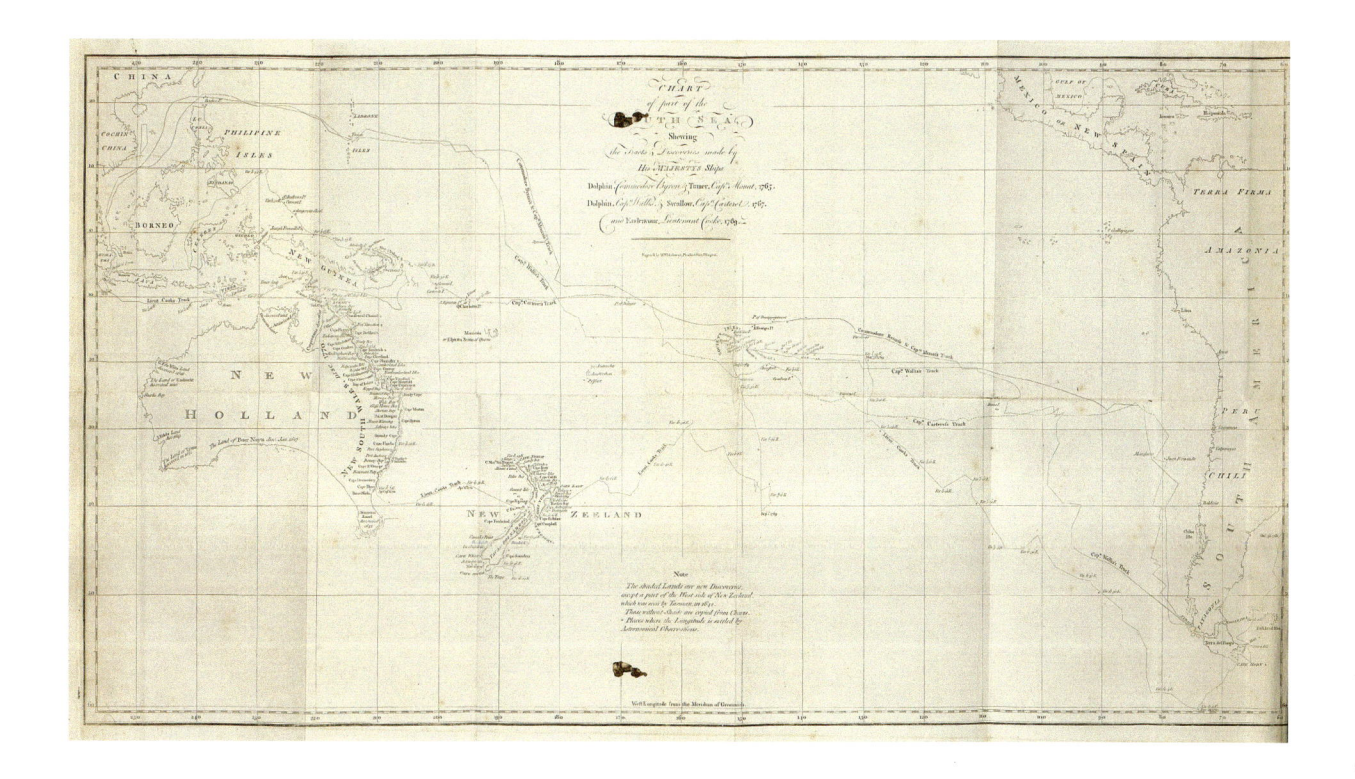

1784, for example, being borrowed 115 times between 1773 and 1775, and 201 times over the whole period.[21] By the end of 1773, a second English edition and a New York edition had appeared. It was translated into French, German, Spanish and Italian, and excerpts were widely published in the periodical press.[22] It also came out in shilling parts entitled *Genuine Voyages to the South Seas, published in sixty weekly numbers*.[23]

Cook's second voyage produced three accounts, partially due to the falling out between the Forsters, the naturalists who accompanied the expedition, and the commanding officer. It appears that Cook, having seen the success of Hawkesworth's efforts (and of whose sensationalizations he did not approve), was determined not to lose out on a potentially lucrative opportunity. He kept several journals during the voyages and there is evidence of some editorial agonizing on his part over the representation of the voyage, a sign that he was thinking about future publication as he wrote his notes. On his return, and with the full support of the Admiralty and assisted by Dr John Douglas (who also worked on the account of the third voyage), he published a two-volume work in May 1777. In a highly successful synthesis of text and image, *Voyage to the South Pole* included 12 charts and 51 monochrome engravings of places, people and artefacts, based mainly on originals produced by the official voyage artist, William Hodges (figs 141–3).

As already touched upon, when he returned to Britain, Hodges was employed by the Admiralty at £250 a year to work up the drawings he had brought home with him in preparation for publication of the voyage, and do other oil paintings for them, of which several small examples were also the basis for engraved plates in this.[24] Whatever its aesthetic and intellectual merits, the book was a roaring commercial success: the first edition sold out in just one day. A second edition appeared the same year, and by 1784 the book was in its fourth edition. This success was even more remarkable given the competition it faced. The Forsters had understood that they would be responsible for the official voyage proceedings, and it appears that Cook's eagerness to publish a narrative came as something of an unwelcome surprise. Nevertheless, they retaliated with vigour. Six weeks before Cook's account appeared, George Forster published a two-volume account. His father, Johann Reinhold Forster, whom he had assisted as botanical draughtsman on the expedition, published his single-volume observations on 'Physical Geography, Natural History and Ethnic Philosophy' in 1778. Although neither this account nor George's contained any illustrations, the fact that they appeared at all gives a sense of the public's appetite for such material.[25]

The book describing Cook's last voyage to the Pacific – Cook and King's *Voyage to the Pacific Ocean* – appeared in June

Fig. 141 (above)
*View of the Island of
New Caledonia*
William Byrne (engraver),
after William Hodges, 1777
National Maritime Museum
(PAH3200)

Fig. 142 (opposite, above)
*The Landing at Mallicolo,
one of the New Hebrides*
James Basire (engraver),
after William Hodges, 1777
National Maritime Museum
(PAI2104)

Fig. 143 (opposite, below)
*The Landing at Erramanga,
one of the New Hebrides*
John Keyse Sherwin (engraver),
after William Hodges, 1777
National Maritime Museum
(PAI2107)

EXHIBITING THE PACIFIC 151

1784 and sold at 4½ guineas (fig. 144). Although it was more the work of its editor, John Douglas, than of Cook or King, in terms of presenting a view of Britain's exploration activities for widespread public consumption, it was an even more impressive operation.[26] The official voyage artist, John Webber, was engaged to illustrate the account, again at a salary of £250 per annum. The project eventually incorporated 63 drawings as copper plates, which necessitated the help of 25 assistants. Perhaps unsurprisingly, it took two and a half years, rather than the 18 months originally estimated by Webber.[27] The Admiralty assumed the production costs for the prints, including Webber's salary and the engravers' fees of about £1000, thereby effectively subsidizing the *Voyage*'s publication by about £2 per copy.[28] Despite the book's delayed appearance and its relative costliness, the results were again impressive: it sold out within three days.[29] One London periodical captured the public mood:

> We remember not a circumstance like what has happened on this occasion. On the third day after publication, a copy was not to be met with in the hands of the bookseller; and to our certain knowledge, six, seven, eight and even ten guineas, have since been offered for a sett.[30]

This interest extended beyond the 'official' accounts: George William Anderson's large folio edition of Cook's voyages was published in 80 weekly parts between 1784 and 1786. The 'Striking Likeness of the Late Captain James Cook' that formed the frontispiece to the collection (fig. 136) was loosely based on Hodges's portrait, of which a better version had already prefaced the official publication of the second voyage (see fig. 234). Despite the somewhat variable nature of the engravings used to illustrate the Anderson compendium, such as the incorrect depiction of a captain's undress uniform in this image, it proved popular with a general public eager for information about the Pacific and the British travellers who had explored it.

The runaway success of Cook's voyages illustrates not only the 'craze for Cook' that proliferated at this time, but the fact that publication outlets of the period brought people closer to exploration and engagement with the wider world. The results were not just limited to the enlightenment and education of armchair travellers like Thomas Pennant: there is evidence that they inspired future voyages. As we have seen, Anson's account accompanied Cook's first expedition as a point of reference and inspiration and, from his posting in India, James Strange (Henry Dundas's future son-in-law) decided to outfit a voyage to the Columbia River on the Pacific coast of America, influenced by 'an attentive perusal of Captain Cook's last voyage'.[31]

Fig. 144 (above)
Title page to James Cook,
A Voyage to the Pacific Ocean
(London: E. Newbery, 1781)
National Maritime Museum
(PBG1185)

Fig. 145 (opposite)
A View of the Island of New Caledonia
William Hodges, 1777–78
National Maritime Museum
(BHC2377)

IMAGES

Just as published descriptions brought exploration to the attention of people in Britain, so visual images also contributed to this expanding universe of knowledge. This included, as we have already seen, work by the artists who went on the voyages, as well as others who never went beyond the shores of Europe, but who exploited interest in exploration for their own commercial and artistic ends.

Within months of setting out with Cook on the second voyage, William Hodges sent pictures of Funchal, Porto Praya and the Cape of Good Hope back to Britain (all apparently from Cape Town) for public display. His work offered a sort of visual travelogue, charting the progress of the voyage for the public at home. Seven paintings by Hodges were exhibited at the Free Society of Artists Exhibition of 1774, more than a year before the *Resolution* returned to Britain.[32] Even in the culturally rarefied atmosphere of the annual Royal Academy exhibition, with its royal visits and classical allusions, the impact of Cook's voyage was felt. In 1774 and 1775, several of the oils painted by Hodges for the Admiralty were exhibited at the Academy.[33] The subject matter was not the usual fare served up to the capital's art critics (fig. 145), but they had to acknowledge the wider public appetite for these sorts of depictions. For example, the critic of the *London Packet*, while generally unfavourable to Hodges's works in the Royal Academy exhibition of 1777, recognized his impact, acknowledging that 'the public are indebted to this artist for giving them some idea of scenes which before they knew little of'.[34] The artist on Cook's third voyage, John Webber, was particularly prolific. Over the course of his later career, Webber exhibited some 29 paintings based on the voyage, and the Royal Academy exhibition of 1784 included three paintings related to it.[35] The range of subject matter, and the excitement and novelty with which Webber managed to imbue his work, contributed to a wider public acceptance of 'exploration' as a legitimate subject for so-called 'high art'. His large picture of walruses being shot for food (fig. 146) was painted after Webber's return to London, and was one of those shown at the Academy in 1784. In 1785 he exhibited pictures of Macao and Krakatoa, as well as his portrait of Poedua, the Raiatean princess (see fig. 82).

Fig. 146
A Party from His Majesty's ships
Resolution & Discovery
Shooting sea-horses [walruses],
Latitude 71 North, 1778
John Webber, 1784
National Maritime Museum
(BHC4212)

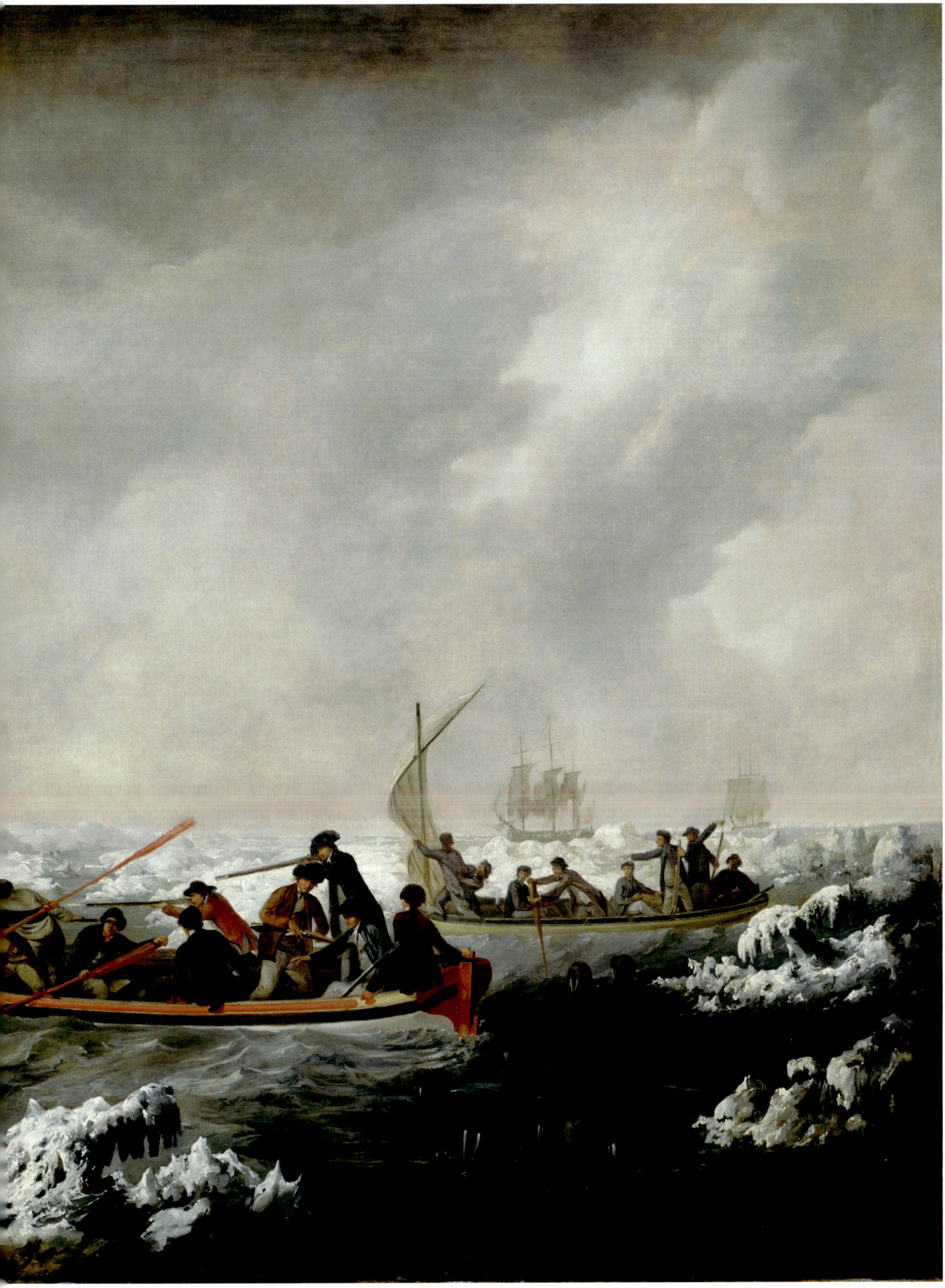

Kamtschatka

of flames, and burn up the neighboring forests; clouds of smoke succeed, and darken the whole atmosphere, till dispersed by showers of cinders and ashes which cover the country for thirty miles round. Earthquakes, thunder, and lightning join to fill the horror of the scenery at land; while at sea the waves rise to an uncommon height, and often divide so as to show the very bottom of the great deep. By an event of this kind was once exposed to sight the chain of submarine mountains which connected the Kuril isles to the end of this great peninsula. I do not learn that they overflow with lava or with water, like the volcanoes of Europe. There are in various parts of the country hot springs, not inferior in warmth to those of Iceland: like them; they in some places form small jets d'eaux, with a great noise, but seldom exceed the height of a foot and a half.

hot springs

Climate

The climate during winter is uncommonly severe, for so low as Bolcherefsk, lat. 52.30 all intercourse between neighbors is stopped. They dare not stir out for fear of being frost bitten. Snow lies on the ground from six to eight feet thick as late as May; and the storms rage with uncommon impetuosity, owing to the subterraneous fires, the sulphureous exhalations, and general volcanic disposition of the country. The prevailing winds are from the west, which passing over

Kamtschatka

It is navigable from its mouth for small vessels two hundred versts or one hundred and fifty miles. On this river the Cossacs made their first settlement; and they, and the russians founded several Ostrogs or forts.

This peninsula, and the country to the west, are inhabited by two nations; the northern parts by the Koriacs, who are divided into the Rein deer or wandering, and the fixed Koriacs; and the southern part by the Kamtschadales, properly so called; the first lead an erratic life in the tract bounded by the Penschinska sea to the south east; the river Kowyma to the west, and the river Anadir to the north. They wander from place to place with their Reindeer, in search of the moss, the food of those animals, their only wealth. They are cruel, squalid, and warlike; the terror of the fixed Koriacs as much as the Tschutschi are of them. They never frequent the sea; nor live on fish. Their habitations are jourts or places half sunk in the earth; they never use balagans, or summer houses elevated on posts like the Kamtschadales: are in their persons lean, and very short; have small heads and black hair which they shave frequently; their faces are oval; nose short; their eyes small, mouth large; beard, black and pointed, but often eradicated.

Rein deer Koriacs

Fixed

The fixed Koriacs are likewise short, but rather taller than the others, and strongly made: they inhabit the north of the

A WOMAN of KAMTSCHATKA.

Many images were not only displayed in the fairly exclusive surroundings of London fine-art exhibitions, but circulated more widely in the form of prints, which were a key medium through which visual information about exploration reached a wider public. Though a little later, James Gillray's famous caricature *Very Slippy Weather* is perhaps the best-known image demonstrating how print-shop windows were themselves popular and free public exhibitions on every subject for which there was such a market – exploration among them. Over the course of his career, John Webber utilized the storehouse of Cook images he had amassed on the third voyage. In 1788, he issued a series of plates of Pacific views based on drawings and studies not included in the official account. By 1792, 16 views by him were for sale by subscription.[36] As an indicator of the range and impact of such prints, Thomas Pennant included several of Webber's in his manuscript compilation describing 'Northern Regions', focusing on Siberia and Russia's frontier in the Far East (figs 147, 148). In 1808, John Boydell published a folio volume of *Views in the South Seas* based on Webber's drawings.[37] The effect of such enterprise in creating visual and cultural expectations can be gauged by the words of George Mortimer, a lieutenant on a later voyage to the Pacific north-west. In 1789,

Figs 147, 148 (opposite)
Pages from Thomas Pennant's *Outlines of the Globe*, 1787–92, including pages from William Coxe's *Account of the Russian Discoveries between Asia and America*, 1780
National Maritime Museum (P/16/4)

Fig. 149 (below)
Views in the South Seas from the Drawings of the Late James Webber [sic]. *View in Queen Charlottes Sound, New Zealand*. Engraving after John Webber, 1809
National Maritime Museum (PAH9570)

EXHIBITING THE PACIFIC   157

Fig. 150 (above)
A View of Huaheine
William Byrne (engraver) after
John Webber, 1777
National Maritime Museum
(PAI4157)

Fig. 151 (opposite, above)
View of Huaheine, one of the Society Islands in the South Seas
Francis Jukes (engraver), after John Cleveley, the Younger, c.1788
National Maritime Museum
(PAI0470)

This is one of the four prints that Cleveley drew aided partly by sketches by his brother James, carpenter of the *Resolution*, but also using other sources.

he described how 'in this and in every other particular [the natives of Oonalashka] exactly resemble the prints of them in Captain Cook's last voyage, taken from the elegant drawings of Mr Webber'.[38] These products of exploration had become so widespread and instantly recognizable that they themselves became a benchmark for assessing the people and places encountered by future expeditions.

As well as being visual documentary records in conjunction with published accounts, the prints had other functions in display and exhibition. They could almost literally domesticate the Pacific, bringing it into British homes in ways that were beyond the published narrative nestling on the library shelf. William Hodges's landscapes, derived from the second voyage, were so popular that they encouraged Matthew Boulton to buy depictions of Tahitian landscapes and have them reproduced using the new mechanical, copperplate method, which imitated oil paintings.[39] When Thomas Pennant received a gift of the prints from Cook's third voyage, he hastened to have 'half a room hung with them'.[40] Illustrations of Cook's voyages providing basic information about the Pacific, and based predominantly on Webber's drawings (figs 149, 150), were published in the great Italian and French costume books that appeared during the first third of the nineteenth century.[41] In 1804–05 the Parisian firm of Joseph Dufour also issued magnificent colour-printed panoramic wallpaper, designed by Jean-Gabriel Charvet, under the evocative product title *Sauvages de la Mer Pacifique*.

The voyage artists were, in fact, well outnumbered by others producing more derivative images for public exhibition and consumption. On his return from the Pacific, Joseph Banks commissioned George Stubbs, the renowned animal painter, to depict two previously unknown animals that the first expedition had encountered. Capitalizing on the heightened public awareness of the voyage, Stubbs exhibited both paintings at the Society of Artists in London in 1773, thereby closely identifying the kangaroo and dingo with the new world of Australia (see figs 10, 152). Banks subsequently hung the paintings, along

Fig. 152 (left)
'Portrait of a Large Dog' [Dingo]
George Stubbs, 1772.
National Maritime Museum
(ZBA5755). Acquired with the assistance of the Heritage Lottery Fund; the Eyal and Marilyn Ofer Foundation; The Art Fund (with a contribution from the Wolfson Foundation) and other donors.

EXHIBITING THE PACIFIC 159

with the iconic portrait of Cook by Nathaniel Dance (see fig. 1), at his residence in Soho Square, which was a meeting place both for savants and for others in his wide circle among the social elite who were interested in 'natural science'. In the absence of live models, Stubbs worked from written and verbal descriptions provided by Banks himself and, in the case of the kangaroo, a small group of pencil sketches made by Sydney Parkinson and a stuffed or inflated pelt (now lost) that was in Banks's possession. By employing such a celebrated specialist painter, renowned for his scientific methods and representational accuracy, Banks helped to give the unfamiliar sights and discoveries of Cook's first voyage the kind of cultural significance that could only be achieved through the public display and dissemination of great works of art. Later the same year, an engraving after the image of the kangaroo was used to illustrate John Hawkesworth's official account of the voyage, thereby confirming the marsupial's place among the dramatis personae of Cook's expeditions.

Stubbs was among the most renowned artists who benefited from the voyages in terms of patronage and exposure for their work, but this was a much wider phenomenon. We have already seen how John Cleveley, the Younger, who was a 'Cook artist' despite not sailing with him, published good prints of the voyages, in terms of his sources, at second hand (fig. 151). Another example was William Parry, who in 1776 painted a historically important group portrait of Sir Joseph Banks, his botanist-librarian Daniel Solander and the young Polynesian generally called Omai.

PEOPLE

One of the most striking ways in which exploration could be exhibited was through the 'display' of people. The exhibition of non-Europeans to audiences in Europe has recently inspired much valuable scholarship and engendered new academic debates.[42] In some cases, it is clear that such people were taken to Europe against their will and were treated deplorably while there; but in other cases, such as that of Omai, discussed below, there was humanity and an element of mutual benefit.

Omai – more properly 'Ma'i' – was the most famous example of an eighteenth-century Pacific Islander in Europe (fig. 153). He was a native of Raiatea in the Society Islands and became the first 'Tahitian' to visit Britain, when he arrived in July 1774 with Tobias Furneaux in the *Adventure*'s premature return from Cook's second voyage. One member of the British press described Omai as a 'wild Indian, that was taken on an island in the South Seas'.[43] Apparently, however, he demonstrated such refinement and good grace that he charmed everyone with whom he came into contact. Omai had enrolled as a supernumerary in the *Adventure* under the pseudonym of Tetuby Homey. He eventually returned to Tahiti on Cook's third voyage. During his two years in Europe, Omai provided elite society with a living example of the 'noble savage' and a focus for discussions about the virtues of natural man as against the artificiality produced by civilization. Fanny Burney praised him as somebody who 'appears in a *new world* like a man [who] all his life studied *the Graces* . . . I think this shews how much more *Nature* can do without art, than *art* with all her refinement, unassisted by Nature.'[44]

Omai was taken under the wing of Joseph Banks who, along with Daniel Solander, presented him to George III at court on 17 July 1774. He was also painted by a number of significant artists, most notably Sir Joshua Reynolds, whose celebrated portrait was exhibited at the Royal Academy in May 1776.[45] In this image, Reynolds depicted him wearing loose robes and a turban, in a pose influenced by the Apollo Belvedere, a classical statue much revered among artists and connoisseurs alike. By freighting the image so heavily with classical allusions, Reynolds's work was a visual argument for the Rousseauesque notion of the nobility of natural man, whatever his origins.

Such Enlightenment ideals are similarly found in John Webber's portrait of Poedua (see fig. 82). Visiting the artist's studio in March 1785, Dorothy Richardson described 'an Oil painting of a woman of Otaheite, ½ length & as large as life, with white flowers for earrings; tho her complexion is copper colour, she is extremely handsome & graceful'.[46] The half-smiling countenance of Poedua, and her Venus-like pose and gestures, further reinforces this idea. Although she was not physically brought to Europe in the same way as Omai, her image similarly sheds light on the Enlightenment attitudes about Pacific people that were circulating at the time.

Fig. 153
Omai, native of the South Pacific
James Caldwall (engraver),
after William Hodges, 1777
National Maritime Museum
(PAI2070)

EXHIBITING THE PACIFIC

OBJECTS

If texts and images formed a key part of exhibiting the fruits of exploration, then tangible objects and material culture were equally important in presenting the voyages of Cook and his colleagues to audiences in Britain. Since publication could increase the aura of such truth, the evidence of the object was potentially even more incontrovertible.[47] The accounts of individual explorers, the images produced by artists and the presence in Britain of people from the Pacific gave readers and audiences a personal connection to the region; but the voyages, and the scientific and cultural results they produced, could also be exhibited in Europe though the objects collected, bought and bartered by those on the expeditions. The return of these artefacts, and their subsequent European display and interpretation (and misinterpretation), is one of the most enduring and tangible legacies of Cook's voyages. Such artefacts were collected as part of the process of exchange and transaction that occurred as European ships sailed from place to place. They provided tangible confirmation of these interactions, offering material proof of the ways of life of the people encountered, and of the resources and the environments in which they lived.[48]

Many objects were also collected because they were perceived to have scientific value, and as another way of gathering and preserving objective data, while other items that were the products of voyage information were in turn collected as ways of using it for a broader educational function. The terrestrial and celestial pocket globe produced by John Miller in 1793 (fig. 154) was advertised as 'a New Year's gift for the instruction and amusement of young ladies and gentlemen'. The terrestrial one displayed trade winds, while monsoons were indicated by arrows, and New Zealand and Australia were drawn 'according to Cook's discoveries'. As was almost mandatory by this stage, the location of Cook's death – 'Owhyhee' – was noted in the middle of the Pacific.

Collecting was an important activity across all sections of eighteenth-century society. The collecting of 'curiosities' was a pastime equally popular among those who sailed with

Fig. 154 (below, left)
Terrestrial and celestial pocket globe
John Miller, 1793
National Maritime Museum, Caird Collection (GLB0009)

Fig. 155 (left)
Terrestrial and celestial pocket globe
West of Soho Square, c.1800
National Maritime Museum, Caird Collection (GLB0010)

Fig. 156 (below, right)
Pocket globe
John Cary the Elder and William Cary, 1791
National Maritime Museum, Caird Collection (GLB0001)

Cook as it was with connoisseurs back in Europe. Cook himself recognized the 'prevailing passion for curiosities'.[49] As a result of the collecting activities of various voyages, and the subsequent exhibition of the fruits of their endeavours, the British public had the opportunity to see, at first hand, the material cultures of various Pacific Island, American and Australasian societies. These 'numerous specimens of the ingenuity of our newly discovered friends', as John Douglas (Cook's best original-voyage publication editor) termed them, were accompanied by a variety of other media that exhibited exploration to the public. All of these offered 'real matter for important reflection' on the nature of Britain's place in the world, the status of the societies encountered on these voyages and the relative merits of each.[50]

Even Cook himself was not immune, promising 'to add to your collection of curiosities' in a letter to Hugh Dalrymple before he set off on his third and final expedition.[51] George Forster made repeated reference to the eagerness with which all members of the ship's company acquired specimens, as 'targets without number were bought by almost every sailor'. He noted that 'not less than ten' of the elaborate mourning dresses of Tahiti 'were purchased by different persons on board and brought back to England'. Officers and sailors built up huge collections of material on their travels, partly because of their desirability and demand in Europe (figs 157, 158). These objects found a willing market among museums and collectors: one mourning dress brought back to England by a seaman on Cook's second voyage was sold for the sum of 'five and twenty guineas', which was just over six months' pay for a naval lieutenant at the time.[52] Similarly, David Samwell, the assistant surgeon on the third voyage, auctioned off his collection in 248 lots in 1781. The whereabouts of more than 2000 items related to the voyages were known at the end of the nineteenth century, and these objects were distributed far and wide.[53] When Cook's ships arrived back in Europe, the London-based merchant, dealer and collector George Humphrey obtained shells that he subsequently sold to purchasers such as the Duchess of Portland and the Literary and Philosophical Society of Danzig.[54]

The most notable collector on Cook's voyages was Joseph Banks, who accompanied the first expedition. On his return to London, his Soho Square home became 'an early "Museum of the South Seas"'. It provided a focus for those seeking information about Cook's voyages, much of which was gleaned from the arrangement and interpretation of objects there. The reaction of one visitor in 1772 offers a glimpse of the sort of impact that material culture collected through voyages of exploration had on British views of the rest of the world. The Reverend William Sheffield, Keeper of Oxford's Ashmolean Museum, could hardly be described as an average visitor, yet he felt 'utmost astonishment' and could 'scarce credit my senses'.[55]

Such material was initially most in demand from educated and scientifically curious audiences or those with a direct connection to the voyages. As early as 1773, Thomas Falconer (classical scholar, antiquary and friend to both Banks and Thomas Pennant) wrote to Joseph Banks expressing genuine interest and curiosity in the indigenous arts of the Pacific: 'I was highly entertained at Oxford with a sight of some curiosities you sent from Otahieta and new Zealand [sic].'[56] By 1783 the *Museum Academicum* in Göttingen had acquired, under the supervision of Johann Friedrich Blumenbach, 'a wonderful collection of curiosities from the South Seas, gathered during the last two voyages of Capt Cook'.[57] In 1780, George Humphrey auctioned his museum, which included 182 'artificial curiosities'. The items for sale from his collection represented the types of objects that could convey the ways of life of these newly explored Pacific spaces: 'the best and most extensive collection of cloths, garments, ornaments, weapons of war, fishing tackle and other singular inventions of the natives of Otaheite, New Zealand and other new discovered islands in the South Seas'.[58] In negotiating his place on the second voyage, which he later declined, Johan Zoffany asked for £1000 and 'a third of all curiosities'.[59] John Webber returned from the third voyage with more than 100 items, many of which he bequeathed to the library in his home town of Bern, though they are now in the museum there. Instead of seeing 'his south Sea drawings', when she visited John Webber's house in March 1781, Fanny Burney spent the morning inspecting the 'curiosities' with which he had returned. The author had a personal connection with Cook's voyages, since her brother, James, had accompanied the second and third voyages and became a notable writer on exploration, as well as an admiral. When Webber and Captain King explained the objects and their significance to his sister, Fanny proclaimed them to be 'extremely well worth seeing'.[60]

The interest in such material – and the impulse to put it on public display for general audiences – went beyond those directly connected with the voyages. Cook's first biographer, Andrew Kippis, acknowledged that 'the curiosities which have been brought from the discovered islands, and which enrich the British Museum and the late Sir Ashton Lever's (now Mr Parkinson's) repository, may be considered as a valuable acquisition to this country, as supplying no small fund of information and entertainment'.[61] Several donations of Cook voyage material were made to the British Museum at regular intervals. In February 1770, Philip Stephens, Secretary to the Admiralty, wrote to its trustees with the 'offer of several Curiosities from the late discovered Islands' in the Pacific.[62] On 3 February 1775, the Lords of the Admiralty again wrote to consign 'a collection from New Zealand and Amsterdam in the South Seas, consisting of 18 articles, Domestic and Military, brought by Captain Furneaux', while on 24 November 1780 the museum's 'Book

VARIOUS ARTICLES, at the SANDWICH ISLANDS.

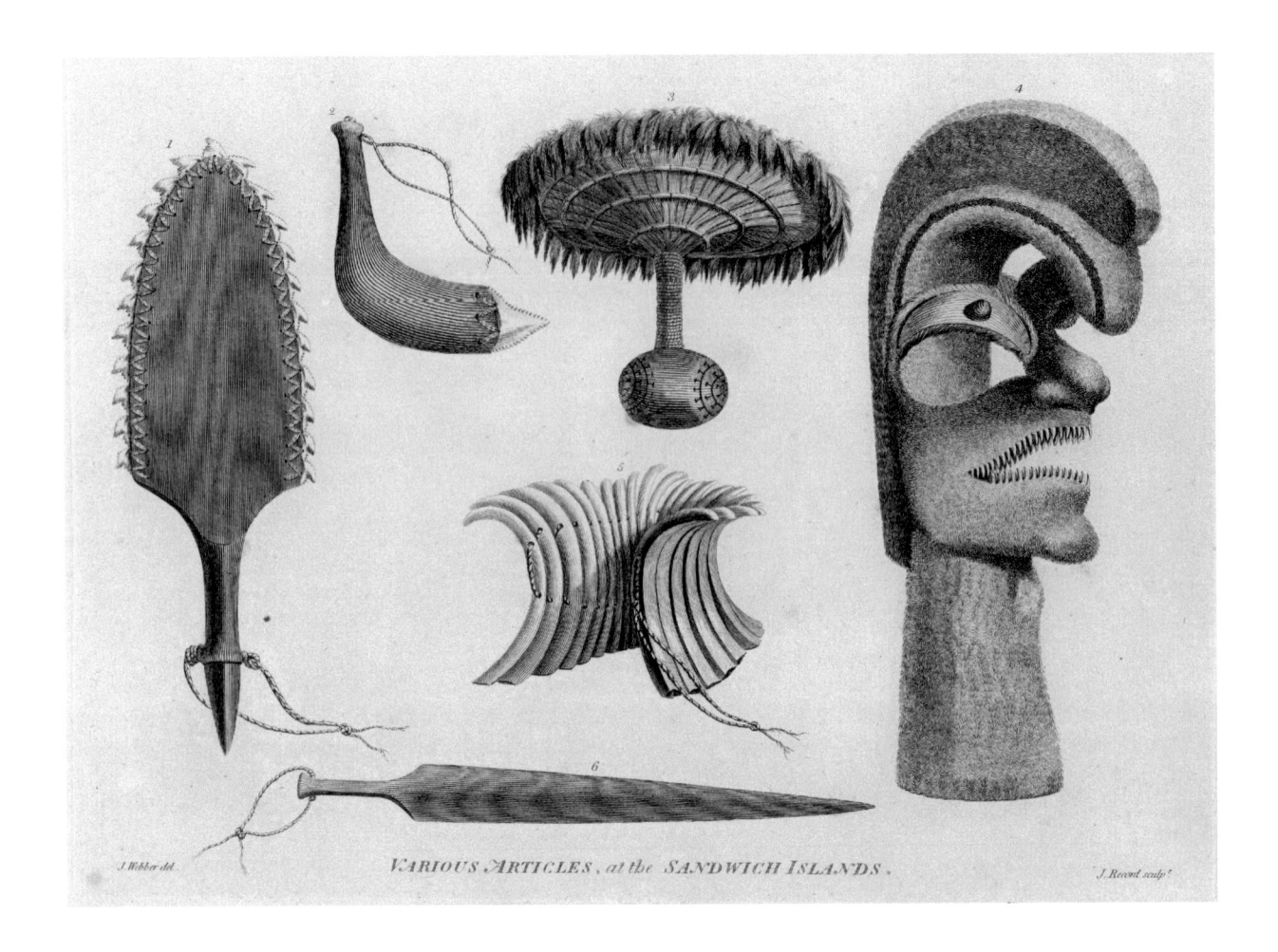

of Presents' records a gift of 'several artificial curiosities from the South Sea from Captain Williamson, Mr John Webber, Mr Cleveley, Mr William Collett, and Mr Alexander Hogg'.[63]

Joseph Banks, who made donations in 1778 and 1780, described the 'several cart loads' of 'arms and curiosities from the South Sea' that he had sent to the museum, which had 'engaged to fit up a room for the sole purpose of receiving such things'.[64] A 'South Seas Room' was planned at the British Museum as early as 1775. In April 1776, the carpenter was reprimanded for his tardiness and instructed to complete the arrangements as a matter of urgency. By March 1778, the room was being papered in a 'neat mosaic pattern'.[65] The display subsequently became a major public attraction. It was the first occasion that the museum had organized a display with specific geographical and cultural references.[66] Daniel Solander, together with some assistants, arranged and labelled everything.[67] In doing so, they had to 'intirely [sic] renew the arrangement'. On 10 August 1781, Solander reported that the display was opened to visitors.[68] When Sophie von la Roche visited the museum in the summer of 1786, she was intrigued by the objects ('all the pots, weapons and clothes from the South Sea islands just recently discovered'), which were 'just as they are shown in the prints illustrating the description of [Cook's] voyage: crowns, helmets and war-masks, state uniforms and mourning'. Even at this stage, the material was inextricably linked to the personal story of Cook: Sophie described the room as being 'devoted to Captain Cook, that luckless, excellent man'.[69] By the early nineteenth century the British Museum's collection of Cook-related material formed 'one of the most conspicuous parts' of that museum.[70] The 'Otaheite and South Sea Rooms' were regarded as among the great sights of London for visitors. James Malcolm described the sight in *London Redivivum*:

> The researches of Captain Cook are well known to the publick; and perhaps no age or country were ever more happy in the choice of a Navigator, and his companions in science and perseverance. The visitor will find in this

164   CAPTAIN COOK AND THE PACIFIC

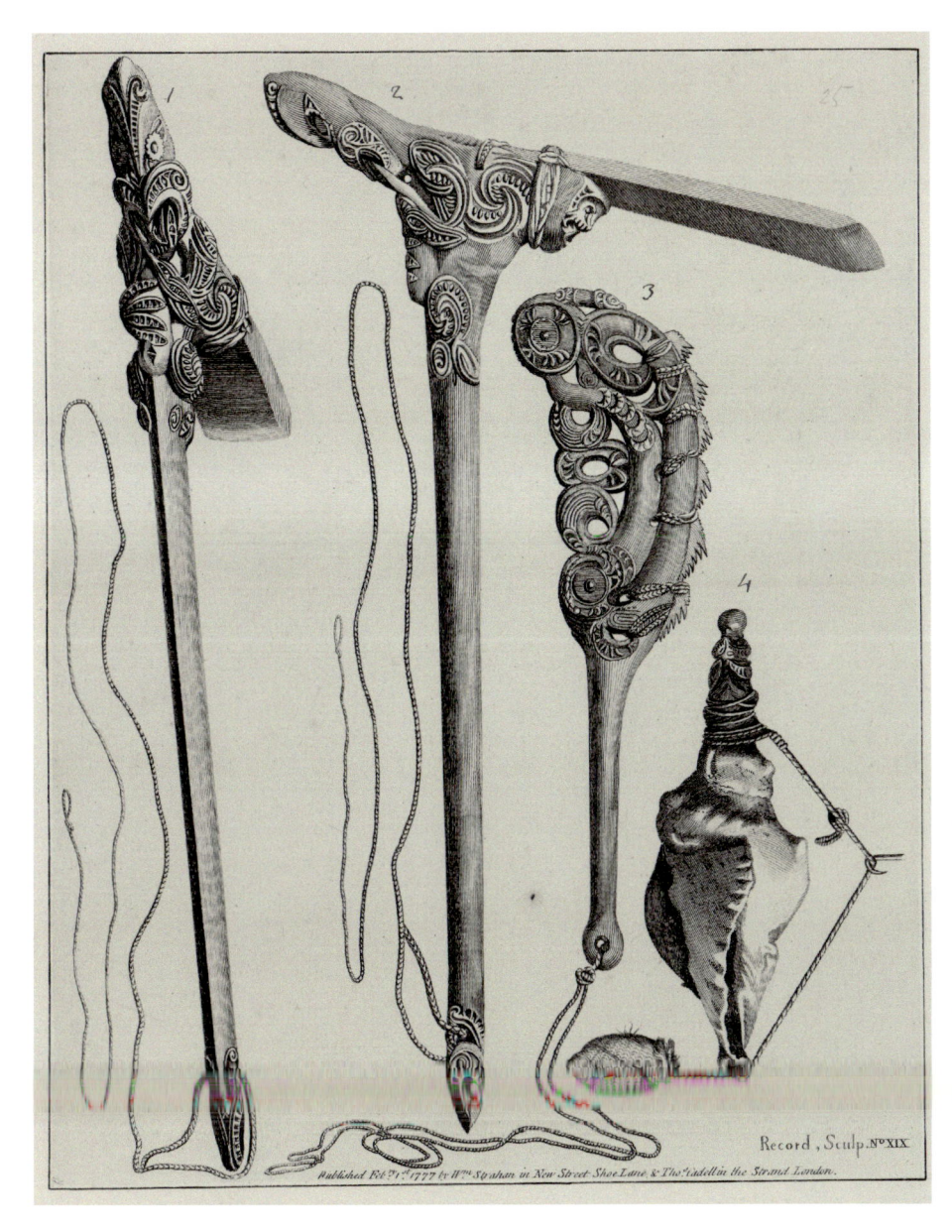

Fig. 157 (opposite)
Various articles, at the Sandwich Islands
James Record (engraver), after John Webber, 1779
National Maritime Museum (PAI4196)

Fig. 158 (left)
Various articles belonging to South Pacific natives, including weapons with heavy carving
James Record (engraver), after William Hodges, 1777
National Maritime Museum (PAI2081)

room the result of years of labour and danger: a fund of information, supported by undoubted authenticity; and a source for poignant regret, that our possession of these treasures led to the unhappy end of this illustrious seaman.[71]

Of course the exhibition of exploration did not happen just in London. Many other public institutions in Britain and across Europe acquired and displayed items and material culture derived from these voyages. As we have seen, John Webber donated objects to Bern. Johann Reinhold Forster presented some of his collection to the University of Oxford, while Anders Sparrman (a Swede who was picked up and dropped back at Cape Town as assistant botanist on the second voyage) left material to the Academy of Sciences in Stockholm.[72] In October 1771, only two months after Cook had transmitted to the Admiralty 'the bulk of the Curiosity's [*sic*]' collected on the first voyage, Lord Sandwich presented Trinity College, Cambridge, with 'a great number of curiosities brought from the newly discovered islands in the south seas', creating considerable interest among visitors to the Wren Library.[73] William Hunter, the great Scottish collector, was another who acquired a variety of 'First Contact'

EXHIBITING THE PACIFIC

pieces from Cook's voyages. These were initially exhibited in his museum in Great Windmill Street, Soho, but by the early nineteenth century the entire collection was on display in Glasgow.[74] Both James Patten, surgeon on the *Resolution*, and James King, second lieutenant on the *Resolution* on the third voyage and later commander of the *Discovery*, left large collections to Trinity College, Dublin. Patten's donation even led the College board to consider how best to exhibit this collection.[75]

Interest in collecting and displaying objects derived from exploration extended beyond metropolitan centres. Samuel Rush Meyrick's collection at Goodrich Court, Herefordshire, had a South Sea Room. In the account of a 'Tour of a German Prince', the room is described as being:

> *filled with the rude weapons, feathered cloaks, etc., of the islanders of the Pacific Ocean. Among these is a war cloak made from the plumage of the tropic bird, brought from the Sandwich Isles by Captain Cooke, and presented by the late Sarah Napier, and a cap with the representation of the whale fishery upon it, from Nootka Sound, engraved in the to his third voyage, and formerly in the Leverian Museum.*[76]

As late as 1865, the museum of the Bedfordshire Architectural and Archaeological Society contained, among other things, 'two paddles from Owhyhee, curiously carved . . . a war club from Tahiti, brought over by Captain Cook – a chieftain's cloak from one of the South-sea islands'.[77]

Sir Ashton Lever's Holophusicon was probably the most remarkable public display phenomenon of the period, and its popularity was at least partly based on the richness of its Pacific collections. Initially opened in Lever's home, Alkrington

Fig. 159
Pages from 'Thomas Pennant's *Outlines of the Globe or 'Imaginary World Tour'*, including two prints after John Webber
Thomas Pennant, 1787–92
National Maritime Museum (P/16/18)

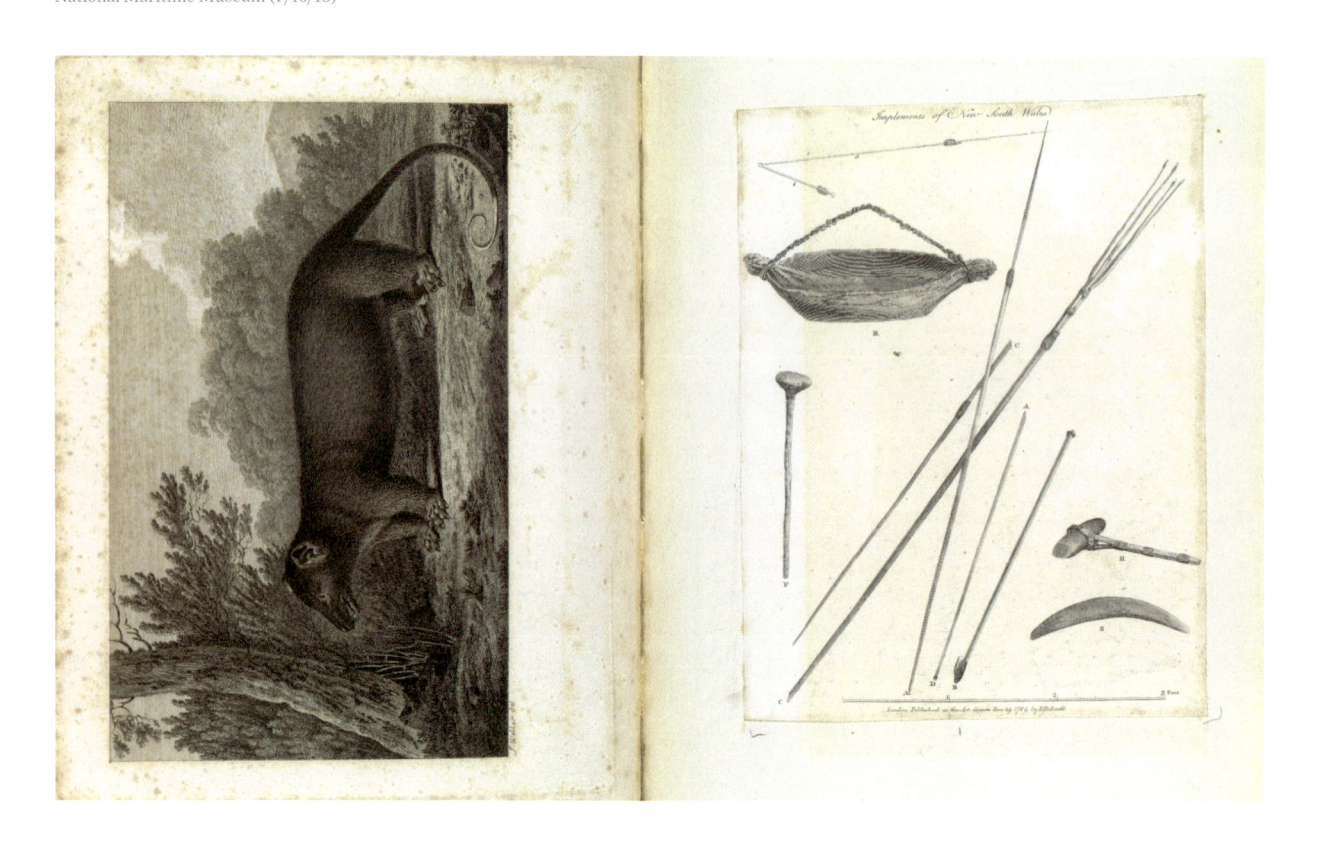

Hall near Manchester, this attempt to exhibit everything produced by nature (*holo* for whole; *phusikon* for natural) was subsequently moved to Leicester House (which dominated the north side of what is now Leicester Square) in London.[78] Objects collected on Cook's second voyage were on public display in the Otaheite Room and Club Room. By 1784, when Lever was in the process of disposing of his collection, he had 1859 Pacific objects, the majority of which must have been derived from Cook's voyages.[79] On 16 July 1778, Susan Burney wrote to her sister, the author Fanny, about a visit to the Holophusicon, where there 'were a great many things from Otaheite' already on display.[80] The Sandwich Room, named for the First Lord of the Admiralty, later opened to accommodate material from the third voyage.[81] In a public announcement dated 31 January 1781, Lever acknowledged 'the patronage and liberality of Lord Sandwich, the particular friendship of Mrs Cook, and the generosity of the officers of the voyage, particularly Captain King and Captain Williamson' in endowing his establishment. He hastened to add that he had made 'many considerable purchases himself' and was 'now in possession of the most capital part of the curiosities brought over by the Resolution and Discovery in the last voyage'. Lever's collection was 'displayed for public inspection: one room, particularly, contains magnificent dresses, helmets, idols, ornaments, instruments, utensils, etc. etc. of those islands never before discovered, which proved so fatal to that able navigator, Captain Cook, whose loss can never be too much regretted'.[82]

The public reception of this material may be partially gauged from a description of Lever's collection that appeared in *The European Magazine and London Review* of 1790, when it was still on display in Leicester House. The author of the piece was in no doubt that 'of all the spectacles contained in this opulent and extensive city, there is not one more worthy the attention of a curious and intelligent person than the Holophusicon'.[83] As well as providing information on the arrangement of the material on public display, the article also gives an insight into the ways in which these objects were interpreted for visitors:

> 15. The next is the Otaheite Room, where are numerous dresses, ornaments, idols, domestic utensils, &c. of the people in the newly discovered islands, which, to an active imagination, convey a forcible idea of them and their manners.
>
> 16. In the Club Room are the warlike weapons of the several savage nations of America. The clubs are many of them curiously carved, and some require prodigious strength to be able to wield with agility.

> 17. The Sandwich Islands Room is a continuation of the subjects in the Otaheite Room, being full of curious Indian dresses, idols, ornaments, bows, &c. which express very strongly the character of the people.[84]

For this author, and for many others besides, objects provided 'a forcible idea' of the strange nations and 'express very strongly the character of the people' and 'their manners'. Their very materiality (the 'curiously carved' clubs that 'require prodigious strength') became a key interpretive device for conveying ideas about the results of British exploration in the Pacific.

The contemporary collecting, exhibiting and interpreting of information and items derived from eighteenth-century voyages of exploration occurred at a time when British responses to the rest of the world were being rapidly forged and reshaped. The enduring appeal of objects, exhibitions and displays relating to Cook, up to the present day, demonstrates the long-lasting impact of these voyages.[85] His death on the shores of Hawai'i, however, meant that exhibiting this history often came to focus on the story of just one man. By the time Sir Ashton Lever's 'Sandwich Room' went on display in Parkinson's Museum (or, similarly, the 'Leverian Museum' at the Rotunda on the south side of Blackfriars Bridge) it was 'dedicated to the immortal memory of Captain Cook'.[86] Exhibition and display were increasingly put to the service of commemoration and celebration. Before examining this issue and its material legacies in the collections of the NMM, it is important to remember that Pacific exploration did not end with Cook's death. Some of his faithful lieutenants had unfinished business in the Pacific, and it is to their stories that we now turn.

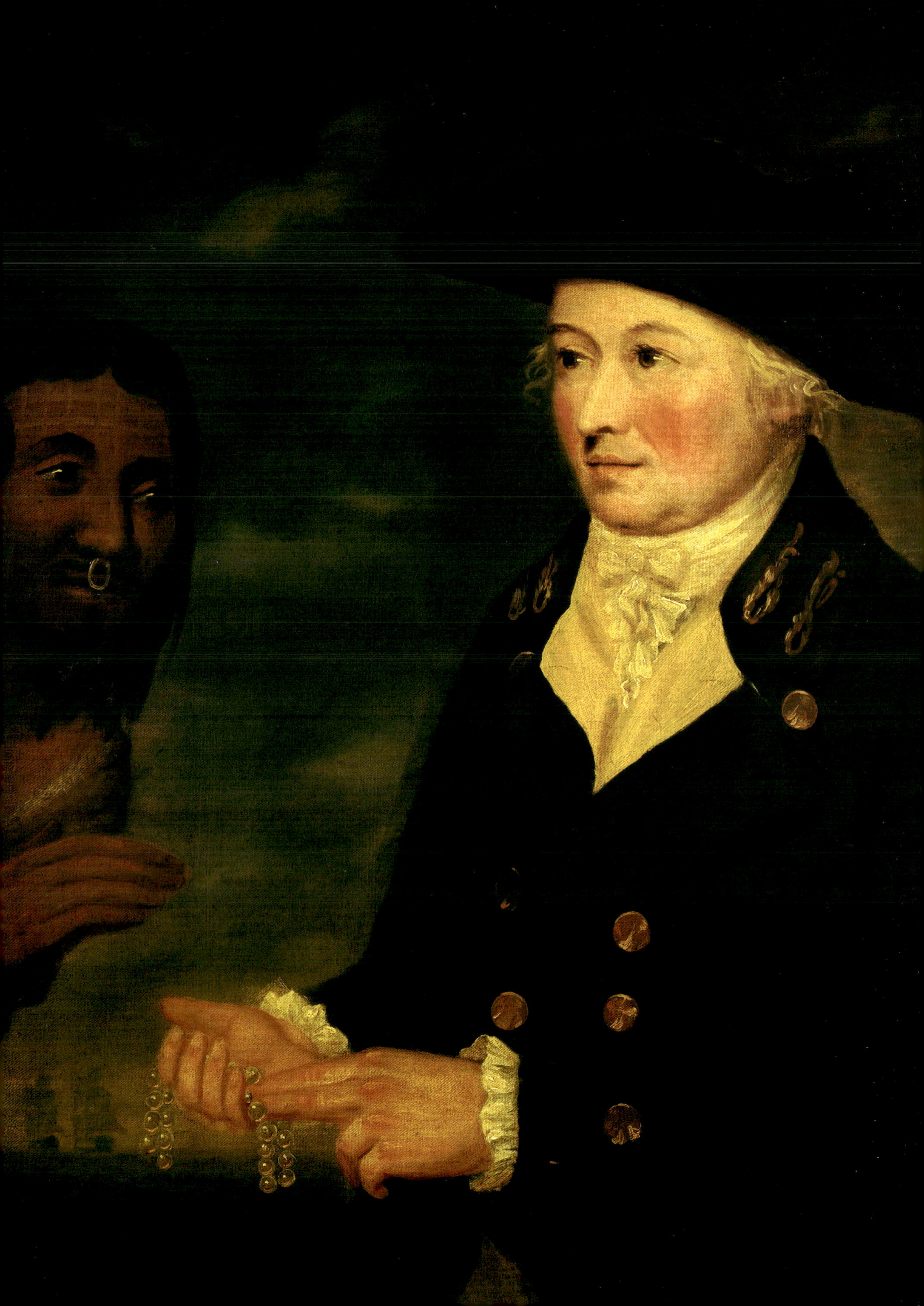

Chapter 7
'Men of Captain Cook'
Pacific Voyages 1785–1803

Nigel Rigby

'But what officers you are! you men of Captain Cook; you rise upon us in every trial'[1]

Cook bestrides the European map of the Pacific like a colossus, as well he might, for his three extraordinary voyages exploded geographical myths, replaced speculation with fact, fixed the positions of many 'wandering islands', 'discovered' others and laid down thousands of miles of coastline with great accuracy and detail. Although Pacific exploration did not die alongside Cook on that beach in Hawai'i, his giant shadow has tended to obscure those who followed him: as Roger Knight has pointed out, historical glory goes to the first explorer and to the first encounters, with little recognition for coming second.[2] Veterans of Cook's voyages, like William Bligh, George Vancouver, Henry Roberts, James Colnett, Nathaniel Portlock (fig. 160), George Dixon and Edward Riou, would all go on to command their own voyages of exploration with distinction. They would in turn train officers who, again, followed them, such as Matthew Flinders and William Broughton. With the exception of Bligh – who is unfortunately remembered for his failings, rather than as both a great navigator and a brave fighting commander – their names are little known today. There were also state sponsored voyages of exploration mounted by France, Spain and Russia, all of which were influenced to greater or lesser extents by the ghost of Cook.

There were still some major gaps in Europe's knowledge of the Pacific after Cook: the Sea of Japan, for example; the insularity or otherwise of Van Diemen's Land; and the north-western coast of America. Building on Cook's map of the Pacific would occupy the navies of Britain, France, Spain, Russia and America for many years to come. Apart from the voyage of La Pérouse, this chapter concentrates largely on the British expeditions that consciously followed Cook into the Pacific during the 20 or so years after his death, looking at their triumphs and tragedies, their contexts and the changing character of scientific exploration.

THE VOYAGE OF NATHANIEL PORTLOCK AND GEORGE DIXON

Cook's last voyage revealed to Europe an unexpected Pacific: a more threatening one, perhaps, as his death testified, but also one with serious commercial promise (fig. 161). An anonymous reviewer of George Dixon's *A Voyage Round the World but More Particularly to the North-West Coast of America* (1789) would neatly sum up the changing emphasis:

> Beside the many valuable discoveries which were made in Captain Cook's last voyage relating to geography, navigation and natural philosophy in general, there was one which, taken in commercial view, seemed to promise a new and inexhaustible mine of wealth to such as chose to be adventurers for it. The prodigious number of those animals called by the Russian discoverers, sea-otters . . . would, it might have been expected, have instantly allured the eye of commerce.[3]

Prior to Cook's third voyage, his explorations had revealed relatively little commercial significance in the Pacific: the huge and populous 'Southern Continent', for which Alexander Dalrymple had held such hope, was discovered to be a cartographical chimera, and the numerous but small islands and groups, and even the big ones like New Zealand, had little more than fresh food to trade. However, when *Resolution* and *Discovery* called at Macau on their long voyage home in 1779 they found that the sea-otter pelts they had bought so inexpensively on the

Fig. 160
Captain Nathaniel Portlock (c.1747–1817) (detail)
British school, *c.*1788
National Maritime Museum, Greenwich Hospital Collection (BHC2959)

The composition is based partly on Webber's engraving of *A Man of Nootka Sound* for the figure on the left.

north-western American coast could be sold for indecently large sums of money, with prime skins selling for 120 Spanish dollars each (£40 sterling). 'The rage', declared Lieutenant James King, captain of *Discovery* since Charles Clerke's death, 'with which our seamen were possessed to return to Cook's River [in Alaska], and by another cargo of skins, to make their fortunes . . . was not far short of mutiny'. King helpfully sketched out the likely costs and returns of such a trading voyage from China, his plan being included in the official voyage narrative published in 1784. Edited from the journals of Cook and his officers by King and the Reverend John Douglas, *A Voyage to the Pacific Ocean* introduced a new and apparently commercially viable ocean to Europe, although rumours about the sea-otter trade had been circulating almost as soon as Cook's ships returned, but were not acted upon immediately.[4] George Dixon, commander of one of the first ships to set out from England on a trading voyage to north-west America, suggested in the introduction to his *A Voyage Round the World* that the cause of the delay was the difficulty of finding men to lead such a voyage who had the necessary 'patience and perseverance' and a 'degree of enterprize' – by happy chance, men rather like himself.[5] David Mackay has shown that the actual reasons for the delay were political and commercial: trading rights were held, theoretically at least, by the East India Company, the South Sea Company and the Hudson Bay Company; and there was some scepticism about the financial and practical viability of such world-girdling commerce.[6]

The exploitation of the sea-otter trade was obviously not a Royal Naval matter, although the first voyage from London clearly involved high-level government support. A merchant named Richard Cadmon Etches formed the King George's Sound Company, purchasing two ships and employing two veterans of Cook's last voyage to command them: Nathaniel Portlock, who had been promoted to lieutenant upon the return of *Resolution* and *Discovery*, was given overall command of the expedition and of the larger ship, *King George* (320 tons, or about the tonnage of *Endeavour*), while Dixon, who had been the *Discovery*'s armourer, commanded the smaller snow *Queen Charlotte* (200 tons).[7] It was not uncommon for naval officers to serve in merchant ships if suitable employment in the Royal Navy was not available, and especially so in peacetime, when the Navy could shrink dramatically. The two ships, as Portlock pointed out in his own book of the voyage, thus had the 'size and burden that Captain Cook, after adequate trials, recommended as the fittest for distant employments'.[8] The ships were commissioned in a stage-managed ceremony at Deptford (fig. 163), attended by George Rose, Secretary to the Treasury; the influential Sir Joseph Banks (the 'common Centre of we discoverers', as James King described him);[9] Lord Mulgrave (the polar explorer Constantine Phipps); and Sir John Dick, an ex-diplomat with some knowledge of Russian affairs and the Commissioner for the Inquiry into Fees in Public Offices.[10] The names of the ships further hinted at royal approval, and indeed Portlock was later given permission to dedicate his book to the king, while Dixon's was dedicated to Banks.

Several 'gentlemen's sons, who had shewn an inclination to engage in a seafaring life,' wrote Portlock, 'were put under

Fig. 161 (left)
Title page to *A voyage to the South Atlantic and round Cape Horn into the Pacific Ocean for the purpose of extending the spermacetti whale fisheries and other objects of commerce.*
James Colnett, 1798
National Maritime Museum
(PBG1181)

Fig. 162 (opposite)
A Sea Otter
Peter Mazell (engraver),
after John Webber, 1778
National Maritime Museum
(PAI3916)

my care', effectively as midshipmen, 'for the purpose of being early initiated in the knowledge of a profession which requires length of service, rather than supereminence of genius'.[11] One of the boys, John Gore, had received a berth at the request of the John Gore who had taken command of Cook's last voyage, been promoted to post-captain on his return and was, by 1785, enjoying the 'ease and retirement' that Cook had relinquished, as one of the four resident Captains of Greenwich Hospital, in order to make his final voyage.[12] Portlock would also receive this mark of favour in later life. Two pupils of Cook's astronomer William Wales, who was by then master of the mathematical school at Christ's Hospital, were employed to accompany and teach the young gentlemen 'the rudiments of navigation' and construct charts of useful harbours.[13] The 'best provisions' were purchased and 'an unremitting attention to the rules observed by Captain Cook' was initiated,[14] including his three-watch system, the regular airing and drying of clothes and quarters, and careful attention to preventing scurvy. The King George's Sound expedition was being 'marketed', we might say today, as 'Cook IV' by using Cook's ships, officers, on-board regime, navigational precision and cartographic expertise, and governmental – even royal – blessing.

The voyage was profitable, but not spectacularly so, in part because the ships arrived on the American coast rather too far north and late in the brief sub-Arctic summer season to acquire a full cargo of furs. They overwintered in Hawai'i and returned to America the following summer, sensibly sailing independently to get a better coverage of the coast, and arrived in Macau with more than 2000 good-quality pelts. Portlock and Dixon complained (inaccurately, argues historian David Mackay)[15] that the profits were lower than expected because, under the terms of the original agreement, all sales had to go through the East India Company, which eventually sold the pelts 'for less than *twenty dollars* each'.[16]

Although it was a commercial voyage dressed in scientific clothing, the geographical results were respectable and added valuable information to Cook's surveys of the American coast. It produced three publications (see note 5), of which Portlock's was based on his journal and filled with navigational and practical information of great interest to his profession, if less so to a general reader; for example, giving a detailed account of Cook's method of salting down pork in the heat of the tropics. Portlock's pilotage notes are clear and reliable, and his charts, if not extensive, are serviceable. In the interests of readability, Portlock does, however, relegate to appendices daily navigational information such as date, time, course, latitude, longitude, compass variation, weather and general remarks, along with a list of the plants, birds and fossils found in Cook's River.

Fig. 163
Model of the Royal Dockyard at Deptford
Thomas Roberts; William Reed, 1772–74
National Maritime Museum
(SLR2906)

Portlock's hesitant investigations into the natural history of north-west America would soon be built upon, for in 1786 Sir Joseph Banks persuaded Etches to employ a naval surgeon trained in botany, Archibald Menzies, on a second fur-trading voyage led by yet another Cook officer, James Colnett, in command of *Prince of Wales* and *Princess Royal*. Banks would later get Menzies appointed as supernumerary naturalist on George Vancouver's survey of the American coast, an untidy confusion of roles with serious consequences that will be discussed later in this chapter. What the commercial voyages lacked, for obvious reasons, was consistent attention to cartography: their priority was finding furs and so they tended to navigate the coast in a series of hops, leaving large gaps between the more detailed charts and sketches of places where they had found good opportunities for trade.

THE TRAGIC VOYAGE OF JEAN-FRANÇOIS DE GALAUP DE LA PÉROUSE

Using Cook and his methods as an organizational touchstone would become commonplace on state-sponsored explorations, although non-British voyages in particular were more obviously poised between admiration of Cook and a desire to maintain a certain distance from his methods. Spain's Alejandro Malaspina firmly pointed out at the beginning of his journal that his voyage of 1789–94 could not be compared to Cook's, for it was not a voyage of discovery, but one to inspect Spain's Pacific empire. Malaspina's recent editors argue convincingly, however, that the frequent references to Cook in his journal show that he was seldom far from Malaspina's thoughts and that 'there is no doubt that Cook's voyages were the yardstick against which Malaspina measured his own'.[17]

Perhaps none of Cook's many disciples followed him as self-consciously as did the experienced and highly regarded

172   CAPTAIN COOK AND THE PACIFIC

French navigator Jean-François de Galaup de la Pérouse, whose voyage between 1785 and 1788 ended tragically with the loss of both his ships on the reefs of Vanikoro in the Santa Cruz Islands – a fate that would not be discovered for another 40 years. The voyage was planned as an equal, natural extension of Cook's third, in which Louis XVI – himself an enthusiastic amateur geographer and keen student of exploration – was actively involved. La Pérouse requested three copies of Hawkesworth's *Voyages Undertaken...*, three of Cook's second voyage account and no fewer than 12 of the third (in English and in translation), asking particularly for the second edition, only just published in 1785, which he believed to be more accurate.[18] One of his civilian supernumeraries, Paul-Antoine de Monneron, was despatched to London to obtain the most appropriate scientific instruments for maritime exploration and to hear the latest thinking on scurvy – ever the scourge of long voyages. Monneron did not advertise the real purpose of his visit, but disguised himself, as John Dunmore has remarked, in true comic-opera tradition as the accredited agent of a Spanish grandee.[19] However, his mission must have been obvious to all, for he spent a total of £4000 on the latest scientific instruments, consulted Joseph Banks and commissioned John Webber to paint his portrait. His purchases included azimuth compasses from George Adams; theodolites, steering compasses, night telescopes, thermometers and sextants from Jesse Ramsden (figs 164, 165); a sextant and pantograph from Troughton's;[20] and barometers from Nairne and Black. Joseph Banks, who since Cook's death had become an increasingly influential voice in the matter of maritime exploration, contributed to La Pérouse's instructions and was generous with his help and advice, even persuading the Board of Longitude to let La Pérouse have two of the Nairne dipping needles used on Cook's last voyage, which Monneron had been unable to buy in London.[21]

The La Pérouse expedition was intended to be scientific voyaging on a grand scale, far exceeding Cook's and even surpassing Joseph Banks's ambitious but unrealistic plans for a large scientific party on Cook's second voyage. La Pérouse's two ships, *Boussole* and *Astrolabe*, eventually left France with a team of 15 civilian scientists and artists (figs 167, 168). For a voyage that was planned as the natural successor to Cook's third, this was a significant departure from the example of the 'great navigator', whose scientific parties had shrunk as his own reputation and influence within the Admiralty had grown.[22] La Pérouse's journey had two overriding objectives: first, to examine the Sea of Japan – 'the only part of the globe that has escaped the attention of the tireless Captain Cook';[23] and second, to survey the coast of north-west America, seek the still-elusive North-West Passage and assess the potential of a trans-Pacific fur trade. La Pérouse's orders, however, gave him considerable latitude and he changed the suggested route, heading north as soon as he entered the Pacific and making for Easter Island, the Hawai'ian islands and reaching the American coast at modern-day Yakutat Bay, Alaska. This left La Pérouse

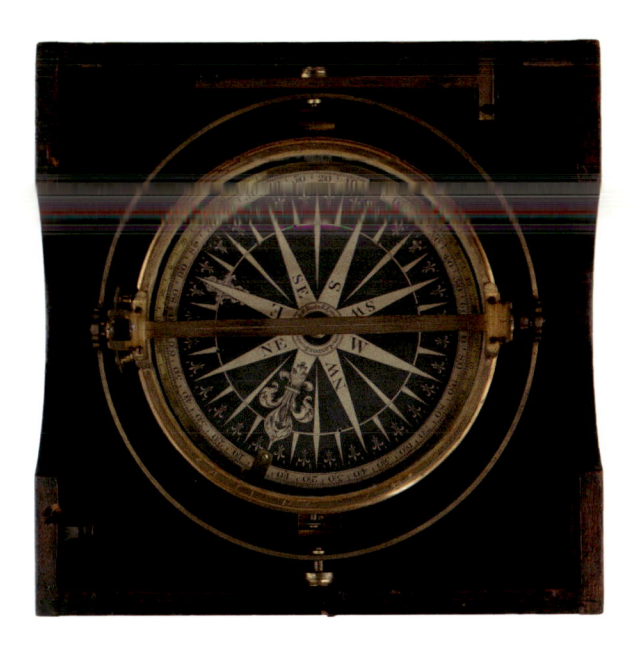

Figs 164, 165
Azimuth compass, Rust & Eyres, c.1770–90 (left), and sextant, Jesse Ramsden, c.1785 (right)
National Maritime Museum (NAV0289) and Adams Collection (NAV1104)

MEN OF CAPTAIN COOK 173

Tableau des Decouvertes du Cap.ne Cook, & de L[...]

1. Hab.ts de Nootka.–2. Hab.ts de la Zelande.–3. Hab.ts de l'Entrée du Prince Guillaume.–4. Hab.ts de l'Ile de Pâques.–5. Hab.ts de la Bay[e]
7. Hab.ts de Tanna.–8. Hab.ts de S.te Christine.–9. Hab.ts de la Baye de Castries.–10. Hab.ts de la Baye ou Port des França[is]
13. Hab.ts de la Baye de Langle.–14. Hab.ts de la Conception.–15. Hab.ts de la Baye des Manilles.–16. Hab.ts des Iles Pele[w]
19. Hab.ts des Iles Marquises.–20. Hab.ts de l'Ile des Amis.–21. Hab.ts de la Nouvelle Caledonie.–22. Hab.ts d'Otaiti.–23. Hab[.ts]

Tout Contrefacteur sera poursuivi d'après la Loi, le depot étant fait à la Bibliothèque Nationale.   l'an — de la République Française.   Par Jacques Gra[...]

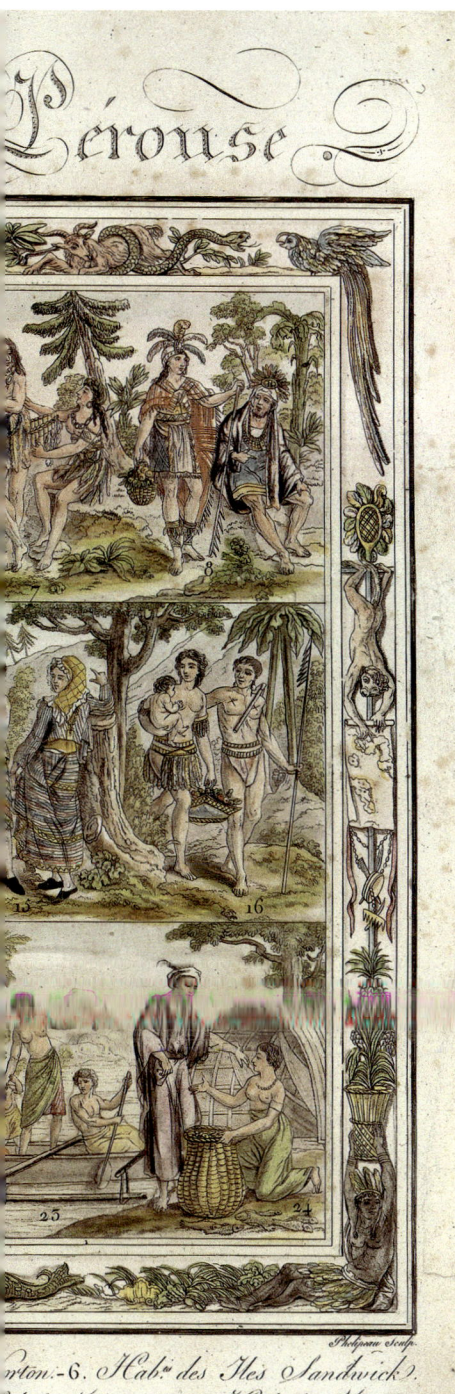

Fig. 166
Tableau des Decouvertes du Capne Cook, & de la Perouse. Figures in costume from many various places surveyed around the world.
Phelipeau (engraver), after Jacques Grasset de Saint Sauveur
National Maritime Museum (PAG7907)

The largely fanciful engraving of the lands and peoples encountered by Cook and La Pérouse was produced on the posthumous publication of La Pérouse's voyage narrative in 1797, edited by M.L.A. Milet-Mureau.

MEN OF CAPTAIN COOK 175

Figs 167, 168 (left)
Medal of Lapérouse's departure, commemorating the expedition of the ships *Boussole* and *Astrolabe*
Pierre Simon Benjamin Duvivier, 1785
National Maritime Museum (MEC0826)

Fig. 169 (opposite, above)
Views in the South Seas ... View in Macao
John Webber, 1809
National Maritime Museum (PAH9588)

with three months to survey this long, endlessly indented and confusing coastline, since the timetable was being driven by the need to arrive in the Sea of Japan early the following summer. The plan was hopelessly optimistic: it would take George Vancouver, admittedly a perfectionist, three surveying seasons.

A few days later, the French expedition entered Lituya Bay, which was not marked on Cook's charts, but was of a size and conformation that offered the hope of its being the mouth of the North-West Passage. The bay, which they named Port des Français, and the coastline either side of it were duly surveyed, but hopes of a route through the American interior were quickly shattered by the discovery of two huge glaciers closing off the bay after little more than five miles. However, they had better success in evaluating the fur trade, for they were 'constantly surrounded by native canoes' offering skins of sea otters and other animals. They found that buying pelts for a handful of glass beads was already a thing of the past, for La Pérouse noted that the locals were 'as skilful in their deals as the ablest buyers in Europe' and were only really interested in iron.[24] *Astrolabe* and *Boussole* sailed south, surveying as much of the coast as possible in the dense fogs and limited time available, and reached the recently established Spanish settlement of Monterey in the middle of September 1786. The ships then crossed the Pacific to arrive in Macau on 1 January 1787 with 1000 furs on board (fig. 169). Unsurprisingly, given their encounters with the Native American traders, they found that, in the seven years since *Resolution* and *Discovery* sold their cargoes, there had already been a number of fur-trading voyages from China and Bombay and the value of the pelts had plummeted in the expectation of a glut in the market.

The French ships left Macau and headed north to begin their examination of the Sea of Japan and continue up to Kamchatka. The area had been little sailed by Europeans and its survey was 'one of the most rewarding parts of the entire voyage', writes La Pérouse's modern editor and biographer, John Dunmore.[25] They were warmly welcomed at Petropavlovsk in remote Kamchatka, a recent extension to Russia's empire, albeit one where visitors were rare beasts. The explorers visited the grave of fellow explorer Charles Clerke, the popular captain of *Discovery* and later of *Resolution*, who had been buried there in 1779, and received permission from the governor to replace, with an engraved copper plate, the carved inscription on Clerke's wooden tombstone. More importantly, they received orders from France to investigate reports that Britain was establishing a colony in Botany Bay, New South Wales. It was at this point that Ferdinand de Lesseps, a young Russian-speaking supernumerary originally employed to assist in encounters with Russian fur traders in America, left the voyage and began a long and arduous overland journey back to Paris, taking with him La Pérouse's reports, papers and journals. Over 40 years later, de Lesseps was able to confirm that relics found on Vanikoro by the Irish trader, Peter Dillon, did indeed come from *Boussole* and *Astrolabe*, which finally solved the mystery of the expedition's later disappearance with all hands.

The ships worked their way south through the Navigators' Islands (Samoa), where 12 men, including the captain of *Astrolabe*, Fleuriot de Langle, and two of the scientific party, were killed in an attack on a watering party. They entered Botany Bay just as Commodore Arthur Phillip was moving his nascent penal colony further up the coast to the more promising location of Port Jackson, which had been seen by Cook, but not entered. French and British socialized briefly and amicably, the latter being particularly appreciative of the French visit to Clerke's grave. A few weeks later in March 1788, *Astrolabe* and *Boussole* sailed from the bay to meet their fate. Once the expedition was officially declared missing in 1791, France launched a search mission under the command of Bruny d'Entrecasteaux, but found no trace of the lost ships (fig. 171). In a moment of sad irony, d'Entrecasteaux sailed past uncharted Vanikoro, one of scores of tiny islands they sighted on the voyage, but did not stop to investigate. It is now believed that some of the survivors could still have been alive there at that time, although this has not been firmly established.

Fig. 170 (left)
Monument to Monr de la Perouse and his companions, erected at Botany Bay
W. Sproat (engraver, printer), after Paul Westmacott, c.1826
National Maritime Museum (PAD2134)

The monument's foundation stone was laid in September 1825 by a descendant of the French explorer Louis-Antoine de Bougainville.

Fig. 171 (overleaf)
Naufrage de l'Astrolabe sur les recifs de l'ile de Vanikoro
Louis Le Breton, c.1843
National Maritime Museum (PAH0750)

An artistic impression of the shipwreck by the surgeon and artist on a later Pacific voyage, Louis Le Breton.

MEN OF CAPTAIN COOK   177

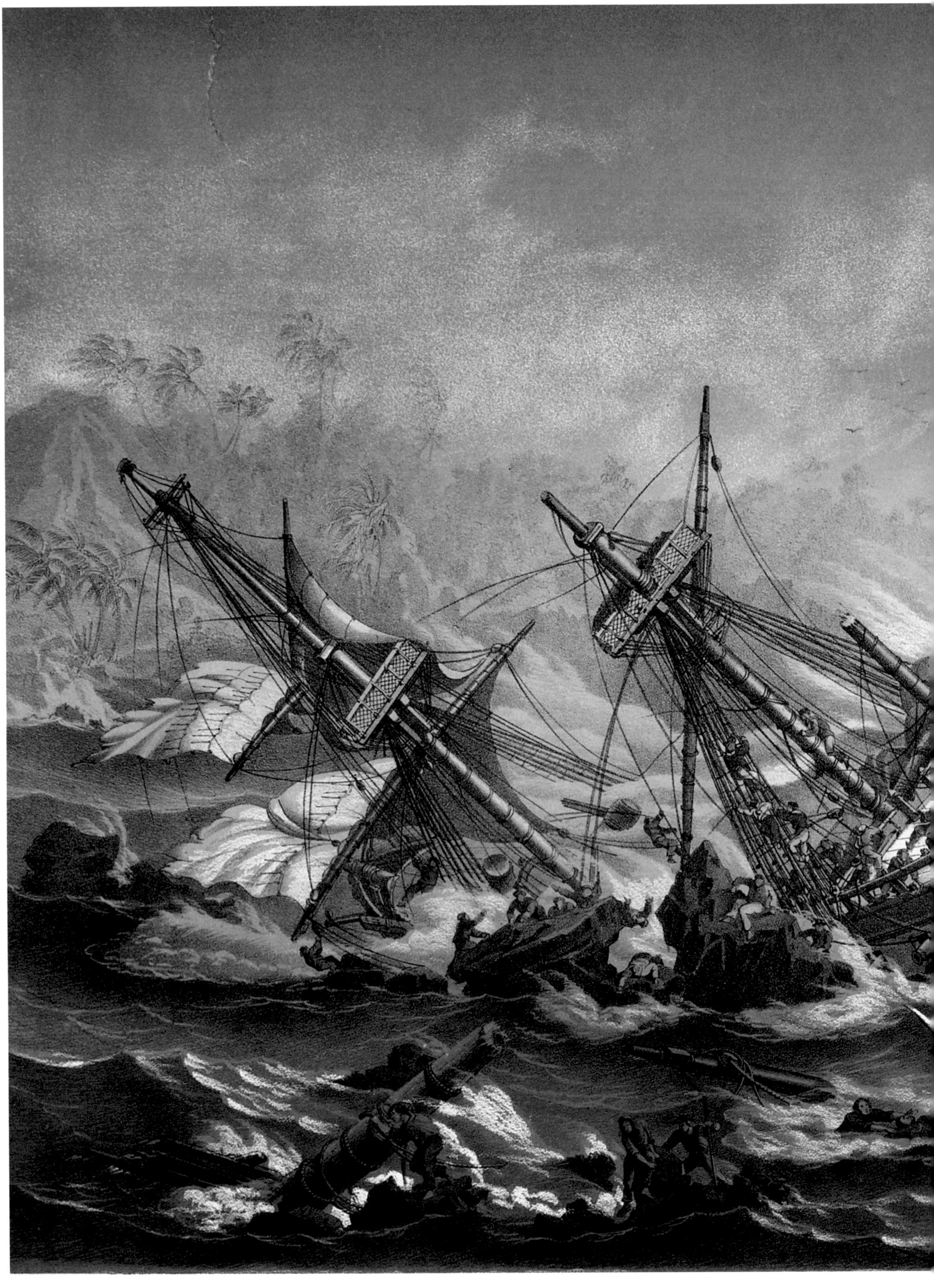

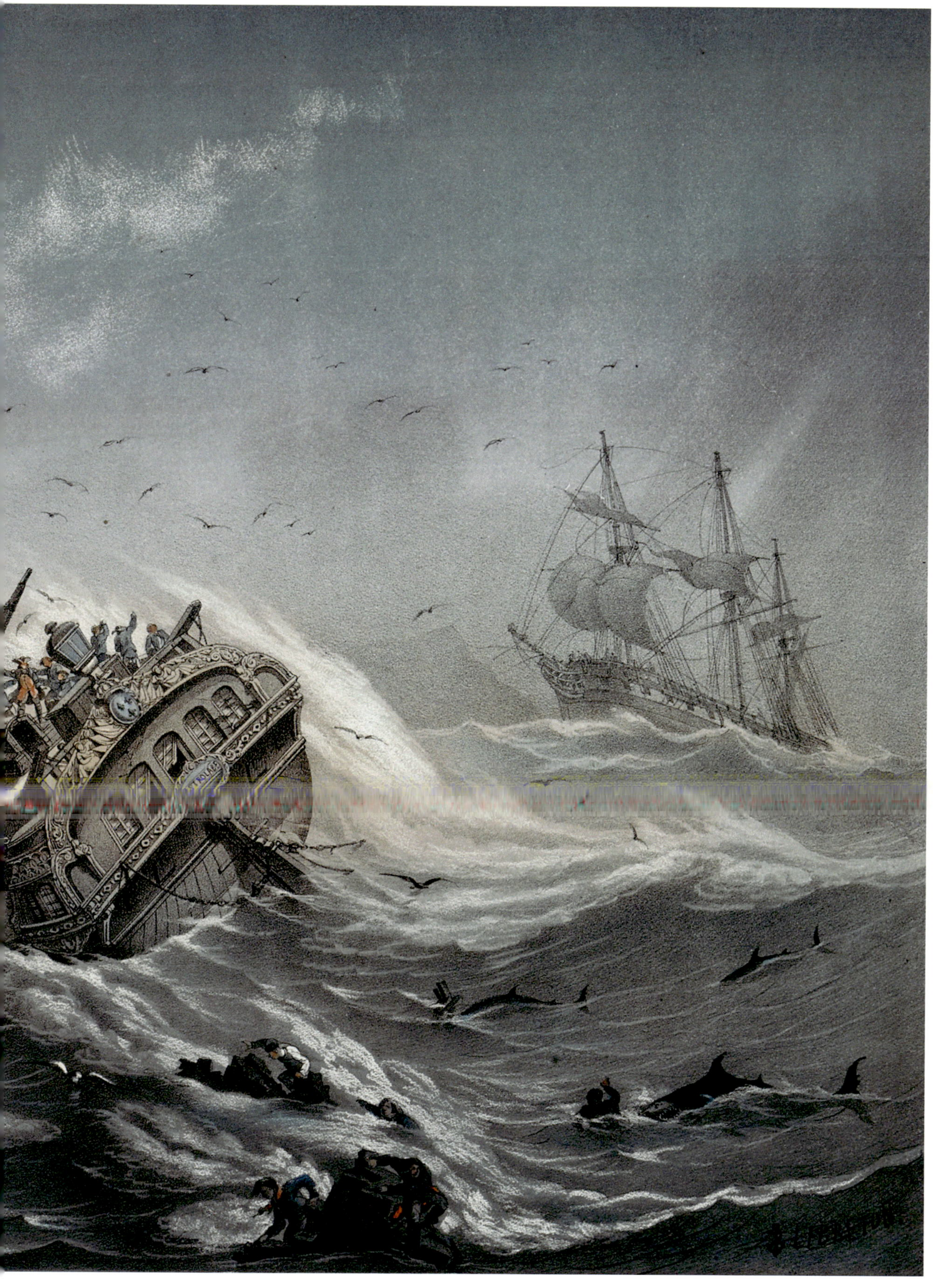

THE TWO VOYAGES OF WILLIAM BLIGH

As *Astrolabe* and *Boussole* sailed out of Botany Bay, another ill-fated ship was rounding the Cape of Good Hope bound for the Pacific: HM armed vessel *Bounty*, commanded by the old sailing master of Cook's *Resolution*, William Bligh, who was no longer a warrant officer, but a commissioned lieutenant since 1781 (fig. 172). His orders were to collect a cargo of breadfruit plants from Tahiti and transport them to the West Indies, where it was hoped they would flourish in the warm, oceanic climate and would in time provide a cheap staple food for plantation slaves. Bligh had originally been ordered to enter the Pacific around Cape Horn or through the Straits of Magellan – since on paper this was the shortest route to the Society Islands – but extremely strong westerly winds and eastbound currents made this impossible, so Bligh sensibly turned around and made for Tahiti via Cape Town and New Zealand instead. Owing to Admiralty delays, *Bounty* had been late leaving England and, combined with the abortive attempt to round Cape Horn, this meant that the expedition reached Tahiti out of the season for successfully transplanting young breadfruit trees (fig. 173). The enforced wait to do so contributed to the disintegration of normal order that followed.

The mutiny on the *Bounty* would become one of the most powerful European stories of Pacific exploration, one that is possibly as well known today as it was at the time. It has been told and retold in books, journals, courts of law, poems, theatre, school packs, museum exhibitions, television and film.[26] There are three main elements to the *Bounty* saga: the breadfruit mission and the mutiny itself; Bligh's subsequent extraordinary open-boat voyage; and, as a coda, the mutineers' return to Tahiti and then to their hideaway – and generally grim fates – on remote Pitcairn Island.

The original plans held by the National Maritime Museum show the alterations that were needed to convert the small merchant ship *Bethia* into the armed vessel *Bounty*, a ship of around 90 feet and 212 tons, smaller than either *Adventure* or *Discovery* and under half the size of *Resolution* (fig. 174). Through Sir Joseph Banks, West Indian merchants had successfully petitioned George III for such a ship to transport a large number of breadfruit plants from Tahiti to the West Indies, although the evidence of the plans suggests that the Admiralty may not have been entirely committed to the project. *Bounty* was far too small for the long, two-year mission, and the plans clearly show the effect that 700 plant pots would have had on the living quarters, which also had to accommodate 46 men. Officers and men were constantly and inescapably on top of each other, and little thought had been given to the inadequate command structure inherent in the manning regulations for a vessel of the small size: for, unlike Cook's experience when he was a lieutenant in the *Endeavour*, normal rules were not sensibly 'bent' in this regard for Bligh, who was the only commissioned officer on board. Overcrowding, ill-matched personalities and divided loyalties would become key factors in the mutiny.

This took place on 28 April 1789, just over three weeks after a fully loaded *Bounty* left Tahiti. The immediate and absurd cause was an argument between Bligh and his acting lieutenant, Fletcher Christian, over missing coconuts, but there were other underlying factors: Bligh's intolerance of inefficiency and poor seamanship (in which he excelled); his verbal abuse of those who fell short of his standards; his readiness to undermine his officers' authority; and his inability to see how this sapped their self-confidence. Once Bligh had delivered his 'punishment', the matter was over as far as he was concerned, but he was never able to appreciate how deeply his outbursts could wound people. His behaviour could also seem inconsistent, swinging violently and unpredictably from affability to fury. As a general rule, British seamen could accept harsh captains, as long as they

Fig. 172 (left)
Captain Bligh, frontispiece to *A Voyage to the South Sea*, 1792
John Condé (engraver), after John Russell
National Maritime Museum (PAD3277)

Fig. 173 (opposite)
A branch of the bread-fruit tree with fruit
John Miller (engraver), after Sydney Parkinson, 1773
National Maritime Museum (PAI3959)

180   CAPTAIN COOK AND THE PACIFIC

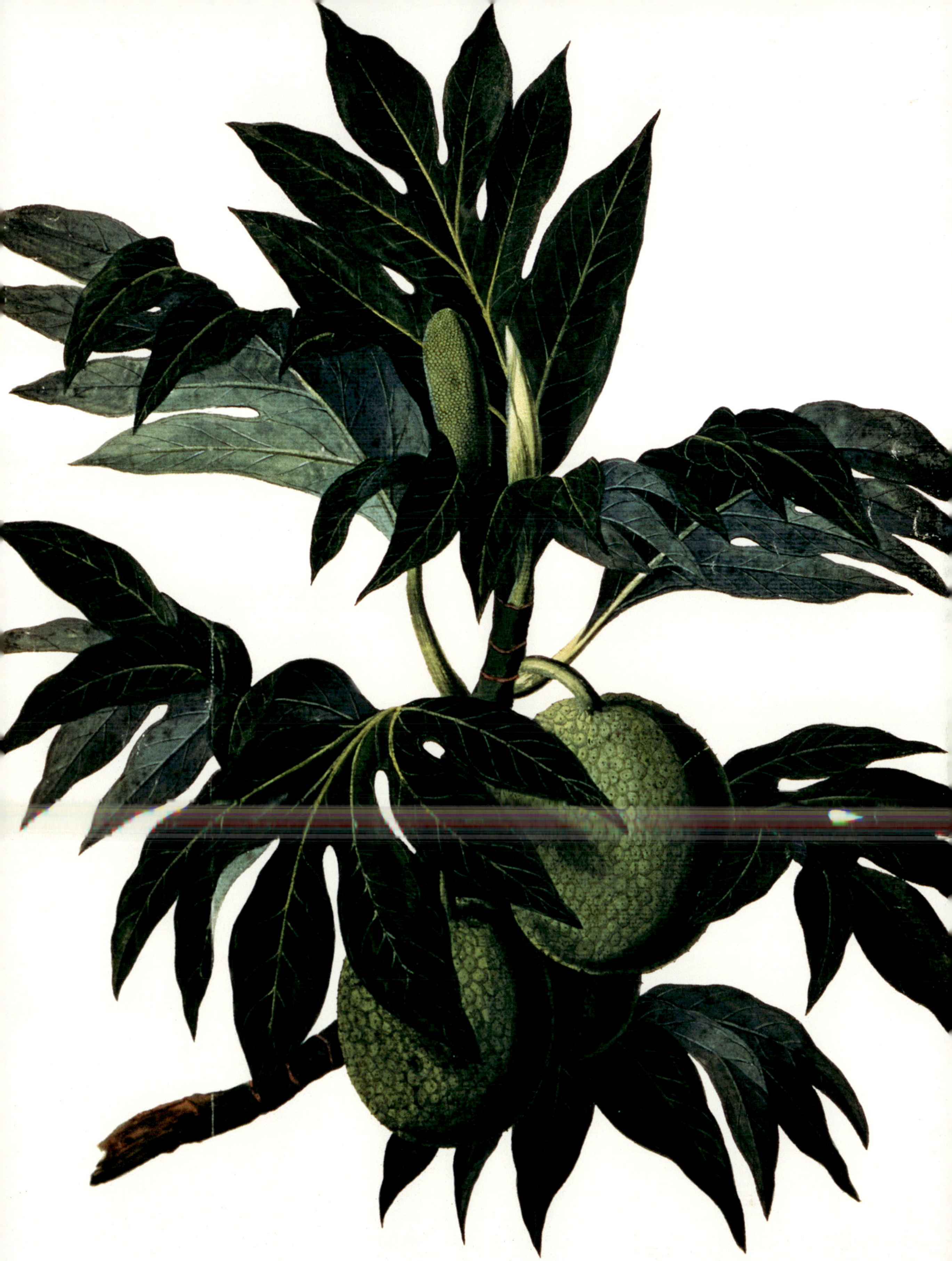

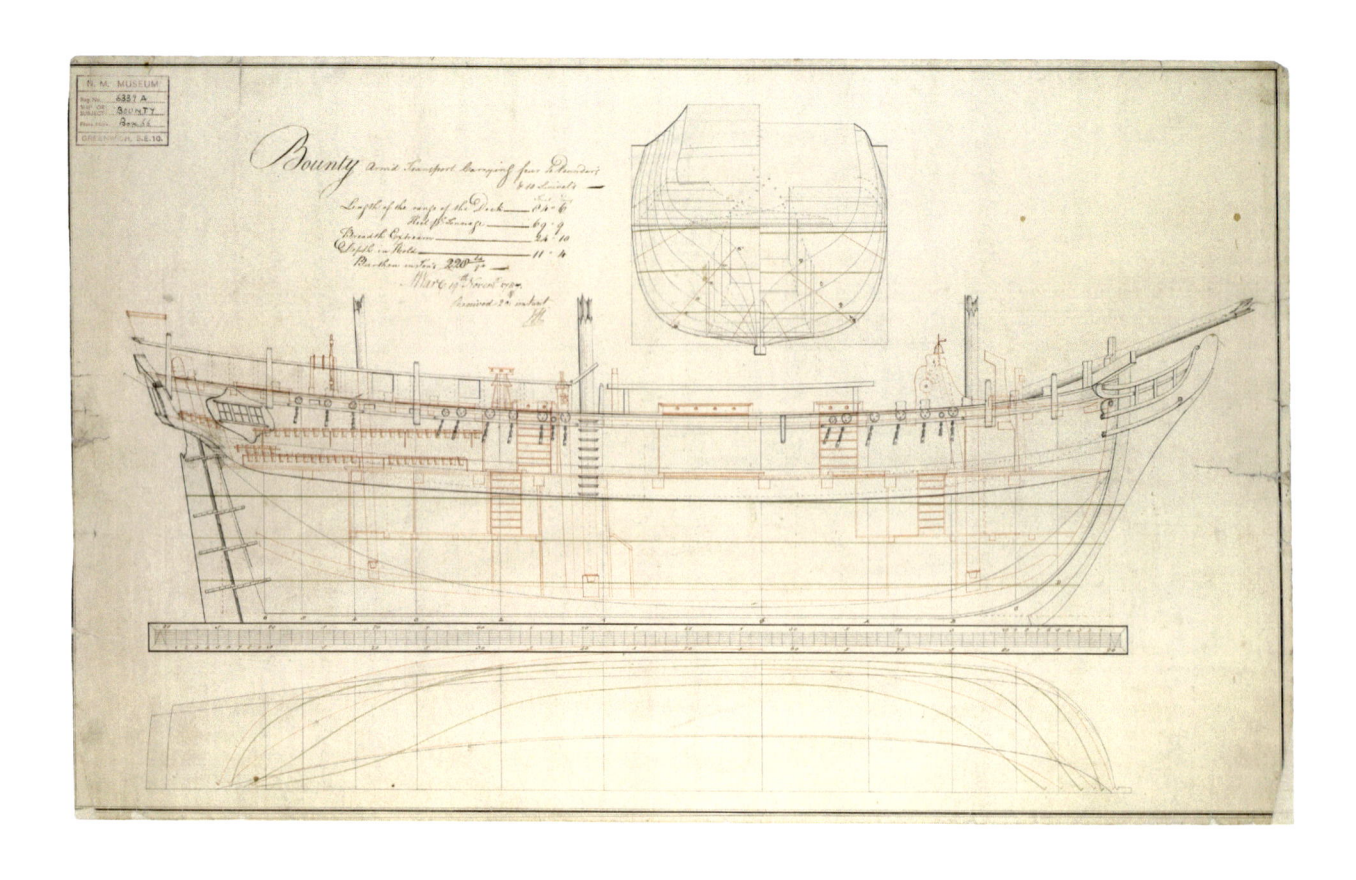

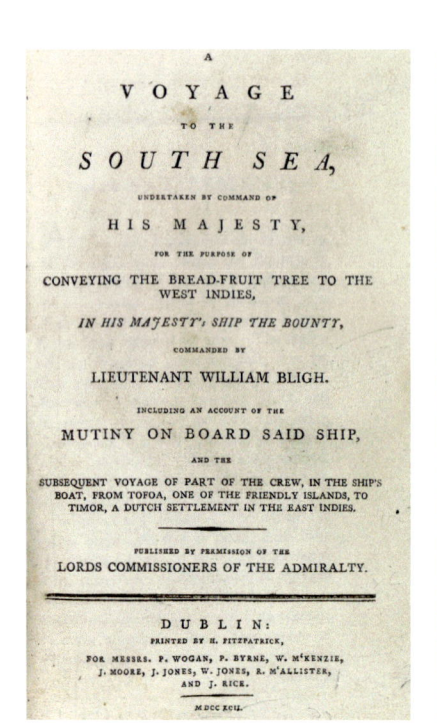

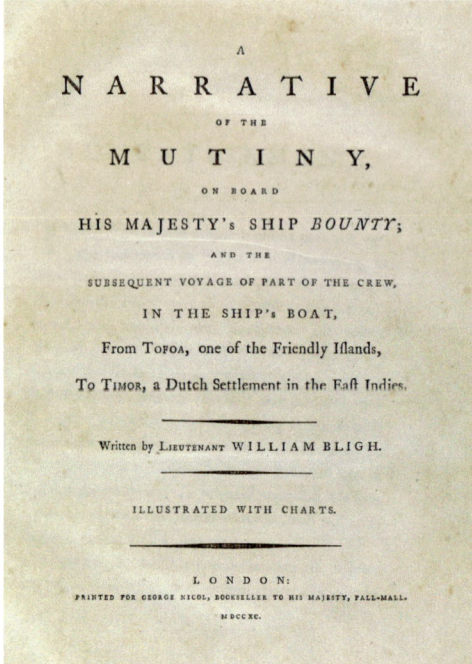

Fig. 174 (above)
Plans of the armed transport *Bounty* (detail)
1787
National Maritime Museum
(ZAZ6668)

Fig. 175 (far left)
Title page to *A Voyage to the South Sea*
William Bligh, 1792
National Maritime Museum
(PBC4219)

Fig. 176 (left)
Title page to *A Narrative of the Mutiny on Board His Majesty's Ship* Bounty
William Bligh, 1790
National Maritime Museum
(PBE3399)

knew where they stood with them. It has also been suggested that Bligh was cheating his crew by manipulating their rations for financial gain, which is very possibly true, although certainly not an exceptional malpractice in the service. What is remarkable about the mutiny is that such a small number of officers and men were able to take the ship: for there was also a strong loyalist element, but not one that was prepared to go beyond urging caution and actually fight for their captain. Bligh had, for whatever reason, 'lost' the ship in more senses than one.

Bligh and 18 loyal men were put into a 23-foot open boat with a week's rations, a sextant and a compass, but no chart or firearms, and were abandoned thousands of miles from the nearest safe refuge (fig. 177). It took them 48 days to reach Timor. That they did so at all is astonishing, and their survival was due almost entirely to Bligh's determination, navigation and seamanship: he measured out the meagre rations, using a pistol ball to weigh the daily allowance of bread, and a beaker for the thrice-daily ration of water (fig. 178). He returned to England the hero of the hour. Although he duly faced a routine court martial for the loss of the *Bounty*, he was honourably acquitted and appointed to the command of a second breadfruit voyage, which left Britain in the summer of 1791 and returned in 1793.

Fig. 177
The Mutineers turning Lieut Bligh and part of the Officers and Crew adrift from His Majesty's Ship the Bounty [29 April 1789]
Robert Dodd, 1790
National Maritime Museum
(PAH9205)

Bligh thus missed the return of Captain Edwards, who had been ordered to the Pacific in the frigate *Pandora* to search for the mutineers in 1790. Edwards captured 14 of the crew who had elected to stay on Tahiti rather than roam the Pacific looking for a safer place to hide, but could find no trace of Christian and the remaining mutineers. Four of those whom he captured died in the wreck of *Pandora* on the Great Barrier Reef, but ten were brought back for trial after Edwards also managed to reach Timor, with the surviving crew, in the ship's boats. Four were acquitted and six sentenced to death, although only three were finally hanged.

After an abortive attempt to find a safe and remote haven 400 miles south of Tahiti on the island of Tubuai, Christian took

Fig. 178 (left)
Coconut bowl made and used by Bligh to eat his ration while adrift; plain horn beaker used by Bligh to measure water rations at a daily rate of 3/4 pint per man; bullet used as a weight to measure out rations during the voyage of the *Bounty*'s launch to Timor (the plate is engraved: 'This bullet 1/25th of a lb was the allowance of bread which supported 18 men for 48 days served to each person 3 times a day') 1789
National Maritime Museum (ZBA2701; ZBA2703; ZBA2702)

Fig. 179 (below)
Kendall's marine timekeeper, K2
Larcum Kendall, 1771
National Maritime Museum, Royal United Services Institution Collection (ZAA0078)

the ship back to Matavai Bay, where he left a few who wanted no further part in the mutiny. These were the men who would be captured by Edwards and tried in England, including William Morrison, the *Bounty*'s gunner (eventually pardoned at the court martial). He left one of the most important accounts of the mutiny and the first description of Tahitian culture and society by someone who had lived there for more than a few weeks, for the mutineers lived among the Tahitians for some 18 months before recapture. A copy of the journal, clearly written by Morrison while waiting for the court martial, was found only recently in an archive of letters acquired by the NMM in 2012, though its content has long been familiar, and published, from another copy in Australia.[27]

The third element of the story relates to the fate of Christian and his party of mutineers, who eventually found the refuge they sought in the small island of Pitcairn. This was first sighted in modern times by a midshipman of that name in the *Swallow*, during Phillip Carteret's circumnavigation in 1767, and its position was published in Hawkesworth's *Voyages Undertaken*. Carteret, however, had plotted it incorrectly some 3° of longitude west of its real position. Cook tried and failed to do so on his second voyage, but had no time to investigate further: Christian did, and perhaps no other realistic option. Within a few years, though, only one mutineer was left alive on Pitcairn: John Adams (who had joined the *Bounty* under the name Alexander Smith). The rest of the *Bounty* rebels and the few male Tahitians with them were mostly killed in a series of sordid and brutal disputes. Pitcairn's population of John Adams, 11 Tahitian women and 23 children would not be discovered for nearly 20 years, first in 1808 by an American whaling captain, Mayhew Folger, whose rapid report to the Admiralty was virtually ignored – as indeed were the reports of Captains Staines and Beechey, who 'found' Pitcairn once more in 1814. Adams gave or sold the *Bounty*'s chronometer (K2; fig. 179) – previously issued to Constantine Phipps in 1773 for his search for the North-East Passage – to Folger, but it was quickly confiscated from him by the Spanish governor of Juan Fernandez Island, disappeared into South America and was only returned to England and presented

Fig. 180
Susan Young, The only surviving Tahitian woman, Pitcairn's [Island], *Augt 1849*
Edward Gennys Fanshawe, 1849
National Maritime Museum
(PAI4615)

to the Royal United Services Institute Museum in London in 1843. When that closed in 1963, it was transferred to the NMM. During the nineteenth century Pitcairn would become almost a mandatory call for Royal Navy ships entering the Pacific. The only surviving sketch of one of the mutineers' wives – Susan Young, wife of Midshipman Ned Young (fig. 180) – was one of three made on Pitcairn in 1849 by Captain (later Admiral) Edward Gennys Fanshawe, which, like a number of other objects in the NMM collections, suggests the enduring fascination that the island held for visitors, both professional and civilian.

There is a coda to the story, one largely ignored at the time and still much less well known today: Bligh's second and successful breadfruit voyage, which sailed for the Pacific in 1791. Once more, Joseph Banks was a great champion of the voyage, as he had been of Bligh from the moment he had reached London in March 1790. This time – with a public point to prove – the Admiralty made no mistakes and Bligh, instead of being given command of an enterprise that he had almost no part in setting up, as was the case with the *Bounty*, was fully involved in the planning. Two well-manned and more appropriate vessels were allocated: the newly built *Providence* (420 tons) and the smaller tender *Assistant* (100 tons). The latter was commanded by the experienced Nathaniel Portlock, already known to Bligh as a fellow veteran of Cook's last voyage, and later described by him as having 'an alertness and attention to duty [that] makes me at all times think of him with regard and esteem'.[28]

The voyage was successful in both cartographical and botanical terms. During his open-boat voyage, Bligh had passed through the southern extremes of the Fiji islands, still unfamiliar to Europeans, but he had literally steered *Bounty*'s small and vulnerable boat well clear of them because of Fiji's reputation among the Polynesians for cannibalism. Voyaging in company, the well-armed *Providence* and *Assistant* were the first European ships to sail through the group, using Cook's methods to conduct a detailed running survey of the southern Fijian islands and those of the Torres Strait (fig. 181). One of the young midshipmen on *Providence* was Matthew Flinders, who would later be the first to circumnavigate and chart much of the coast of Australia. Flinders would subsequently acknowledge the importance of the breadfruit voyage to his career, and particularly the technical guidance on surveying that he received from Bligh.[29] Having said that, tensions arose between the two when Bligh followed the usual Admiralty orders to gather up all private journals and charts created on the voyage. In Flinders's opinion, this removed any possibility of building a professional reputation and gaining financial reward from the *Providence* voyage.[30] He would, however, get his revenge many years later when he declined Bligh's request that he dedicate his *Voyage to Terra Australis* (1814) to his old captain.

As for the breadfruit, Bligh was able to write to Banks from Port Royal, Jamaica, with the good news that 'I now give you joy of my Voyage being completed, and I hope you will think it as fully done as human attention & industry could accomplish', detailing his delivery of 690 live saplings to St Helena, St Vincent and Jamaica. The cheerful optimism did not last long, for six months later Bligh was writing once more to Banks complaining that his second-in-command, Portlock, had been granted an audience with the First Lord of the Admiralty (John Pitt, Earl of Chatham) while Bligh had not, despite calling several times 'without Success'.[31] Bligh was experiencing fallout from the trial of the *Bounty* mutineers, which had been held during his absence on the second voyage. He had been heavily criticized during their court martial and was no longer the hero of the hour, but something of an embarrassment to his former supporters. The book of his second breadfruit voyage had not been published, either, and Bligh replied to a correspondent that 'I cannot possibly say when my voyage will be printed. At present books of voyages sell so slow that they do not defray the expense of publishing.'[32] There was some truth in his observation, for the official account of George Vancouver's voyage to the north-west coast of America, published posthumously in 1798, had been criticized for its dauntingly inaccessible style. Voyages continued to be published, however, and while in terms of sales they never reached the contested heights of Hawkesworth's *Voyages Undertaken*, it is more likely that the Admiralty was not prepared to offer expensive financial support to an unpopular public figure when all its resources were being stretched by the French Revolutionary War, which Britain joined in February 1793. This also denied Bligh much of the public credit he would probably have received, had his second breadfruit voyage returned home earlier. By the time it did so in August that year, war – not botany and exploration – dominated public concerns, and his journal of the expedition remained unpublished until the late twentieth century.

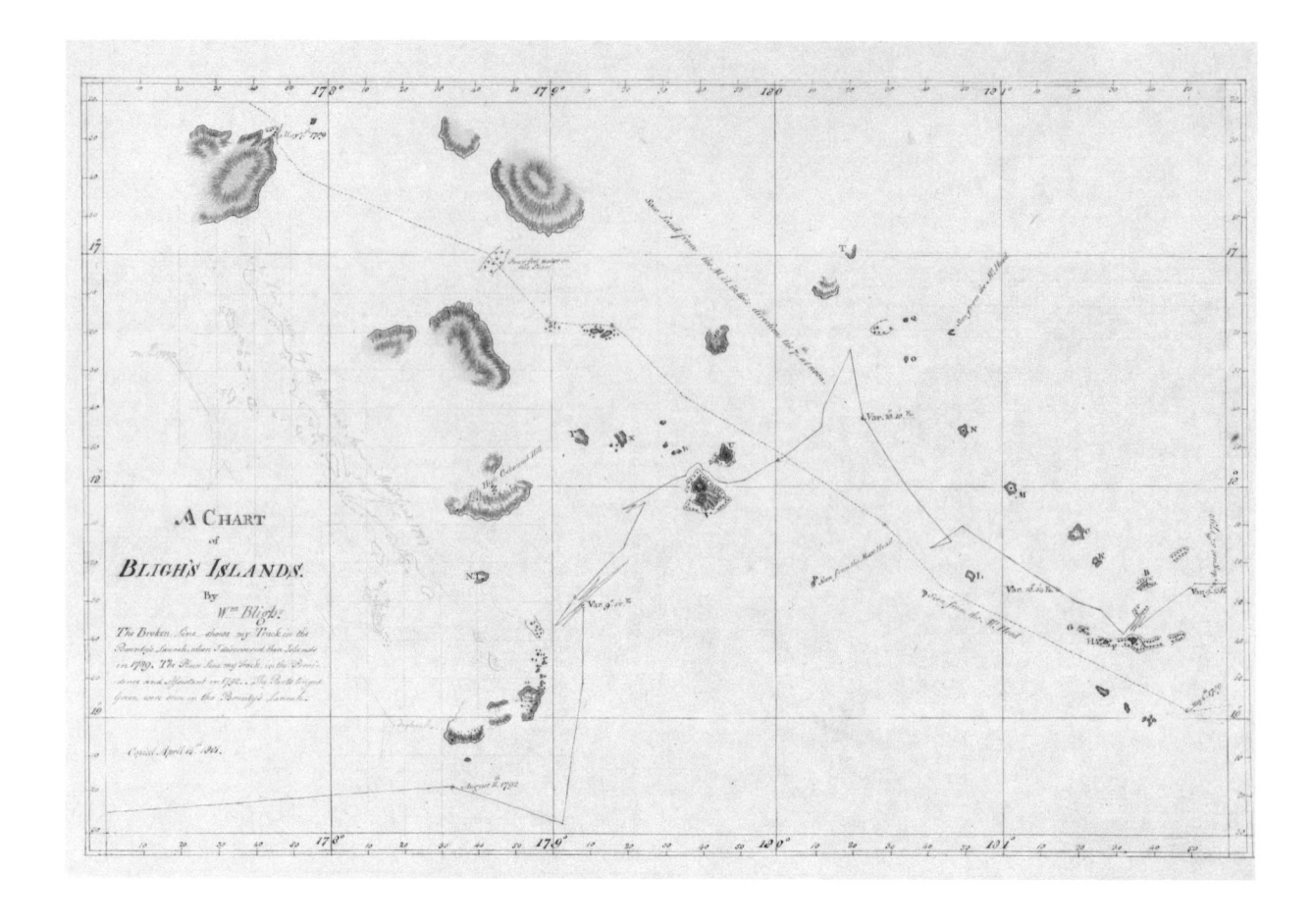

Fig. 181
Copy of a chart of Bligh's Islands [Fiji] by William Bligh, showing the track of the launch from *Bounty* when the islands were discovered in 1789, the track of *Providence* and *Assistant* in 1792
Captain Matthew Flinders, 1801
National Maritime Museum (FLI/15/23)

GEORGE VANCOUVER AND THE EXPLORATION OF THE NORTH-WEST COAST OF AMERICA

It soon became clear, after the series of trading voyages to the north-west coast of America, that its safe and productive navigation posed considerable challenges. 'A Chart of the Interior Part of North America' was published in *Voyages Made in the Years 1788 and 1789 . . .* (1790) by John Meares, a naval officer who, like Dixon and Portlock, had left the Navy to command two trading voyages from India and China to America between 1786 and 1788, although unlike Dixon and Portlock he had never served with Cook. The chart draws on Cook's survey, Russian charts and those of the traders who had followed him, without crediting them, and reveals the still-speculative state of geographical knowledge of this potentially profitable coastline after some ten fur-trading expeditions (fig. 182). However, Meares makes the reasonable point in the introduction to his *Voyages Made* that his were 'Voyages of commerce, and not of discovery', so any new knowledge gained would inevitably be 'an incidental part of a commercial undertaking' rather than a primary objective.[33] This was partly a jibe directed at Portlock and Dixon who, as we have seen, made much of their service with Cook to market their voyages as both scientific and commercial. Meares was nevertheless right to point out that there had still been no consistent and consecutive survey of the coast, and that there would not be one without significant state involvement. That sensible observation was moderated by his more speculative espousal of the search for a continental North-West Passage from the coast to the Great Lakes, with the chart making the bold claim of 'the Very Good Probability of an Inland Navigation from Hudson's Bay to the West Coast'. Here he was following the lead of Alexander Dalrymple, by now the hydrographer of the East India Company, who was campaigning as vigorously for the North-West Passage as he had for the Southern Continent.[34]

Resolving the issue of the North-West Passage and conducting a more detailed survey of the coast fell to another of Cook's former officers, George Vancouver. Vancouver's voyage of 1791–95 was further charged with on-the-ground negotiation of a political agreement that had been hammered out between Spain and Britain to settle what was called the 'Nootka Crisis' – a territorial dispute between Britain and Spain on the trading rights on the Spanish-American coast. The incident also involved Meares, for some of his ships (sailing under the Portuguese flag to circumvent the East India Company's monopoly on British trade within the Pacific) had been arrested by Spanish authorities and a trading post demolished. In one sense, the Nootka Crisis was little more than a footnote to history, but its political and commercial overtones were considerable and very nearly brought the two countries to war in 1789.

George Vancouver had sailed as a midshipman on Cook's second and third voyages. In 1789 he was appointed first lieutenant of *Discovery* (fig. 183), a merchant vessel of 330 tons bought by the Admiralty from the Rotherhithe shipbuilders Randall and Brent, and launched and named after Captain Clerke's former ship. *Discovery* was converted for exploration and put under the command of another experienced Cook officer, Henry Roberts (fig. 184). The original plan had been for *Discovery* to conduct a survey in the South Atlantic and Pacific in support of the Southern Whale Fishery, which had also expanded, based on Cook's reports of the large numbers of whales in the Southern Ocean. The survey was suspended during mobilization for the Nootka Crisis, but reinstated when a ship was needed to

Fig. 182
A chart of the Northern Pacific Ocean containing the northeast coast of Asia and northwest coast of America explored in 1778 & 1779 by Captain Cook and farther explored in 1788 & 1789 by John Meares
William I. Palmer (engraver), 1790
National Maritime Museum (G266:2/7)

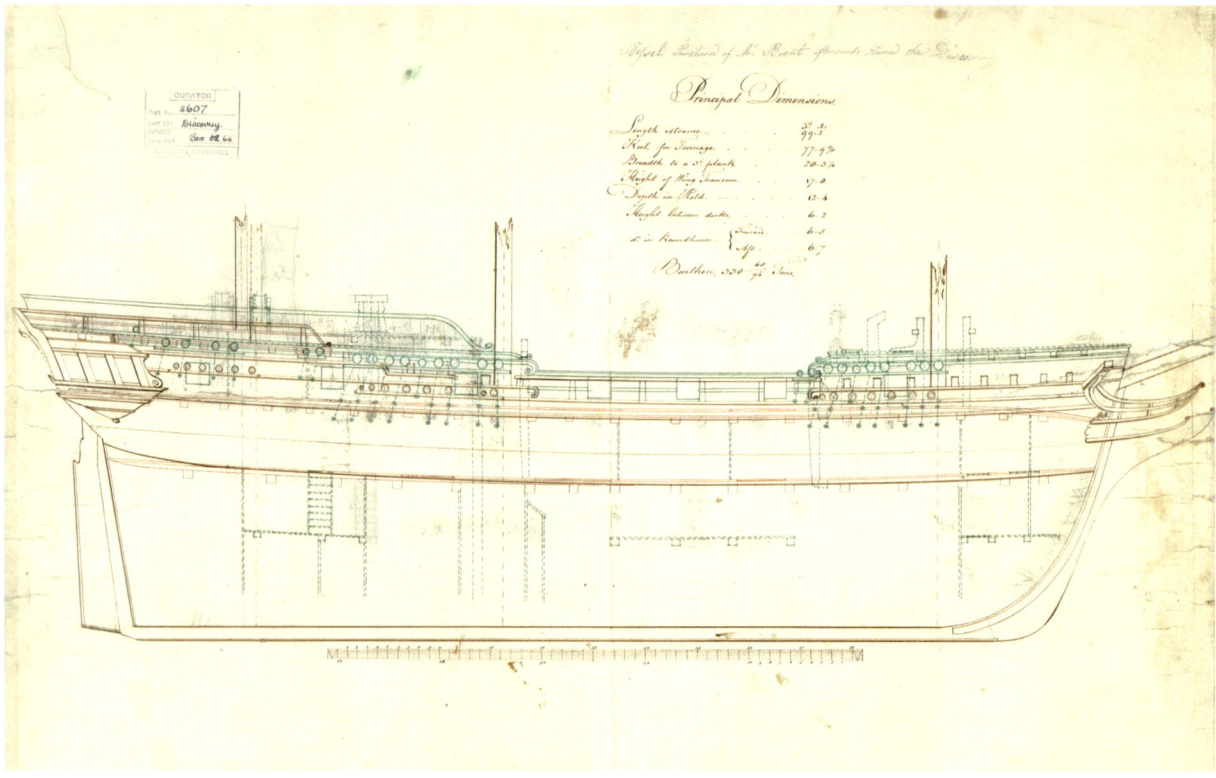

Fig. 183 (above)
George Vancouver's Discovery
Unknown artist, *c.*1789
National Maritime Museum
(ZBA4268)

Fig. 184 (below)
Plans of the *Discovery* showing modifications for her fitting as Vancouver's exploration vessel *c.*1789
National Maritime Museum
(ZAZ7834)

Fig. 185 (opposite)
Kendall's chronometer, K3
Larcum Kendall, 1774
National Maritime Museum
(ZAA0111)

188   CAPTAIN COOK AND THE PACIFIC

implement the eventual political agreement, the Nootka Convention. Roberts was relieved of his command – the reason is not absolutely clear – and Vancouver was promoted to captain in his stead.

A watercolour sketch shows *Discovery* recently coppered and undergoing final fitting out in Deptford Dockyard. Although undated, it was drawn prior to the installation of a substantial wood-and-glass plant cabin on the quarterdeck, ordered in December 1789 at the request of Sir Joseph Banks (who, as already mentioned, also ensured the appointment of a supernumerary botanist, the naval surgeon Archibald Menzies). The ship was well appointed with the usual antiscorbutics, as well as 42 bottles of rob of orange and of lemon, which Cook had considered useful, but too expensive for widespread use.[35] Kendall's third chronometer, K3 (fig. 185), and a large number of other scientific instruments were assigned to Vancouver, as were the latest Pacific publications: Cook's second and third voyage accounts, along with Portlock's, Dixon's, Meares's and de Bougainville's. Vancouver's request for additional trade goods, including calico, 20,000 copper nails, 200 pounds of vermilion and 20 pieces of baize, was valued at £473 and agreed.[36] Following the now-familiar practice for voyages to remote shores, a smaller vessel was also commissioned: this was a brig of 130 tons named *Chatham*, after the First Lord of the Admiralty (Lord Chatham), and commanded by Lieutenant William Broughton.

The two ships made their way to the Pacific via the Cape of Good Hope, the south-west tip of Australia (which Vancouver named King George III's Sound) and Dusky Bay, New Zealand. On the way, Broughton discovered the isolated group of low-lying islands south of New Zealand called The Snares, and was the first European to sight the Chatham Islands. After what were by now routine calls at Tahiti and the Hawai'ian islands, *Discovery* and *Chatham* reached the survey's starting point at the Strait of Juan de Fuca, which, according to a sixteenth-century Spanish account, led towards an extensive passage to America's Atlantic coast. Cook had missed the strait, but Vancouver's ships entered it in April 1792, working their way along the continental shore and taking a slow and systematic approach to the search for the Passage. Here, too, Vancouver began to appreciate that his survey would have to be done largely by the ships' boats, so intricate and indented was the coastline: 'one such survey', writes Glyn Williams, 'took twenty-three days, and charted 700 miles of shoreline, but advanced knowledge of the mainland coast by only sixty miles'.[37] Constantly drenched by heavy rain and chilled by the frequent fogs, the boat crews endured a miserable first season, although tents, boat covers and dry storage boxes were made for the second, which improved matters to an extent.

While making their steady way north and west on this painstaking task, the British ships were surprised to meet two Spanish naval vessels, *Sutil* and *Mexicana*, part of Malaspina's voyage of imperial inspection, which were engaged on the same job. Vancouver gathered from them that his Spanish counterpart, Juan Francisco de la Bodega y Quadra, was already in Nootka Sound awaiting the arrival of the British ships to conclude the Convention. At Quadra's request, what is now Vancouver Island was named Quadra's and Vancouver's Island, to mark the cordial professional relationship that ensued between the two men, despite their inability to agree on the terms of the Convention. The map was published in Vancouver's *A Voyage of Discovery*, although the Spanish part of its name did not survive later British colonialism (fig. 186).

The ships left Nootka Sound once the political impasse prevented further progress, and sailed first to Monterey, where the courteous Quadra arranged passage across New Spain for Broughton, who was being sent back to London for further advice on the Convention. Broughton never managed to re-join Vancouver, as by the time he returned to the Pacific he learnt that Vancouver was already on his way home. However, this produced an unexpected bonus; for Broughton, given command of Bligh's old ship *Providence*, decided instead to conduct a survey of the Sea of Japan. Although *Providence* was lost during the voyage, he produced a valuable survey of a strategically important area.

Discovery and *Chatham* wintered in Hawai'i before returning in the spring of 1793 to the point where the previous year's survey had finished, and worked their way north along an increasingly challenging coast. The following season, Vancouver sailed directly to Cook River, which his orders stipulated was to be the northerly limit of the survey, before heading south, finally finishing at Port Conclusion in the late summer of 1794. Despite a short surveying season, an extremely convoluted coast and onerous political considerations, the ships had charted 5000 miles of America's north-western shore to 'a degree of minuteness far exceeding the letter of [his] commission', as a weary, probably very ill, but satisfied Vancouver was able to claim. It had also proved beyond reasonable doubt that a transcontinental passage between the Pacific and the Atlantic did not exist. Like Bligh, however, Vancouver's significance as a surveyor, a major contributor to the map of the Pacific and, indeed, to the formal establishment of Britain's rights to

Fig. 186 (left)
A chart showing part of the coast of N.W. America, with the tracks of His Majesty's Sloop Discovery and Armed Tender Chatham; commanded by George Vancouver Esqr. and prepared under his immediate inspection by Lieut. Joseph Baker
Warner (engraver); James Baker (surveyor) for Captain George Vancouver, 1798
National Maritime Museum (G278:1/1)

Chart of the area around Vancouver Island surveyed in 1792, from George Vancouver's *A voyage of discovery to the north Pacific Ocean 1790 to 1795*

Fig. 187 (opposite)
The Caneing [sic] *in Conduit Street: Dedicated to the Flag Officers of the British Navy*
James Gillray, 1796
Private collection

This satirical engraving shows George Vancouver being accosted by the ex-midshipman Thomas Pitt, Lord Camelford. The 'South Seas Fur Warehouse' (right, background) has sea-otter furs hanging in the window, referring to the accusations that Vancouver had used his position to profit from the trade, while the scroll tumbling out of his pocket lists the excessive punishments of his officers and crew. Camelford is abusing Vancouver, demanding satisfaction for his being punished on the voyage. Although Camelford seems to be winning, in actual fact Vancouver and his brother laid some telling blows in return.

navigate and trade in the region was little recognized at the time; and outside north-west America, Canada and Alaska, he is not well known today. This is partly because his dogged recording of the coast had little of the grand, romantic appeal held by Cook's huge oceanic sweeps, and partly because, while his meticulous survey was of great importance to Britain's sea-otter trade, the practical issues of carrying on such a traffic on the other side of the world meant that its long-term success could only ever be a chimera if a North-West Passage remained undiscovered, as Cook had sagely noted. Within a few years Britain's involvement in it had virtually ceased. Vancouver – like Bligh on his *Providence* voyage – also set off in peacetime but returned in war, when exploration was undeniably far less newsworthy.

Vancouver's anonymity was also due to deeper issues of personality. He was a great navigator but a difficult man, who by the end of the voyage had alienated many of his officers and had put the naturalist Archibald Menzies under arrest – thus further alienating the powerful Banks, who had been irritated by Vancouver's high-handed attitude before the voyage even began. Back in England, Vancouver became involved in a public street brawl with his ex-midshipman Thomas Pitt (who had become Lord Camelford on the death of his father), but few supported him or contradicted malicious gossip circulating in London that he had been using government trade goods to buy sea-otter furs for his own profit. This action diminished him on two counts: first, the fraudulent exploitation of his employers; and second, because trade did not befit an officer and a gentleman – an accusation that sat uncomfortably with the enthusiasm of Cook's officers and men for selling sea-otter furs in Macau a few years earlier. Sadly, the enduring public image of Vancouver's survey became not his charts, but the satirical cartoon by James Gillray, *The Caneing in Conduit Street*, which repeated the gossip brilliantly but cruelly (fig. 187). Vancouver would die three years after his return, leaving the completion of his voyage account to his brother Charles, and was soon forgotten by the British public.

MEN OF CAPTAIN COOK

MATTHEW FLINDERS AND THE FIRST CIRCUMNAVIGATION OF AUSTRALIA

Joseph Banks's interest in promoting voyages of Pacific exploration continued when he supported Matthew Flinders's plan for a circumnavigation of Australia from 1801 to 1803. Although Britain had in 1788 established a penal colony at Port Jackson (present-day Sydney), rather than nearby Botany Bay as originally intended, development of better-informed knowledge of the country had hardly begun by the end of the century.

The theory that Terra Australis was not one country but, like New Zealand, two or possibly even three, divided north and south from the Gulf of Carpentaria to the south coast by a huge and as-yet-undiscovered strait, had its champions, not least of whom were Banks and Alexander Dalrymple.[38] It was still not known whether Van Diemen's Land (Tasmania) was an island or part of the main, and the existing Dutch charts were, with some justification, considered to be limited in extent and accuracy. The most up-to-date survey had been carried out during the French search mission for La Pérouse between 1791 and his death in 1793 by Bruny d'Entrecasteaux in the ships *Recherche* and *Espérance*. D'Entrecasteaux's surveys were drawn to a very high standard by the civilian engineer Charles-François Beautemps-Beaupré, whose manual on surveying was later translated into English and was for many years the definitive professional work on the subject. However, d'Entrecasteaux's search had to cover a huge geographical area, and Australia was only one of a number of places that might have held the secret of La Pérouse's disappearance: as a result, his only detailed surveys were limited to Tasmania in the south-east and to the south-west coast. Large tracts of coastline remained untouched and the question of Tasmania's insularity was still not resolved.

Matthew Flinders made his name in the history of early colonial Australia when he arrived at Port Jackson as a master's mate in HM ship *Reliance*, which in 1794 had been ordered to take out the second governor of New South Wales, John Hunter. In between Flinders's time with Bligh on *Providence* and his posting to *Reliance* he had served on *Bellerophon* at the Battle of the Glorious First of June 1794, leaving a well-observed account of it, together with a plan of its various stages. He also witnessed his patron Admiral Thomas Pasley's loss of a leg there to a cannonball. Although shocked by the injury to his patron, Flinders also realized that Pasley would hardly be in a position to help his career for some considerable time, so when he was given the opportunity to join *Reliance*, which offered a very different form of sea service, Flinders took it immediately. Together with *Reliance*'s surgeon, George Bass, and with Governor Hunter's steady support and encouragement, Flinders conducted a series of exploratory surveys of the coast north and south of Port Jackson in a variety of small vessels. The most significant voyage discovered and charted a navigable channel between Tasmania and the Australian mainland, which Hunter named Bass's Strait, while the large island at its entrance became Flinders Island. This discovery enabled ships sailing the faster southern route from Cape Town to Port Jackson to avoid the necessity of making a lengthy and time-consuming detour south of Tasmania. Such knowledge became increasingly important as the new colony grew and sought to build trade, which inevitably had to come by sea. The key route linking Port Jackson with India, China and the Dutch East Indies, for example, skirted

Fig. 188 (above)
Captain Flinders R.N. Autograph copy of Parole on his release from six years Captivity in the Isle of Mauritius
Hargrove Saunders, late 19th–early 20th century
On loan to the National Maritime Museum, from a private collection
(PAF3511)

Figs 189, 190 (opposite)
Plans of *Investigator*
1801
National Maritime Museum
(ZAZ6539 and ZAZ6540)

Xenophon was renamed *Investigator* by Sir Joseph Banks, and adapted for exploration at the Royal Dockyard at Sheerness.

192   CAPTAIN COOK AND THE PACIFIC

the reefs lining the north and east coasts of Australia, which, as Cook had discovered, were fraught with navigational dangers and were a significant discouragement to trade.

Matthew Flinders saw that the future of the colony would depend on the production of accurate charts to make its coasts safer for shipping. This was science in the humanitarian spirit of the Enlightenment as well as in the service of empire, but Flinders was equally clear-sighted in trying to turn his own expertise as a surveyor to his financial and professional advantage. He was a man quite refreshingly honest about his ambitions and the role that exploration could play in supporting them: 'Sea; I am thy servant, but thy wages must afford me more than a bare subsistence,' he wrote, anticipating that the publication of a narrative of the voyage and his charts could bring considerable reward (fig. 188).[39]

Flinders put his idea for a major voyage of exploration to Joseph Banks, whom he had met through the *Providence* voyage. Unsurprisingly, Flinders found that he was pushing at an open door, for Banks had long been trying to persuade the Admiralty that it should build on his brief but productive

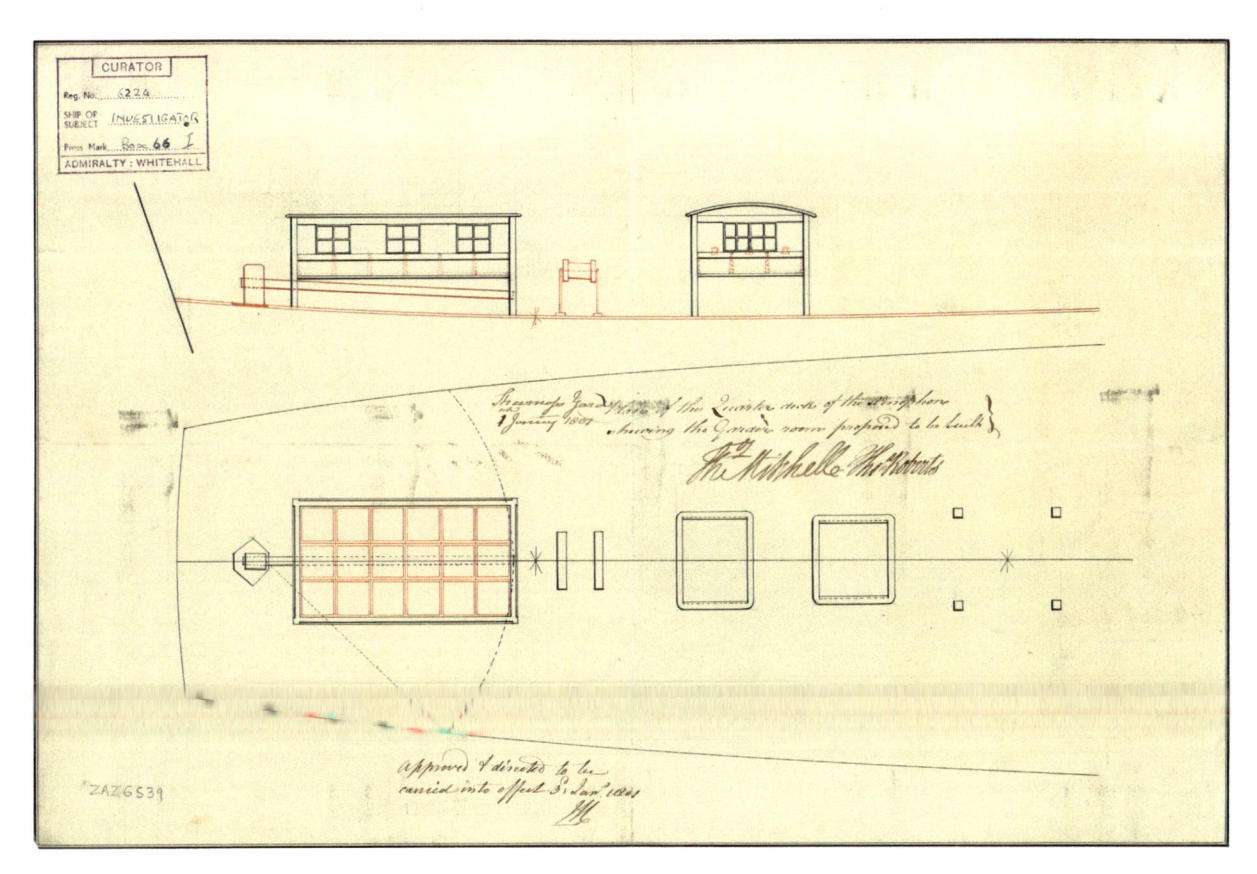

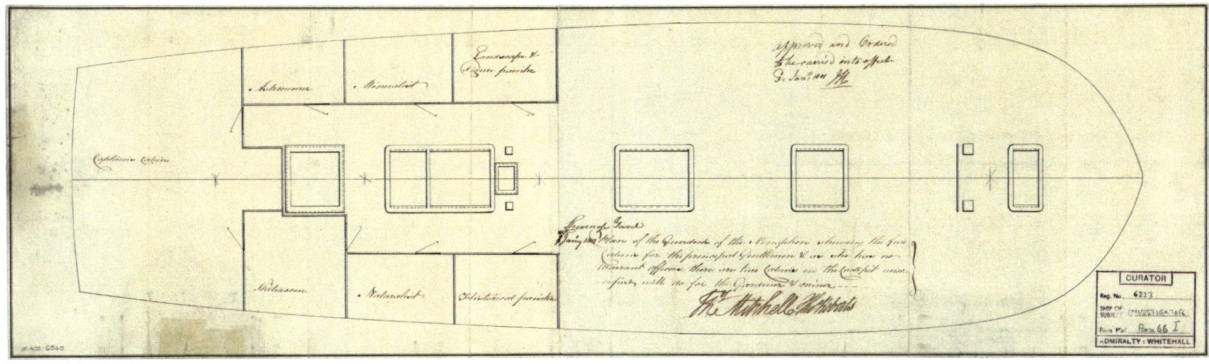

MEN OF CAPTAIN COOK 193

contact with the continent in 1770. After some initial setbacks, the Admiralty approved the voyage, giving Banks what amounted to a free hand in its planning, and Flinders promotion and command of a collier that had been bought on the stocks by the Navy as an escort vessel and named *Xenophon* (figs 189, 190). It had the same tonnage as Cook's *Endeavour* and, as Flinders underlined in the book of the voyage, 'in form, nearly resembled the description of a vessel recommended by Captain Cook as best calculated for voyages of discovery'.[40] A second vessel, *Lady Nelson*, built with a patented sliding keel (that is, a retractable keel), was to meet them at Port Jackson to assist in the close inshore work. To Banks, *Investigator* (as the ship was renamed) represented the voyage of scientific discovery from which he had withdrawn in *Resolution*. Although Banks had no intention of sailing on Flinders's expedition himself, his appointments reflected the broad spread of his scientific ambitions: a botanical artist, landscape painter, botanist, gardener and miner. The only civilian scientist he did not directly select was the astronomer John Crosley, who was appointed by the Astronomer Royal. Space was a political as well as practical issue on board ship: *Investigator*'s original plans show that the 'principal gentlemen' were berthed just forward of the captain's cabin, leaving no room for the ship's senior executive officers, who presumably had to berth in the cockpit.

There was some debate between Flinders, Banks and the Admiralty about the order in which he was to conduct the coastal survey. Flinders finally decided to ignore the Admiralty's preference and first survey the south coast 'minutely' in an easterly direction, before refitting at Port Jackson and heading north to cover the east coast in greater detail, then search for a channel through the reef that had so nearly ended Cook's first voyage, and finally examine the Gulf of Carpentaria as well as north and north-west Australia. Flinders's ambition was 'to make such a minute investigation of this extensive and very interesting country' that no other would be needed for many years.[41] 'Minute' had been a key word in Vancouver's vocabulary as well, suggesting a changing, more confident and possessive relationship between exploration and empire, one in which detailed knowledge produced an imaginative as well as practical control of remote lands.

Whichever way round the survey was to be done, it would have been impossible to complete in a single voyage. It started promisingly enough at the large, natural harbour that Vancouver had found and named King George III's Sound on the south-west point of Australia, where the town of Albany was later established. Here he landed on 9 December 1801 to begin the series of sightings that would give him an accurate position from which to begin the survey. These were carried out by Flinders himself and his brother Samuel, who had been appointed second lieutenant of *Investigator* and given special responsibility for the chronometers and astronomical matters, after the astronomer John Crosley resigned from the voyage due to ill health. Here, too, they had the voyage's longest sustained period of contact with Australian Aborigines, which was good-humoured on both sides, and Flinders discovered that their language was completely different from that of the east-coast peoples he knew from Port Jackson and its environs.

Robert Brown, the botanist, began his collection in King George's Sound. Over the four years that he and the botanical artist Ferdinand Bauer would spend in Australia, more than 4000 plants were collected, some 727 of which were from west Australia, 500 specifically from the Albany area and a further 100 from the Recherche Archipelago, to which *Investigator* then sailed (fig. 191).[42] Brown collected Australia's extraordinary fauna as well as its flora, although his interests lay so firmly with the latter that he did not make specimens available to scientists until the 1820s, by which time the large collections accumulated on Nicolas Baudin's voyage of exploration, which Brown and Flinders would soon meet, had already been catalogued and named.[43]

After leaving King George III's Sound, adverse winds prevented Flinders from starting his survey of the south coast, so he extended d'Entrecasteaux's survey of the Recherche Archipelago, before following the Great Australian Bight eastwards, landing where he thought it feasible. This was not often enough to please the scientific party and created a tension by now familiar on voyages of exploration, which was fed back to an unsympathetic Banks by Robert Brown. On 8 April 1802, off what is now Encounter Bay, the *Investigator* met Nicolas Baudin's ship *Géographe*, also charting the south-Australian coast, but in the opposite direction. The exchanges between Flinders and Baudin were cordial and professional, if not particularly warm, neither realizing that the Peace of Amiens had briefly halted the long war between their respective countries. The two men exchanged information about their surveys, before parting the following day. Flinders arrived in Port Jackson a month later on 9 May, where he was joined after a few weeks by a weakened and scurvy-ridden *Géographe*, which, after meeting him, had sailed as far west as the Nuyts Archipelago, a small group of islands off the coast of South Australia. Baudin's accompanying vessel, *Naturaliste*, which had been sailing separately, arrived a few days later. Flinders and Baudin took the opportunity to compare their charts and agree on where the credit should lie for first discovery, a negotiation much eased by arrival of news of the Peace of Amiens.

Flinders also met his survey vessel, the 60-ton cutter *Lady Nelson*, which was to serve under his command until it was no longer of use in completing the work. The ship had an experimental design by Lieutenant John Schank that should have made it ideal for inshore work: it was flat-bottomed, as the traditional long keel running the full length of the ship was replaced by three sliding keels that could be raised or lowered as required. An original model shows the housing for *Lady Nelson*'s keels on the deck by the main hatch, the deckhouse and windlass (fig. 192). The principle is little different from the centreboard or lifting keel of yachts and dinghies today, and offered the same advantages of being able to sail in much shallower water than standard ships and to stay upright if it had to 'take the ground'. In the event, Flinders found his consort frustratingly slow, easily damaged if it did go aground (which it often did) and its commander, Lieutenant Murray, unused to close inshore work. Flinders was soon convinced that *Lady Nelson* was more of a hindrance than a help and he eventually sent it back to Port Jackson, once they had completed the next leg of the voyage, which was the survey of the east coast.

The Admiralty had not considered this stretch a priority. Flinders's orders directed him to concentrate on the south coast, which he had now done, before charting the crucial but complex waters of the north coast, including the Torres Strait and the Gulf of Carpentaria. Cook's survey of the east coast was thought to be detailed enough, but Flinders felt that, since

Fig. 191
A bay on the south coast of New Holland, January 1802
William Westall
(ZBA7939)

After a trying passage through the dangerous waters of the Recherche Archipelago in January 1801, Flinders named his timely discovery Lucky Bay.

MEN OF CAPTAIN COOK 195

Cook did not have a chronometer on his first voyage, his longitudes were likely to be inaccurate; so he decided to combine the voyage to the north with a more detailed survey of the east. Although Flinders's concerns about Cook's longitudes were justified, Kenneth Morgan is surely right that the ambitious Flinders also welcomed the chance to 'improve' on the work of the great man.[44] The implication of Flinders's decision, right or wrong, was that his survey of the eastern seaboard of Australia considerably delayed his arrival on the northern coast. The further implication – that it committed him to returning to Port Jackson the long way round – was not apparent at the time and only became obvious when *Investigator* was actually on the north coast and leaking once more at an alarming rate. Flinders ordered a survey, which found that the ship was in no condition to fight the strong seasonal winds likely to be experienced by retracing its route south and east back to Port Jackson. This frustrating and disappointing news was effectively the end of his survey, although the passage back to Port Jackson made *Investigator* the first ship to circumnavigate Australia, which would have been solace of some sort.

Investigator was nursed back to Port Jackson via the west and southern coasts of Australia, the crew suffering heavily from diseases that killed eight men, including the popular gardener, Peter Good. Once in Port Jackson, it was obvious that *Investigator* was in no fit state to continue the survey and, as no suitable replacement vessel was available, Flinders decided to take passage back to Britain to persuade the Admiralty to give him another ship in order to complete the job. Robert Brown and Ferdinand Bauer elected to remain in Australia to continue their researches into its natural history: they did not return to Britain until 1805, in a crazily cut-down, rotting and repaired *Investigator*. Ill fortune continued to dog Flinders: the ship carrying him, his crew and most of the supernumeraries home foundered on an uncharted reef, an event that would both shock and inspire the expedition's artist, William Westall, to create one of his most haunting paintings (fig. 193). Flinders then sailed some 700 miles back to Port Jackson in an open boat to get help. The governor of New South Wales, by now Phillip Parker King, provided a small schooner, the *Cumberland*, to take Flinders back to the wreck and then, with a small crew, to make the long voyage across the Indian Ocean and back to London, with the remaining survivors taken to Canton by a merchant ship. But Flinders's troubles were still not over: for he called at Île de France (Mauritius) to discover that Britain and France were at war once more; that his passport, which had been made out in the name of *Investigator* rather than Flinders, was consequently invalid; and that the French governor, Decaen, was entitled to place him in detention. Flinders would remain a prisoner for more than six years, before being exchanged in 1810 and returning to London in ill health. He did, however, just live to see the publication of his book, *A Voyage to Terra Australis* (1814) (fig. 194).[45] His main claim to fame outside Australia is probably that he was the first to use the name 'Australia' rather than the unwieldy Terra Australis. In actual fact he was not quite the first, but he fully deserves the accolade.

The character of maritime exploration was changing by Flinders's time. The large parties of independent scientists sent out to the Pacific with Cook, Flinders, La Pérouse, d'Entrecasteaux and Baudin produced exceptional work, having mapped the flora, fauna, rocks, terrain, cultures and customs they saw, and engrossed Europe with their results. However, they came with their own sets of expectations and priorities, which did not always sit happily with the practicalities of navigation in unknown waters. Although Flinders did not see it, Baudin's voyage was racked by internal disputes between the 25 members of the scientific party and the seamen, dissensions that were far worse than the murmurings on board *Investigator*. The comparison is not perfect, for the social turmoil of Revolu-

Fig. 192 (opposite)
Model of the *Lady Nelson*
c.1799
National Maritime Museum,
Royal Naval Museum Greenwich
Collection (SLR0601)

The sliding keels used in this experimental vessel were an invention of Lt John Schank.

Fig. 193 (above)
Wreck Reef Bank, taken at low water, August 1803, from Flinders's voyage to King George III's Sound
William Westall
National Maritime Museum
(ZBA7935)

tionary France had its own unsettling impact on Baudin's ships. However, by 1816 France would commission Louis-Claude de Freycinet – a veteran of Baudin's voyage – to lead another Pacific expedition with no civilian scientists on board at all, confident that science would still be conducted on the voyage, but by naval officers and subject to naval discipline.[th]

The Admiralty did not react quite so dramatically, but the establishment of the Royal Navy's Hydrographic Office began to have a profound effect on the character of British voyages of exploration. Adrian Webb has shown that in its formative period between 1808 and 1829 (when Francis Beaufort, arguably the most effective Hydrographer of the Navy in the nineteenth century, was appointed to the post; fig. 195), the Hydrographic Office gradually took control of all elements of surveying, from the commissioning of voyages and the actual surveying itself, to the production and sale of charts and the relationship between the Navy and scientific communities: 'the Hydrographer [of the Navy]', Webb writes, 'also found himself creating his own policies, serving as Secretary to the Board of Longitude, being a consultant on navigational matters, taking responsibility for the acquisition, supply and maintenance of chronometers for the Navy, as well as being a focal point for issues concerning pay, promotion and manning for surveying specialists'.[47] Within a few years of Flinders's voyage, Pacific exploration was no longer being driven or overseen by the influential figure of Joseph Banks (who died in 1820), or conducted by a cadre of officers who had served with Cook or been trained by those who had. It was, instead, led by a single department of the Admiralty with clear priorities and responsibilities, and considerable powers and resources. The defeat of Napoleon in 1815 ended Royal Naval warfare on the grand scale for a century, severely limiting the opportunities for advancement through fighting service. Partly as a result, exploration increasingly became a conscious career choice for scientifically-minded nineteenth-century naval officers, as it had been for Flinders. The map of the Pacific left by the men of Cook, together with their expertise, practices and publications, would form an integral part of the professionalization of British maritime exploration and surveying during the late-Georgian and Victorian *pax Britannica*.

Fig. 194 (above)
Chart of Terra Australis by M. Flinders Cmmr of HM Sloop Investigator, from Flinders' *A voyage to Terra Australis*, 1814
HM Admiralty; Hurd; Commander Captain Matthew Flinders, 1814
National Maritime Museum (G262:14/2(1))

Fig. 195 (opposite)
Rear-Admiral Sir Francis Beaufort, 1774–1857
Stephen Pearce, 1855–56
National Maritime Museum, Greenwich Hospital Collection (BHC2541)

This version of his portrait was presented to the Naval Gallery at Greenwich to mark his retirement as Hydrographer of the Navy and commemorate his importance in the role.

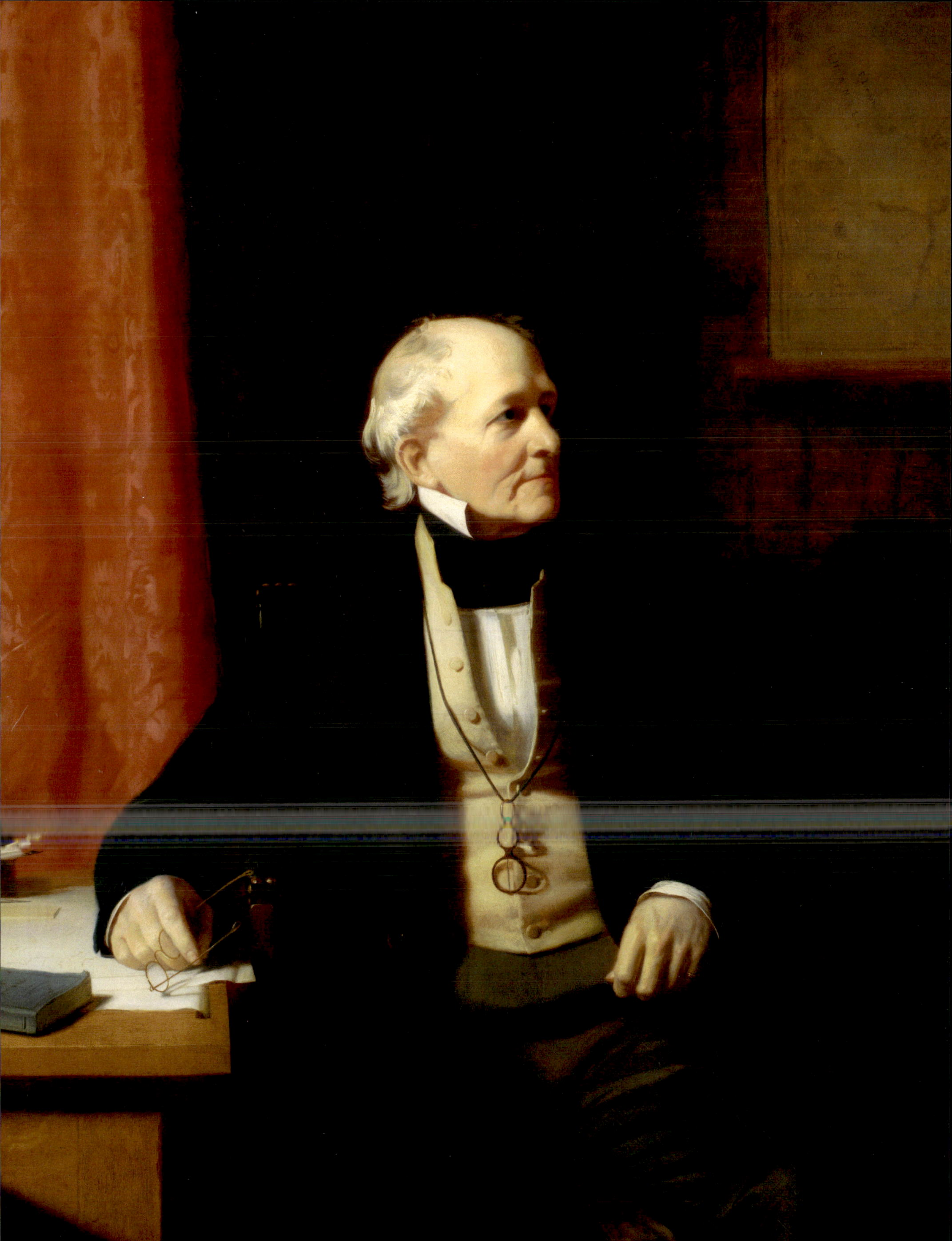

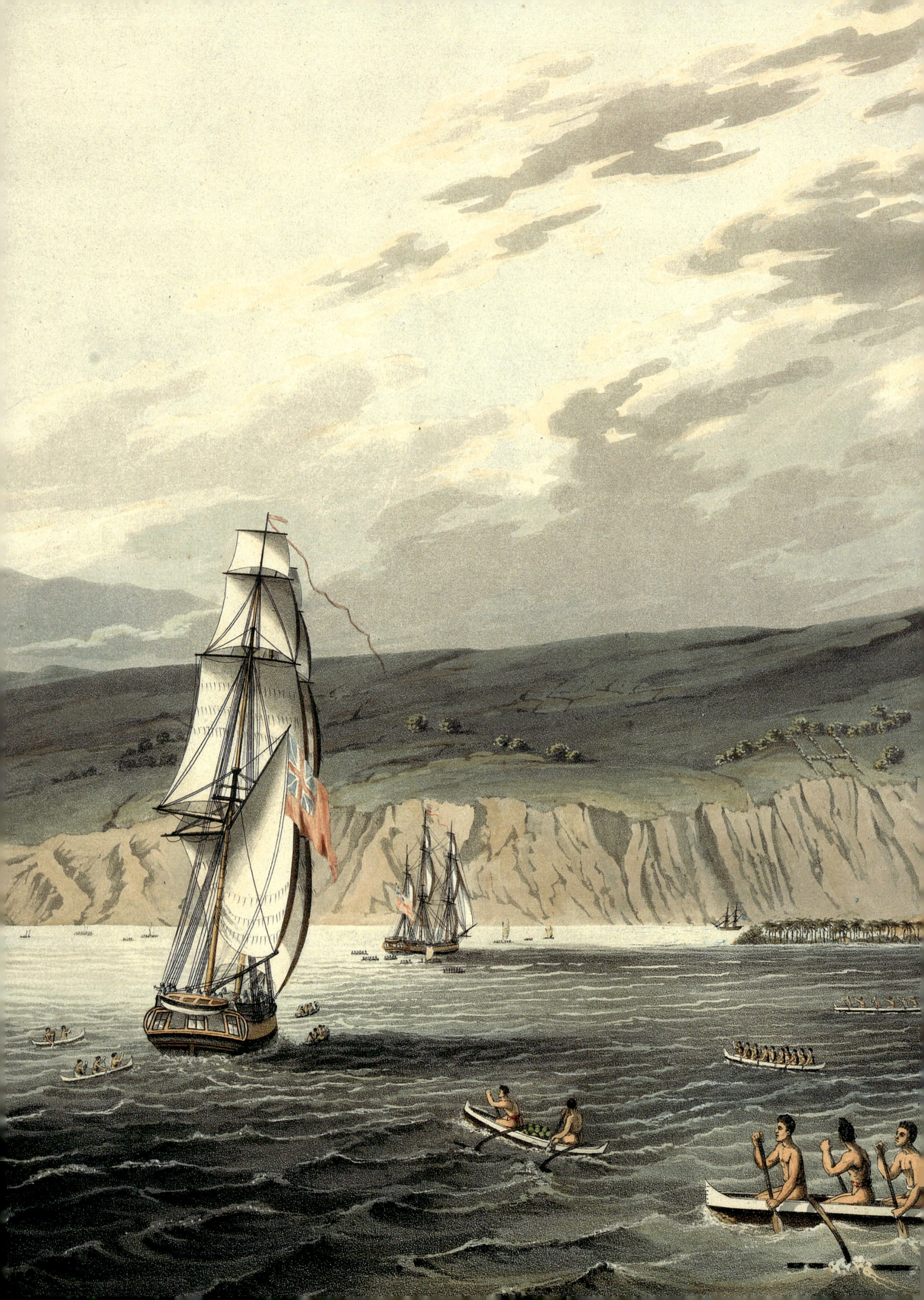

Chapter 8
The Strange Afterlives of Captain Cook
Representations and Commemorations

John McAleer

Even before his death, James Cook was a celebrity. As the leader of three highly publicized voyages around the globe, he was personally associated with extending the frontiers of European geographical and scientific knowledge and, as we have seen, the cultural and artistic legacies of these expeditions were no less profound. Despite the fact that many people – both on the ships themselves and on the various shores at which they called – contributed to the achievements of these voyages, it was Cook who seemed to embody them most closely. Indeed, in the last decade of his life and for centuries thereafter, his name was to become synonymous with the expeditions he led.

The association between the voyages and the man only became amplified after his death on 14 February 1779 at Kealakekua Bay (sometimes spelled 'Karakakoa Bay'), Hawai'i (figs 196, 197). Here was a figure prepared to lay down his life for science, for his colleagues and, by implication, for the country that sent him on such expeditions. Cook's death gave impetus to commemoration and celebration of the voyages, which were increasingly refracted through that of the man Cook assumed an almost mythic status, influencing the ways in which people in Britain thought about scientific exploration and its place in British encounters with the wider world.[1] Crucially, Cook quickly came to stand for a great deal more than the events of his own life and for a lot more than the voyages he led.[2] His life and career were invested with meaning and significance beyond the mere historical facts. Both the man and his missions became shorthand ciphers, deployed to tell a triumphant tale of exploration and empire. The voyages, and the British engagement with the Pacific they represented, were soon intimately connected with larger narratives of British and imperial identities, expansion and purpose, as well as personal and naval heroism. From the beginning, all of this was achieved with the aid of material culture, objects and images. The monuments, memorials and markers that ostensibly commemorated Cook were employed by a range of interested groups for a variety of political ends.[3]

This chapter explores some of the ways in which this transformation from man to myth took place, and how the collections of the National Maritime Museum chart the afterlife of one of Britain's most celebrated maritime heroes. Medals, prints and paintings tell the story of Cook's tragic death, his commitment to science and his status as a true British hero – often, indeed, at one and the same time. These objects remind us that Cook's life and his voyages, along with his death and afterlife and the intersections between these, were crucial in situating this story at the heart of Britain's national museum of maritime history.

IN LIFE

While the work of transforming Cook into a universal icon of exploration started in earnest after his death, it is important to remember that even in his lifetime he was being honoured and commemorated. Along with his naval contemporaries, he personified a spirit of adventure that endured long after his demise and shaped subsequent British attitudes to exploration. W. Fraser Rae observed, for example, that 'when Lord Anson and Captain Cook sailed round the world, they were regarded as heroes of an adventure as daring and wonderful as any which can be found in the histories and poems of Greece and Rome'.[4] Of course, interest in the narratives of his voyages or in the prints derived from them was one facet of this; but there were other, more specific instances of recognition being accorded to Cook. For example, in a mark of its esteem, the Royal Society awarded him its prestigious Copley Medal after he returned

Fig. 196
Karakakoa Bay, Owhyee
(Hawai'i) (detail)
M. Dubourg (engraver),
after Captain Thomas
Heddington, 1814
National Maritime Museum
(PAI0469)

from the second voyage.[5] This medal was (and is) given for outstanding achievements in research in any branch of science. In presenting the medal to Mrs Cook in 1776 (her husband had already departed on his ill-fated third voyage), the Society's president, Sir John Pringle, remarked that Cook had 'fixed the bounds of the habitable earth, as well as those of the navigable ocean'.[6] This was high praise indeed from an august body at the heart of London's intellectual and scientific life, but Cook's influence also spread much further afield.

In 1772, a 'Captain Cook spectacular' was planned by Sir James Caldwell of Castle Caldwell. Only a year after the return of the first expedition, Caldwell's extravaganza was to take place on Lough Erne, County Fermanagh. As a stickler for detail, Sir James even went so far as to enlist 50 labourers from his estate whom he deemed most suitable to represent New Zealand 'savages' in one vessel, and an equal number to play Captain Cook and his crew in another.[7] It is difficult to imagine a starker contrast to the learned discussions that would animate the Royal Society only a few years later as its fellows discussed the awarding of the Copley Medal, and yet both instances indicate just how deeply the expeditions and their commander had penetrated popular consciousness in the 1770s.

Cook and the results of the voyages became a common cultural reference point that represented something fundamental about the processes of exploration and encounter. In summer 1779, Gabriel Beranger, a Huguenot artist born in Rotterdam who moved to Ireland in about 1750, travelled to the west of the country to make sketches of antiquities and monuments for William Burton Conyngham, founder of the Hibernian Antiquarian Society. Beranger's views of the people he encountered were heavily influenced by the ideas of primitivism and the noble savage evoked in the published travels of Cook. He had evidently read Cook's reports: 'I fancied myself to be at Otaheite, since we found here the same good nature, but accompanied by modesty in the sex, who grants plenty of innocent embraces, since they could not enjoy our conversation.'[8]

The influence of the Pacific on views of Britain's own ancient primitives was evident in the rise in antiquarian ethnography. The English naturalist George Pearson used remnants of ancient Roman and Saxon weapons found in Lincolnshire to make extensive comparisons between the state of British society at the time of the Roman conquest and 'that in which our late discoverers found the natives of the South Sea islands'. In the early nineteenth century, the Reverend Frederick Clark, a professor of mineralogy at Cambridge, reported to Joseph Banks that diggers had found an ancient weapon 'exactingly resembling the Stone Hatchets of the South Seas'.[9] Cook's voyages, therefore, became a common intellectual currency, but it was only through his death that Cook could enter the ultimate pantheon of heroes, taking his scientific and exploration missions with him.

IN DEATH

The circumstances of Cook's demise – pored over in detail in contemporary press reports – offered the first opportunity for

Fig. 197 (left)
Karakakoa Bay, Owhyee (Hawai'i)
M. Dubourg (engraver), after Captain Thomas Heddington, 1814
National Maritime Museum (PAI0469)

Figs 198, 199 (opposite)
Medal commemorating Captain James Cook (1728–79)
Lewis Pingo, 1784
National Maritime Museum (MEC1317)

debates about his legacy. The tone was immediately set by his crew. They seemed willing to present their erstwhile commander as an object of saintly and sacrificial veneration. They made a relic for his widow: a coffin-shaped 'ditty box' containing a lock of Cook's hair and a small watercolour scene of his death. This extraordinary object is surmounted by a silver plate, inscribed with the words 'Lono and the seaman's idol', curiously conjoining indigenous religious belief and professional admiration.[10]

Cook's death mirrored and magnified his popularity in life. Regardless of the actual details of the event, his achievements and his end were immortalized in the decades that followed in numerous ways. They were commemorated by poets and playwrights. Presumably hoping to capitalize on the appetite the public had already shown for his previous expeditions, a welter of unofficial prose accounts of Cook's last voyage were published, providing plenty of detail about his last hours and moments. John Rickman's *Journal of Captain Cook's Last Voyage to the Pacific Ocean* and Heinrich Zimmermann's account both appeared in 1781. William Ellis's two-volume *An Authentic Narrative of a Voyage performed by Captain Cook and Captain Clerke* followed in 1782, with John Ledyard's *A Journal of Captain Cook's Last Voyage to the Pacific Ocean* published in the newly independent United States in 1783, and David Samwell's *A Narrative of the Death of Captain James Cook* in 1786. A raft of objects also appeared, intended to exhibit Cook's death (or sacrifice) and his noble legacy in the same way as they had earlier exhibited his exploration endeavours. Medals were quickly struck, for instance.

It was not just those keen to cash in on a celebrity story who contributed to the situation. Cook's earliest biographer, Andrew Kippis, remarked that the Royal Society of London could not lose such a member of their body without being 'anxious to honour his name and memory by a particular mark of respect'. They resolved to do this in time-honoured fashion by having a medal struck. A voluntary subscription was opened for this purpose: 'To such of the fellows of the society as subscribed twenty guineas, a gold medal was appropriated; silver medals were assigned to those who contributed a smaller sum; and to each of the other members one in bronze were given.' Gold medals were also given to members of the British and other royal families.[11] Louis Antoine de Bougainville, a great explorer of the Pacific in his own right, wrote to Joseph Banks immediately:

> *Since I have learnt from the daily papers that the Royal Society has struck a medal in honour of the immortal Cook and that one is to be given to every member, I request that you should act as go-between to the Society and make known how strongly I would like, on this occasion, to stress that in 1755 I was elected a member. I dare say that few of its members have more right than I to share in this gift of a medal dedicated to the memory of a navigator so similar to Magellan. The steps I have tried to take in the same career have taught me to appreciate his efforts and I have the right to scatter flowers on his tomb.*[12]

Writing from Munich, Sir Benjamin Thompson, Count von Rumford, was equally covetous: 'Have I any chance of getting one?' Fortunately, he remembered that his employer, Karl Theodor, Elector Palatine of Bavaria, would also be keen:

> *And will it be possible to procure one for the Elector. I know H[is] S[erene] H[ighness] would be very happy to have one. He often converses on the subject of Captain Cook and his discoveries.*[13]

The Empress of Russia was apparently so moved by 'the sense of the value of the present' that 'she had caused it to be forthwith deposited in the Museum of the Imperial Academy of

THE STRANGE AFTERLIVES OF CAPTAIN COOK   203

Fig. 200
Death of Cook
Francis Jukes, after John Cleveley, the Younger, 1787–88
National Maritime Museum (PAI2627)

The formal title of this print, which is the fourth completing the Cleveley set (see also figs 46; 133; 134; and 151) is 'View of Owhyhee, one of the Sandwich Islands'.

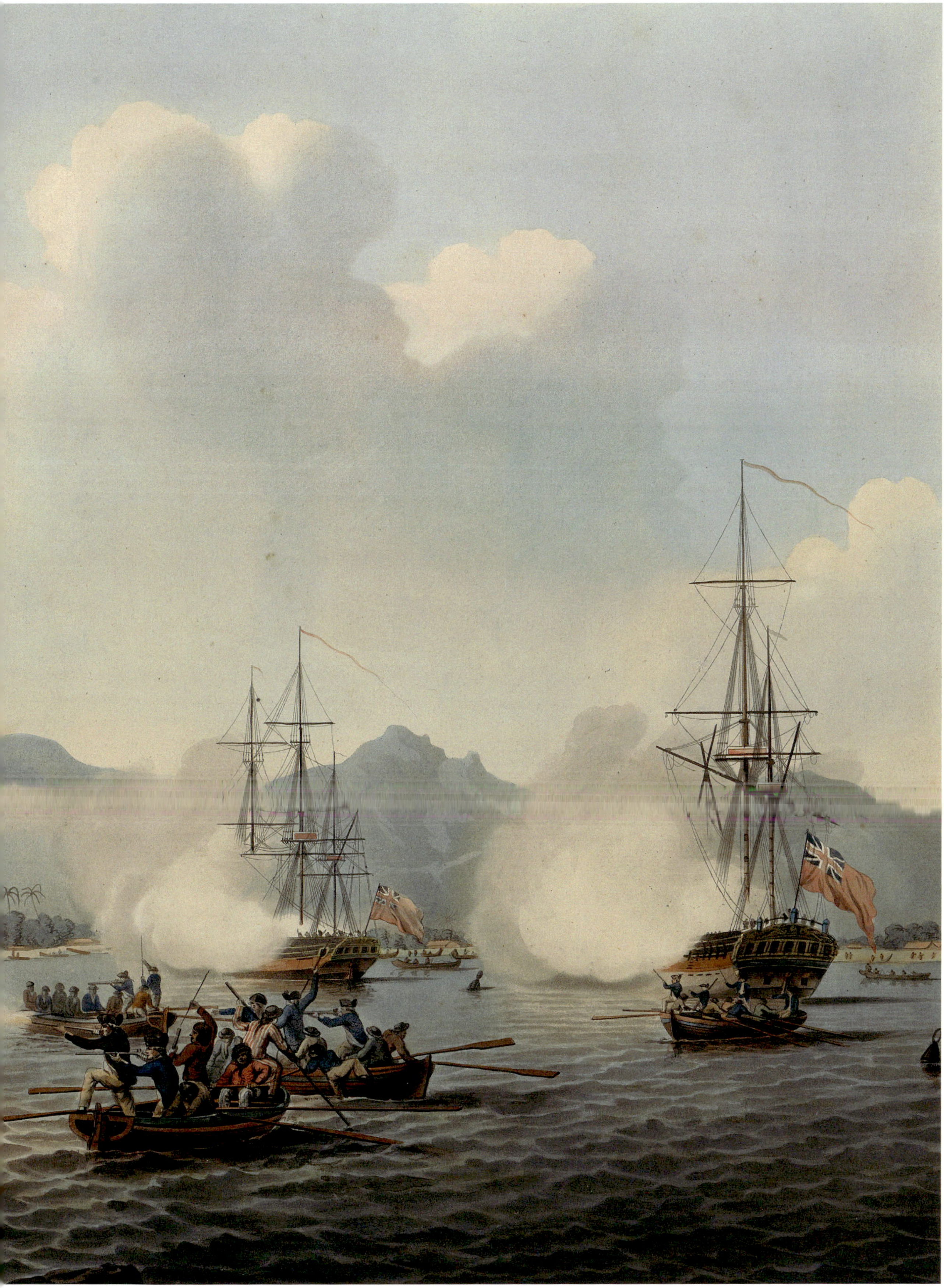

Fig. 201 (above)
Death of Captain Cook
Francesco Bartolozzi
(engraver),
after John Webber, 1785
National Maritime Museum
(PAF6417)

Fig. 202 (opposite)
The Death of Captain James Cook, 14 February 1779
Johan Zoffany, c.1798
National Maritime Museum, Greenwich Hospital Collection (BHC0424)

Sciences'.[14] The medal visually encapsulates some of the views circulating about Cook in the immediate aftermath of his death (figs 198, 199). Designed by Lewis Pingo, the obverse depicts a bust of the explorer with a legend in Latin, which translates as 'The most intrepid investigator of the seas', putting Cook on a parallel with a Roman emperor or contemporary monarch. On the reverse, a draped female figure personifying Fortune leans against a column. She holds a spear in her left hand, while the rudder in her right rests on a globe. A shield bearing the Union flag leans against the column on the right. All of this is under the legend 'Our men have left nothing unattempted'.[15] The rich iconographic detail of the medal, with every item and inscription expressing pride and admiration for Cook, signalled the way he would be remembered for many years.

Although the Royal Society medal offered an opportunity to laud Cook's scientific achievements, the circumstances of his death seemed to call for a larger canvas – both literally and metaphorically – to depict the hero.

As a result, visual representations of Cook's death were quickly and widely circulated. John Webber (fig. 201), D.P. Dodd, John Cleveley, the Younger (fig. 200) and George Carter were among the first who painted versions of the explorer's death at the hands of an incensed Hawai'ian crowd. Works such as Johan Zoffany's image in the NMM collection (fig. 202) went a long way towards reinforcing Cook's rapidly developing status as a hero of exploration through its rich visual iconography. Zoffany travelled extensively and had experience of non-European scenes from his time in India in the 1780s. Indeed, he might even have accompanied the second voyage with Banks, had the latter's original intentions been fulfilled. Nevertheless, like all of the other depictions of Cook's death, this unfinished painting is not an eyewitness account. Rather, it is a deeply mediated work, drawing on other images and accounts of Cook's death and filtering them through Zoffany's classical training and distinctive artistic style. For example, it reflects Zoffany's love of the theatre, as well as the ways in which the events in Hawai'i percolated into British culture. On his return from India, Zoffany went to see *The Death of Cook* at Covent Garden. The composition of the painting, its dramatic gestures and its close attention to detail certainly suggest a theatrical inspiration for the image. It is also clear, however, that Zoffany relied on the work of John Webber to inspire his interpretation of the scene. Although Webber was the official artist on the third voyage, he himself did not see Cook die. His painting of the event, now in Australia, became well known when engraved in 1782–83. Although he drew on all of these sources, Zoffany produced a unique and striking image that attempted to convey the gravity of the moment, and to emphasize the dignity and heroism of Cook.

Zoffany's positioning of Cook is at odds with eyewitness accounts of his death, and is done for aesthetic and narrative effect. The pose of the dying hero was a popular motif in late-eighteenth-century history painting and Zoffany had the obvious example of Benjamin West's *Death of General Wolfe* as a reference point (fig. 203). In the central drama of the image, the doomed Cook and the idealized 'savage' who murders him confront each other. They are both revealed as noble, in Zoffany's moment of classical tragedy. The artist seeks to produce a monumental history composition in the grand manner. He portrays Cook as an ideal tragic hero, but in a realistic setting and in contemporary dress, much as West had done some

THE STRANGE AFTERLIVES OF CAPTAIN COOK

208   CAPTAIN COOK AND THE PACIFIC

years earlier for Wolfe. This fusion of the ideal with the actual relies heavily on classical sculptural precedents: Cook is shown in the manner of the 'Ludovisi Gladiator', while his assailant assumes the pose of the 'Discobulus' (discus thrower) from Charles Townley's renowned collection. The classical allusions are enhanced by the chieftain's headgear, which is similar to the helmet worn by the ancient Greeks. The prime unanswered question is why Zoffany never finished the work. Perhaps he realized that the theatrically 'savage' elements of the composition stood fair, if completed, to overpower the prone nobility of the ostensible hero of the piece (as they had done in life), with the opposite effect on his audience than that intended.

George Carter's *Death of Cook* does not have the same aesthetic pedigree as Zoffany's, even in its unfinished state, but it too offers an insight into the kinds of parallels being drawn to explain Cook's life and achievements (fig. 204). Relying even more explicitly on near-contemporary art, Carter envisaged his image as a pendant for West's famous image of the *Death of Wolfe*.[16] In other words, Carter made a direct visual and intellectual connection between the sacrifices of the two men, as well as well as placing them on equal status, in artistic terms. Wolfe, a great military leader who lost his life at the moment of victory, and whose actions were perceived as laying the foundations for Britain's ultimate triumph in the Seven Years War, was an almost archetypal hero. To compare Cook with Wolfe was praise indeed for the explorer and said a lot for the perceived value of Cook's achievements for the nation, as well as for humanity in general. Indeed, so successful and iconic was this depiction of the explorer's demise that it became the basis for representations of other similar events, most notably the death of the missionary John Williams, at the hands of locals in the New Hebrides in 1839.[17]

As with the other cultural products that resulted from the voyages, images of Cook's death also circulated widely, finding markets and making an impact far and wide. As we have seen, a number were exhibited at the Royal Academy, though Cook's fame in death extended far beyond British shores. In 1796, an image described as *The Death of Captain Cook* was sold to Gopi Mohen Baboo of the Tagore dynasty and presumably ended up adorning one of his residences in Bengal.[18]

These images of Cook's death not only aimed to illustrate the circumstances of the event, but also attempted to say something more profound about his achievements. In doing so, they contributed to emerging narratives about Cook's role in advancing science and as a harbinger of civilization. For example, P.J. de Loutherbourg's image of *The Apotheosis of Captain Cook* circulated widely in print (fig. 205). As its title suggests, this was no matter-of-fact illustration. Instead, it conveyed the

Fig. 203 (opposite, above)
The Death of General Wolfe
William Woollett (engraver), after Benjamin West, 1776
National Maritime Museum
(PAH7700)

West's painting (the original is now in Canada) set the model for heroic 'modern history' painting well into the 19th century.

Fig. 204 (opposite, below)
The Death of Captain James Cook by the Indians of Owhyee, one of the Sandwich Islands
J. Hall and J. Thornthwaite (engravers), after George Carter, 1784
National Maritime Museum
(PAH7773)

Fig. 205 (right)
The Apotheosis of Captain Cook
P.J. de Loutherbourg, after John Webber, 1794
National Maritime Museum
(PAD2895)

developing views of Cook's role and achievements in the Pacific. Loutherbourg's image was partly based on his connection with John Webber. It was originally intended to accompany a stage performance, with Cook descending on to the stage while the cast, attired in the native costumes of all the places he had visited, from Alaska to New Zealand, sang:

> He came, and he saw, not to conquer but to save;
> The Caesar of Britain was he;
> Who scorn'd the ambition of making a slave
> While Britons themselves are so free.
> Now the Genius of Britain forbids us to grieve,
> Since Cook, ever honoured, immortal shall live.[19]

Loutherbourg shows Cook being raised by two angels into heaven. Unlike other representations of this sort, normally reserved for fallen military leaders, this image shows him holding a sextant rather than a sword. Webber assisted by painting the view of Keleakakua Bay, the location of Cook's death, which formed the backdrop of the composition. In lauding Cook's achievements and his efforts to 'save' the Pacific islanders from their benighted heathenism, the image and the play from which it was derived captured only one element of the potent Cook myth. For just as Cook was regarded as a 'saviour' – bringing civilization to the Pacific and its inhabitants – so his journeys also served Britain and its burgeoning empire.

This can be seen in a work such as James Barry's *Commerce, or the Triumph of the Thames*, painted for the Royal Society for the Encouragement of Arts, Manufactures and Commerce between 1777 and 1784 (and revised in 1801). This was the fourth in a series of six paintings that the artist was commissioned to produce on the subject of 'The Progress of Human Knowledge and Culture'. Unlike the previous three paintings, it depicts a modern scene, albeit in a characteristically classical way, by personifying the river as a benevolent deity. A majestic Father Thames, seated on a throne, steering himself with one hand and holding a compass in the other, is surrounded by a full cast of classical actors. Above him is the Roman deity Mercury, a god of trade and particularly important to travellers and communicators. For our purposes, however, the most interesting aspects of the picture are the navigators Barry depicts on the right of the picture, carrying Father Thames out to sea. James Cook is in esteemed company, as he joins such heroes of the country's maritime past as Francis Drake, Walter Raleigh and John Cabot. Just as these historical seafarers facilitated the rise of English sea power in the Elizabethan age, so Cook contributed to its renewal and extension in the late eighteenth century.

The visual representation of Cook's memory, and its usefulness in cementing dearly held national narratives, went beyond the two-dimensional. There was an assumption that Cook, like many other eighteenth-century naval and imperial heroes, would be commemorated by a grand monument or sculptural scheme. In her 'Elegy to Captain Cook' (1780), Anna Seward was confident that a grateful Britannia 'to his virtues just, / Twines the bright wreath, and rears th' immortal bust'.[20] Eight years later, Helen Maria Williams, in her elegiac ode *The Morai* (1788), imagined the nation raising 'immortal wreaths' to Cook's memory in the shape of 'the marble tomb, the trophied bust'. The Royal Academy exhibition of 1780 even included a design by Thomas Banks for a 'National Monument to the Memory of Captain Cook', but nothing came of it.[21] The design was never executed and there was no public monument to Cook until a memorial obelisk was erected on Easby Moor in North Yorkshire in 1827 – and even that was built at the expense of the local lord of the manor, Robert Campion, rather than the king or the nation at large.[22]

Nevertheless, Cook had no lack of admirers prepared to adorn their private estates with monuments. A naval colleague and early supporter, Admiral Sir Hugh Palliser, erected a monu-

Figs 206, 207
Pair of wine coasters commemorating the voyages of Cook
Paul Storr; John Bridge, 1800–01
National Maritime Museum (PLT0031 and PLT0032)

The insides are engraved with a map of the eastern and western hemispheres showing Cook's track; the outsides are engraved with the 12 signs of the zodiac.

CAPTAIN COOK AND THE PACIFIC

ment to Cook at Vache Park, near Chalfont St Giles in Buckinghamshire, where he also appears to have been the first or perhaps second owner of Hodges's striking oil study of him (fig. 234). The legend that the monument bore – 'The ablest and most renowned navigator this or any country hath produced' – spoke volumes for Palliser's pride in his former protégé and in the country that had produced such a worthy seaman.[23] In commemorating 'this great master of his profession, whose skill and labours have enlarged natural philosophy [and] extended nautical science', Palliser's monument touched on some of the key themes that would characterize the way Cook's life, voyages and achievements would be represented over the coming centuries.[24] Not to be outdone, Earl Temple erected a monument at his country seat in Stowe (also in Buckinghamshire) celebrating the exploratory endeavours of Cook and elevating him to the status of the other 'British Worthies' remembered there.[25]

Just as the narratives of exploration were published in their pirated and abridged editions, or the oil paintings circulated more widely translated as prints, so the commemoration of Cook came in less majestic forms. Grand sculptural monuments to Cook were only one part of a vibrant culture of commemoration that was fuelled by all sorts of objects. For example, a pair of coasters traced the route of Cook's voyages in the eastern and western hemispheres respectively (figs 206, 207), mirroring the kinds of narratives portrayed in the hand-held globes that were such a popular accoutrement at the time (see figs 154–6). The fact that the Bridge family crest appears on the objects suggests that these might have been personal items made by, or for, the silversmith John Bridge as a way of recording his interest in Cook's voyages. Indeed, the continuing popular appeal and enduring fascination with Cook can be measured by the number of objects associated with his death – in all shapes and forms – that have made their way into major museums.

The collections at Greenwich, for example, include a red sandstone rock from the main outcrop of Grassy Hill at Cooktown, in northern Queensland (fig. 208). The label suggests that it originated 'almost on the spot where Capt. James Cook is reputed to have stood, when he landed at the Endeavour River on his first voyage, 12th June 1770'. There is also a fragment of a tree with an attached card declaring that it is a 'Piece of the Tree to which Capt James Cook tied his Bark "Endeavour" Cooktown N.Q. 17.6.1770' (fig. 209). Meanwhile, the Australian Museum in Sydney possesses what is perhaps the most bizarre of these secular relics: an arrow said to be made of Cook's leg bone. Even as late as 2003, an early-nineteenth-century gold-mounted walking stick sold at auction in Edinburgh for £153,000 (rather than the average £500–800) because it had an engraved collar with the legend 'made of the spear which killed Captain Cook'.[26] Cook's aura, it seems, endures.

Fig. 208 (above)
Piece of red sandstone reputedly from the spot where Cook stood when landing at the Endeavour River, 1770
National Maritime Museum (REL0038)

Fig. 209 (right)
Fragment of tree: an attached card reads: 'Piece of the Tree to which Capt James Cook tied his Bark "Endeavour" Cooktown N.Q. 17.6.1770'
National Maritime Museum (REL0802)

THE STRANGE AFTERLIVES OF CAPTAIN COOK

AFTERLIVES

The interest in Cook's voyages, evident in the publications and collections derived from them, meant that the outpouring of national grief following the explorer's death was not unexpected. Cook's legacy lived on, as we have seen, and his ghost settled down to an active afterlife. His name and personality were to become synonymous with exploration, and specifically British exploration. A Staffordshire domestic earthenware figure of Cook produced in the 1840s, based on the Nathaniel Dance portrait (see fig. 1), demonstrates just what a permanent feature of the cultural landscape he had become (fig. 212). A terrestrial floor globe made by Edward Stanford in 1880 (fig. 210) included a note on the nomenclature used in the Pacific Ocean: 'The original names assigned to these islands by their illustrious discoverer James Cook in 1778 have been adopted in this globe'. Perhaps the most immediate legacy, however, was Cook's impact on other Pacific explorers and travellers.

In the account of his own second voyage, the French explorer Jules Dumont d'Urville admitted that he was 'haunted' by Cook. Successors to Cook, whatever their nationality, were faced with the fact that his three voyages had determined the main features of the Pacific Ocean and shaped the way the wider public would react and respond to future travellers' accounts. Even when the eighteenth century's 'golden' age of European exploration of the Pacific had passed, Cook's name still acted as a lodestar for others to follow, but never overtake.[27] Russian explorers in the region in the nineteenth century, for example, regarded Cook as 'a kind of Napoleon, and the voyage a kind of battlefield' where Adam Johann von Krusenstern and others could make their reputations.[28] In the last decade of the century, the American painter John La Farge visited Tahiti. Once again, Cook's legacy framed the traveller's experiences. La Farge admitted that his 'impressions of today become confused and connected with these old printed records of the last century, until I seem to be treading the very turf that the First Discoverers walked in, and to be shaded by the very trees'.[29] Visitors to Hawai'i in the same period could purchase guidebooks on which the fateful bay where Cook met his end was prominently marked.[30]

Cook's posthumous reputation and status affected more than just near contemporaries or those who followed in his immediate expeditionary footsteps. The story of his life and voyages seemed to offer the possibility of combining scientific achievement, naval derring-do and civilizing imperial mission, and made it a perennial favourite with publishers into the twentieth century (figs 211, 213). The indefatigable nineteenth-century naval traveller Basil Hall was approached by Archibald Constable in 1825 to write a life of his eighteenth-century predecessor. Hall was taken by the prospect of the project and promised to undertake it 'with heart and soul to produce a worthy work'.[31] The book would have been based on surviving documents, which Hall hoped to source from 'old Mrs Cook' who was still alive, as well as from Lady Banks (the widow of Sir Joseph) and from Lord Melville.[32] Although the book never materialized, there was certainly support for the project. A few months later, Hall told the publisher that he had seen Lord Melville with regard to 'trying my hand at Cook' and Melville offered 'all that the Admiralty has on the subject'.[33]

Fig. 210 (opposite)
Terrestrial floor globe
Edward Stanford, 1886
National Maritime Museum
(GLB0160)

Fig. 211 (above)
Title page of
*The Life and Voyages
of Captain James Cook*
by George Young
1836
National Maritime Museum
(PBC4747)

Fig. 212 (right)
Staffordshire earthenware figure of Cook based on the portrait by Nathaniel Dance
Unknown maker, c.1840
National Maritime Museum
(AAA5953)

One of many Staffordshire decorative pieces commemorating national figures, mass-produced for a wide popular market rather than the social elite.

THE STRANGE AFTERLIVES OF CAPTAIN COOK   213

Walter Besant, writing at the turn of the twentieth century, acknowledged that the story of Cook's voyages was a well-known one (fig. 214). In fact, the narrative had been recounted so many times that Besant was confident that 'every boy has read Cook's *Voyages*':

> Not only every library, but almost every house with a row of bookshelves contains some account of them; there are cheap and popular editions, there are illustrated editions; they have been abridged, condensed, and castigated for the use of the young; they have served for lectures, illustrated by the magic lantern; they are known, in scraps, by everybody.[34]

In making such remarks, Besant might almost have been thinking about the passage in Charles Dickens's semi-autobiographical novel *David Copperfield*, where the eponymous hero recalls his early life at Chatham among the 'small collection of books in a little room upstairs':

> I had a greedy relish for a few volumes of voyages and travels – I forget what now – that were on those shelves; and for days and days I remember to have gone about my region of our house armed with the centre-piece of an old set of boot-trees – the perfect realisation of Captain Somebody, of the Royal British Navy, in danger of being beset by savages, and resolved to sell his life at a great price.[35]

Fig. 213 (opposite)
Title page of *A narrative of the voyages round the world: performed by Captain James Cook, with an account of his life, during the previous and intervening periods*
by Andrew Kippis
1883
National Maritime Museum
(PBP0439)

Fig. 214 (above)
Title page of *Captain Cook*
by Walter Besant
1904
National Maritime Museum
(PBD5402)

Fig. 215 (left)
Title page of *Captain Cook's voyages of discovery*
c.1906
National Maritime Museum
(PBC4746)

THE STRANGE AFTERLIVES OF CAPTAIN COOK   215

While Cook is not named, the basic structure of the tale is recognizable, and this mirrors Besant's views on the impact of the Cook story. He readily acknowledged that 'few of us would sit down to pass an examination on the subject' of Cook's voyages, but he firmly believed that their broad outline was a familiar one:

> We all know in general terms that Cook surveyed the coasts of New Zealand and New Holland, penetrated the southern ocean, traversed the Pacific in every direction, and was finally murdered at the island which some of us still, faithful to tradition, call Owhyhee.[36]

For many, particularly in Britain, Cook's legacy was a straightforward one: a brave and benevolent mission, bringing civilization and expanding the boundaries of human scientific knowledge. This was certainly the view of one of those who followed in his footsteps in the Pacific, Matthew Flinders. Famous for his charting work around the Australian coast and for his captivity on Mauritius, Flinders wrote to his French contemporary, Charles Baudin, extolling the virtues of disinterested scientific endeavour:

> The labours of Newton and Cook were beneficial, whilst those of Alexander and Caesar desolated mankind. Would that our two nations were convinced of this truth, and act accordingly, then might we hope the animosity which makes it a duty for one man to destroy another, would become an honourable emulation for excellence in the useful arts and sciences.[37]

Among the British colonies in the places that Cook explored and encountered most (namely Australia and New Zealand), a similar pattern of celebration can be identified. If, as Walter Besant contended, there was basic familiarity in Britain, then Captain Cook has always loomed large in public history in its antipodes. His visits have been understood there as foundational moments: indeed, for some they marked the true beginnings of each nation's history. As a result, the voyages have been extensively marked through monuments and place names. They have been celebrated in a plethora of popular historical works, poems, plays, films, texts for children and souvenir objects, and the landings themselves were (and continue to be) frequently re-enacted. Their commemoration has sometimes been subject to protest, and criticism of Cook has occasionally been aired. As we will see below, by the later twentieth century, more complex reactions to his representation and commemoration had emerged across Oceania and the Pacific, but his popularity has been surprisingly resilient.[38]

In 1822, members of the Philosophical Society of Australasia marked the spot at Botany Bay where they thought Cook and Banks had landed in 1770. Judge Barron Field published a sonnet on the occasion:

> Here fix the tablet. This must be the place
> Where our Columbus of the South did land;
> He saw the Indian village on that sand,
> And on this rock first met the simple race
> Of Australasia, who presum'd to face
> With lance and spear his musquet. Close at hand
> Is the clear stream, from which his vent'rous band
> Refresh'd their ship; and thence a little space
> Lies Sutherland, their shipmate; for the sound
> Of Christian burial better did proclaim
> Possession, than the flag, in England's name.
> These were the commelinae Banks first found;
> But where's the tree with the ship's wood-carv'd fame?
> Fix then th'Ephesian brass. 'Tis classic ground.[39]

The reason these Australians latched on to Cook might be partially explained by the fact that he was a less ambiguous figure than Governor Arthur Phillip as a symbol of European beginnings. He was not associated with the convict past of New South Wales and, as a result, was popular for his apparently unproblematic and positive associations. Indeed, so dominant did Cook become that, in 1888, many thought they were celebrating the centenary of Cook's landing. (In reality, the commemoration marked the arrival of the First Fleet, commanded by Arthur Phillip, in January 1788.) Even George Belcher, parliamentarian and Geelong pioneer, and a man full of a sense of history and occasion, wrote in his diary for 26 January 1888: '100 years ago today since Capt. Cook landed at Sydney and planted the British flag on the Australian Continent'.[40]

Here again, material culture and images were at the heart of this process of representation. One of those who contributed to the nineteenth-century myth of Cook in Australia was John Alexander Gilfillan. In 1866, he depicted 'Captain Cook taking possession of the Australian continent on behalf of the British Crown, AD 1770', a grand and formal scene set in an idealized Botany Bay.[41] Cook's charting of the eastern seaboard was also marked by the unveiling of a bronze statue of the explorer in Sydney. According to the *Sydney Morning Herald*, upwards of 70,000 people turned out on 25 February 1879 to witness the 'grandest spectacle in Australian history'. The unveiling of Thomas Woolner's statue of Captain Cook in Sydney's Hyde Park was done, according to another report, 'amid great pomp and enthusiasm' and greeted with 'cordial cheering' all round.[42] Plans for the 13-foot figure of Cook, standing on a 22-foot-high granite column, were first mooted in 1874 and some £4400

Fig. 216
Orient Line poster
Albert E. Cox, *c.*1925
National Maritime Museum
(ZBA3073)

THE STRANGE AFTERLIVES OF CAPTAIN COOK

was raised over the next four years. The imposing result represented the explorer in naval uniform, carrying a telescope in his left hand. Cook, it was said, belonged to 'the greater Britain of this hemisphere'.[43] In 1901, following the federation of the Australian colonies as the Commonwealth of Australia, Emmanuel Phillips Fox produced *The Landing of Captain Cook at Botany Bay, 1770* (1902, National Gallery of Victoria). It shows Cook's outstretched arm restraining his men from firing on the gathered Aborigines. Images such as these presented a powerfully persuasive and positive view of Cook's engagement with Australia and would prove to be almost as enduring as the explorer himself.

Across the Tasman Sea, in New Zealand, this process of commemoration is also evident in the statue unveiled by Lord Bledisloe, the Governor General, on 10 August 1932. The inscription on the base of the figure, designed by a local sculptor, William Trethewey, celebrates 'James Cook, Captain, Royal Navy, circumnavigator who first hoisted the British flag in New Zealand and explored her seas and coasts, 1769–70, 1773–4, 1777'. Further down it proudly announces Cook to be 'Oceani investigator accerimus' ('the most intrepid investigator of the seas').[44] This interpretation of Cook and his story by those involved in unveiling the monument is revealing. In his speech, Lord Bledisloe asserted that:

> *In the early history of New Zealand there are three outstanding landmarks – its effective discovery by James Cook, then a lieutenant in the Royal Navy, in 1769; its Christianisation, commencing with the arrival of Samuel Marsden in 1814; and its inclusion in the British Empire under the Treaty of Waitangi in 1840. But for the first, the second would have been improbable and the third impossible.*[45]

The Governor General spoke 'without fear of contradiction' when he said that none would command 'more worldwide veneration or more general assent than that of James Cook. Unsurpassed as a navigator, the variety of Cook's greatness is nevertheless just as astonishing as his supreme greatness in his special sphere.'[46] The Mayor of Christchurch, Mr D.G. Sullivan, a Labour politician, also recruited Cook's memory to support his version of history, referring to Cook's humble origins as the 'son of a Yorkshire farm labourer' who had risen through the ranks: 'Not only a great navigator, a great organiser, and a distinguished scientist, but a man of great humanitarian instincts and one who always had the love and veneration of the crews under his control.'[47] In the first half of the twentieth century, therefore, Cook provided a 'congenial theme', reminding colonial subjects of 'the great things in our own past, and in that of the race to which we are proud to belong'.[48]

MUSEUMS AND OBJECTS IN THE NINETEENTH AND TWENTIETH CENTURIES

In considering the ways in which material culture, images and objects have been used to preserve, shape and challenge Cook's legacies, museums are a vital resource. Much has been written about the way in which these institutions influence our understanding of the past. They select objects and the stories and personalities that accompany them, thereby filtering and mediating our responses to historical events and people. But museums and the displays they present to the public can also play a major part in remoulding our perceptions of the past. The next chapter will take the specific example of the NMM at Greenwich and consider how its response to the Cook story has changed over the years. Here we will look at other cases.

Despite the presence of Cook and his story in many museums and exhibitions, their appearance was often a cameo, with little contextual or comprehensive detail to explain the significance (or otherwise) of the objects on display. While the British Museum and others exhibited material collected by Cook and his crew (see chapter 6), the broader context of these voyages was invariably omitted. Even in places where one might have expected more detail, the Cook-related material was often relegated to the status of quirky relics supporting bigger narratives about the success of the Royal Navy or the advancements of British science. For example, the Royal United Services Institute in Whitehall Yard – formed in 1830 to receive the contributions of returned military and naval officers and including a library and museum – became 'a microcosm of British Military and naval history'.[49] Among the many personal relics in its collection were Cook's punch bowl and chronometer. The great Royal Naval exhibition of 1891, in the grounds of Chelsea Hospital, included a gallery entitled 'The Cook Gallery – Navigation'. As its name implies, the material on display was selected in order to explain 'what modern science has done for the art of navigation' rather than necessarily focusing on the acts of exploration and encounter themselves. Indeed, although it was named after the explorer, only a few of the seven sections in the gallery had any direct relationship with Cook. Visitors, it appears, had to make do with 'Captain Cook's waistcoat' and his 'speaking trumpet'.[50] At the conclusion of the exhibition, the log of the first voyage and the gold medal conferred by the Royal Society went to the British Museum. However, in a sign of Cook's increasing attractiveness and importance to the fledgling colonies that would become Australia, the greater part of the relics preserved were sent to the Colonial Government Museum in Sydney.[51]

When the NMM opened its doors in 1937, Cook 'relics' were also displayed in Greenwich. Once more, the material was more closely connected with his personal story than with the wider geographical, political and social contexts in which his travels were taking place.⁵² He found himself sharing Caird Gallery VI with the story of the War of American Independence. A number of important paintings by Hodges, and the Dance and Zoffany pictures (transferred as part of the contents of the former Naval Gallery in the Painted Hall), were displayed, but the majority of the objects exhibited fell into the category of navigational instruments on the one hand or a hotchpotch of miscellaneous items on the other.⁵³ In the first category, the newly opened museum displayed 'Captain Cook's compass . . . made by Thomas Graydon, an officer of the Royal Engineers in 1768'. It also incorporated a dip circle made by Edward Nairne of Holborn, which 'is almost certain to have accompanied him [Cook] on at least some of his voyages' (see figs 239, 240). The 'two telescopes with octagonal shaped wooden bodies' have an even more tenuous connection: although they did not actually belong to Cook, visitors were reassured that they were 'contemporary'.⁵⁴ The items presented by 'H.A. Baron, Esq.' were much more squarely located within Cook's domestic sphere (or, more specifically, that of Mrs Cook, who survived him by more than half a century) (figs 218–21). They included a pair of silver candlesticks, plated snuffers and a tray, a silver pap-boat, a snuff box, a box of ivory chessmen, a pair of dice boxes with dice, a mahogany wardrobe, a bureau bookcase, a writing table and chest of drawers, four painted rush-bottom chairs, a tea tray, a china punch bowl, teapot, sugar bowl and two drinking mugs.⁵⁵

If museums offered a context for the interpretation of objects connected with Cook's voyages, so the various anniversaries connected with them also inspired the production of objects and, in some cases, the challenging of 'traditional' representations. As we have seen, there were some important nineteenth-century commemorations of Cook's voyages, but the twentieth century heralded the heyday of such occasions in the Pacific, in the settler colonies and in London. These events, and the objects used to mark them, offer some excellent insights into the ways in which Cook's legacy has endured over two and a half centuries, but also the way in which the story of his voyages has been refined and increasingly challenged.

The bicentenary of Cook's embarkation in 1768 was commemorated by a raft of material mementos. Wedgwood produced 200 copies of a blue jasper portrait medallion of Cook, based on an original by John Flaxman. Meanwhile, the Australian Hydrographic Office issued charts depicting the course of Cook's first voyage (fig. 225). The two insets of the map show the entrance to the Endeavour River and Botany Bay respectively. In many ways, these objects projected a very traditional view of Cook, as a single, individual explorer and the harbinger of scientific advancement. Over the following decade, a plethora of anniversaries were marked by similar

Fig. 217
Captain James Cook OB 1779
Nathaniel Dance, c.1779
National Maritime Museum
(PAD4387)

'commemoratives'. However, these increasingly took account of the changed circumstances and political developments in the post-colonial world of the late twentieth century. In 1969, a medal produced for the Royal Numismatic Society of New Zealand commemorated Cook's time there, but bearing a legend noting his 'rediscovery of New Zealand' (figs 226, 227). The use of that term underlined the increasing awareness of Maori navigational and cultural achievements. In 1978, a medal was struck to commemorate Cook's first calling at Hawai'i. The obverse preserves something of the traditional narrative: it shows a bust of Cook with the *Resolution* under sail and a map of the islands, and bears a legend reading 'Captain James

THE STRANGE AFTERLIVES OF CAPTAIN COOK

Figs 218, 219 (left)
Pair of candlesticks belonging to Elizabeth Cook, engraved with Cook's crest
John Parsons & Co., 1784–85
National Maritime Museum (PLT0026–27)

Figs 220 (left, below)
Candle snuffers belonging to Captain James Cook and his wife, Elizabeth Cook
Unknown maker, 18th century
National Maritime Museum (PLT0029–30)

Figs 221 (left, bottom)
Silver pear-shaped pap-boat, belonging to Captain James Cook and his wife, Elizabeth
Unknown maker, 1762–63
National Maritime Museum (PLT0028)

Figs 222, 223 (below)
Coin commemorating the discovery of Hawaii by James Cook, 1778
Unknown maker, 1928
National Maritime Museum (MEC2587)

220   CAPTAIN COOK AND THE PACIFIC

Cook discoverer of Hawaii' (figs 228, 229). The reverse depicts a Hawai'ian chief with his right arm raised and a staff in his left hand, with a boat and shoreline of palm trees in the background. The legend here reads: 'Kamehameha the great founder of the Hawaiian kingdom'. No longer would Cook stand alone as the single most significant figure in the 'opening up' of the Pacific: in future, he would have to share his pedestal with those who came before him. A similar example appears in a Canadian medal issued by the city of Victoria, British Columbia, to commemorate his time in Nootka Sound (figs 230, 231). Here the reverse shows Cook shaking hands with a member of the First Nations.

These objects eschew the violence and power imbalances that frequently characterized encounters in the period, but all of them are in the collections at Greenwich (and in other museums) and are, therefore, available for display to the general public. By incorporating other historical actors alongside Cook, they represent changing perceptions over time and at least nod to new ways of representing him. The next chapter will discuss some of those ways in which Greenwich has chosen to use its rich collections to tell and retell the Cook story over the years.

Fig. 224 (above, right)
Ingot depicting the *Endeavour*
Franklin Mint, 1976
National Maritime Museum,
courtesy of Sequential Brands
Group, Inc.
(MEC2831)

THE STRANGE AFTERLIVES OF CAPTAIN COOK   221

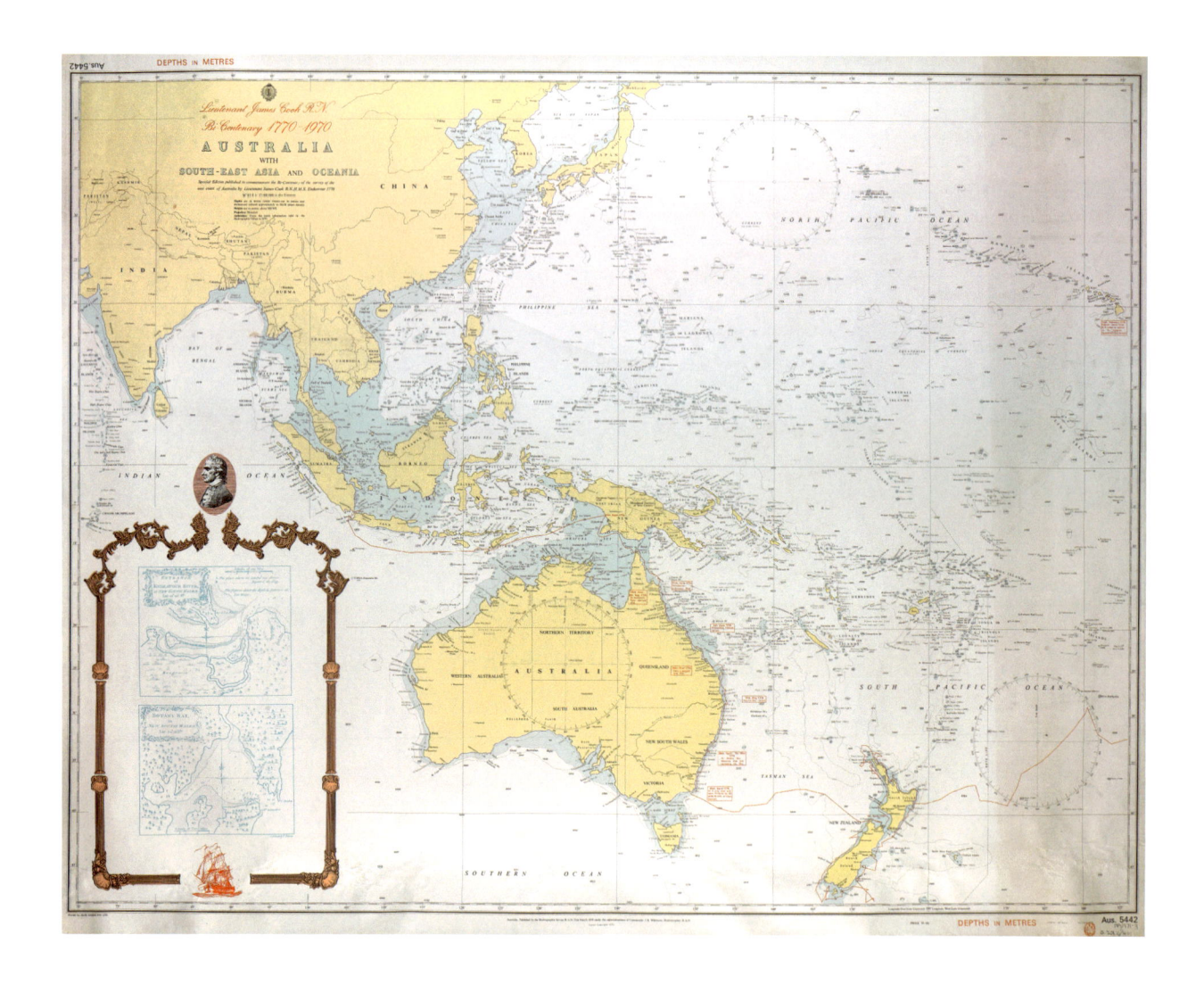

Fig. 225
Australia with S.E. Asia and Oceania special edition to commemorate the bicentenary of the survey of the east coast of Australia by Lieutenant James Cook
Australian Hydrographic Service, 1970
National Maritime Museum (G201:6/4(1))

Figs 226, 227
Medal commemorating the bicentenary of James Cook's rediscovery of New Zealand
Issued by the Royal Numismatic Society of New Zealand Inc.; designed by James Berry; minted by Royal Australian Mint, 1969
National Maritime Museum (MEC2684)

Figs 228, 229
Medal commemorating the discovery of Hawaii by Captain James Cook (1728–79)
Issued by the Hawaiian Trading Company; engraved by E.W. and designed by Jim Dean, 1978
National Maritime Museum (MEC2117)

Figs 230, 231
Medal commemorating Captain James Cook (1728–79) at Nootka Sound
Issued by Greater Victoria Information Centre; designed by Tom Seymonsbergen; minted by The Sherritt Mint, Alberta, 1978
National Maritime Museum (MEC2118)

Figs 232, 233
Medal commemorating the bicentenary (1778–1978) of Captain James Cook (1728–79) meeting Chief Maquinna
Issued by local merchants and Alberni Valley Stamp Club; designed by Bob Eyford; minted by Charles Roynon & Co., 1978
National Maritime Museum (MEC2119)

THE STRANGE AFTERLIVES OF CAPTAIN COOK

Capt.ⁿ James Cook
of the Endeavour.

Chapter 9
Cook on Display
The National Maritime Museum's Cook Galleries and Exhibitions, 1937–2000

Nigel Rigby

EARLY DAYS

The National Maritime Museum was opened on 27 April 1937 by George VI (fig. 236), who marked the occasion by presenting the Museum with what have become two of its most valued objects: manuscript journals of James Cook's first and second voyages to the Pacific in *Endeavour* and *Resolution*. These enriched a small but already growing collection of Cook-related material, much of which the first Director of the Museum, Sir Geoffrey Callender, had installed on the ground floor of the West Wing; Cook displays would become virtually ever-present over the next 60 years. Callender's remarkable feat of establishing a new national museum from scratch with a small and largely inexperienced staff had been achieved in a little under three years, although the Greenwich site, as we shall see, already had a long prehistory as a theatre of naval memory.

With energy, vision, single-minded determination, the help of some outstanding collections – not least those already on site in the Royal Naval College – a small but dedicated staff, and the over-generous support of Sir James Caird, the Museum's first and greatest benefactor, Callender created a largely chronological suite of displays that told the story of British naval history and navigation. Beginning with the Tudors and early Stuarts in the Queen's House, visitors could proceed through 12 further rooms comprising the 'Caird Galleries' of the Museum's West Wings, ending with the Battle of Trafalgar and the death of Nelson. The story, in effect, followed the same didactic approach as Callender's three-volume textbook of naval history, *Sea Kings of Britain* (1907–11), written when he discovered, in his first teaching job at the Royal Naval College, Osborne, that no such book yet existed. In 1922, Callender was appointed the Navy's first Chair of Naval History and English at the Royal Naval College, Greenwich, and from then on he played a leading role in the long campaign to establish what would become the National Maritime Museum. Despite being a man of lively wit and charm in private, and a born teacher, Callender was somewhat aloof and autocratic as a public figure, deeply knowledgeable about naval history and antiquities – an outstanding speaker and a familiar figure in the salerooms. He built a set of galleries that defined the Museum for years after his sudden death in 1946; fittingly, perhaps, this happened in the Museum where, although already 70 years old, Callender

Fig. 234 (opposite)
Captain James Cook, 1728–79
William Hodges, 1775–76
National Maritime Museum.
Acquired with the assistance of the National Heritage Memorial Fund
(BHC4227)

When acquired in 1987, this was the Museum's most expensive painting acquisition to that date.

Fig. 235
The Naval Gallery in the Painted Hall of Greenwich Hospital (the Royal Naval College, Greenwich, from 1872) Before 1936
National Maritime Museum
(C9138-007)

225

was still in command and rebuilding, following the disruptions of the Second World War.[1]

Like Callender's *Sea Kings*, the galleries told their stories through the deeds of great men: Francis Drake, John Hawkins, George Anson, Edward Vernon, James Cook, George Brydges Rodney, Richard Howe, Adam Duncan and Horatio Nelson, though only Cook and Nelson, and to a less extent Drake, have really survived and prospered as important figures in British popular maritime history. Cook had not actually been one of Callender's 'Sea Kings', since the textbook told a predominantly military history, and he got the merest aside in Callender's later and better-known work, *The Naval Side of British History* (1924). He was, however, a figure with national value and international recognition, and he was accordingly accommodated within the overarching naval narrative, if a little awkwardly, in the brief outbreak of peace between the Seven Years War and the War of American Independence. In terms of the quality and significance of objects, Cook amply deserved his place in the Museum: the focal point of the display was Nathaniel Dance's iconic portrait of Cook at the height of his powers and reputation, between the second and third Pacific voyages (see fig. 1). A bust of Sir Joseph Banks by Peter Turnerelli pointed towards the voyages' scientific achievements, as Callender's *Catalogue of the National Maritime Museum* explains: 'Banks accompanied Cook on his first voyage, making valuable natural history collections. He was for many years president of the Royal Society.'[2]

Johan Zoffany's equally imposing, if unfinished oil painting *The Death of Captain James Cook* (see fig. 202) brought the Cook display to a logical end.

An experienced member of staff would later comment that Callender 'took very little interest in the technical development of the ship or the history of the methods used by navigators. He chose pictures, models or instruments because he considered them works of art and things of beauty.'[3] Callender's idea, said another, was 'for the visitor to proceed, General Guide in hand, to the threshold of each room, and suitably priming himself, enter it with the ringing phrases of "Sea Kings" in his head. That meant minimum labelling to pictures and models, and plain dates above the doorway.'[4] Although a little uncharitable, this was probably accurate enough: it was generally acknowledged that no one could weave a story around Callender's displays as well as Callender himself could. That he could actually carry through this aesthetically driven and minimalist plan was only possible because he already had high-quality and extensive private and public collections at his disposal, which had been acquired or earmarked for the Museum through gift, transfer or purchase well before it opened:

> *Thanks to the wealth of the Painted Hall and Royal Naval Museum collections and the generosity of Sir James Caird in the Sale Rooms, Sir Geoffrey ... was able to hang an admiral, painted by a reputable artist,*

Fig. 236 (right)
The royal opening of the National Maritime Museum by George VI, 27 April 1937
National Maritime Museum
(C9138_002)

Sir Geoffrey Callender is on the far left, behind George VI, and Sir James Caird walks behind Queen Mary, the Queen Mother. It was the first public engagement attended by Princess Elizabeth (the future Elizabeth II, then aged just 11).

Fig. 237 (left)
A sketch of the vestibule of
the Painted Hall, with Chelsea
and Greenwich Pensioners
discussing J.M.W. Turner's
Battle of Trafalgar (right)
John Burnet, after 1829
National Maritime Museum
(PAH3983)

Cook's is the portrait on the left,
by the stairs, in the early years
of the Hall's use as the Naval
Gallery at Greenwich Hospital.

Fig. 238 (right)
The Turnerelli bust of
Sir Joseph Banks on display
in Gallery VI
1968
National Maritime Museum
(C7674-026)

COOK ON DISPLAY   227

alongside his naval victory, ditto, and place opposite the Dockyard scale model of his flagship. In this way we 'spotlighted' so to speak the immortal memories.[5]

For the Cook display, as for many of the inaugural galleries, arguably the most significant of these collections would be the 248 oil paintings and 20 pieces of sculpture held by the Royal Hospital for Seamen at Greenwich, known more familiarly as the Naval Gallery, or after 1936 as the Greenwich Hospital Collection (fig. 235).[6] The Hospital had accumulated (mainly through gifts) just over 30 paintings throughout the eighteenth century, the most significant being a full-length portrait of John Montagu, 4th Earl of Sandwich and First Lord of the Admiralty, by Thomas Gainsborough. Its Secretary, Edward Hawke Locker, made these the core of a permanent 'National Gallery of Naval Art', which was the first such public 'national' gallery in Britain. By the time it opened in April 1824, thanks to the assiduous solicitations of Locker, it had already acquired a wealth of naval portraits, and more than 30 of these had been presented by George IV, who added J.M.W. Turner's great *Battle of Trafalgar* as one of his last contributions in 1829. Further gifts, including works by Joshua Reynolds, Benjamin West and other celebrated names, were quickly accumulated, with the overarching aim of commemorating 'those men, by whose prowess Great Britain had won, and had victoriously maintained, the dominion of the sea'.[7] The collection, whose modest entry charge raised money for the Hospital's charitable activities, played a key role in maintaining its public profile. It was a major public attraction,

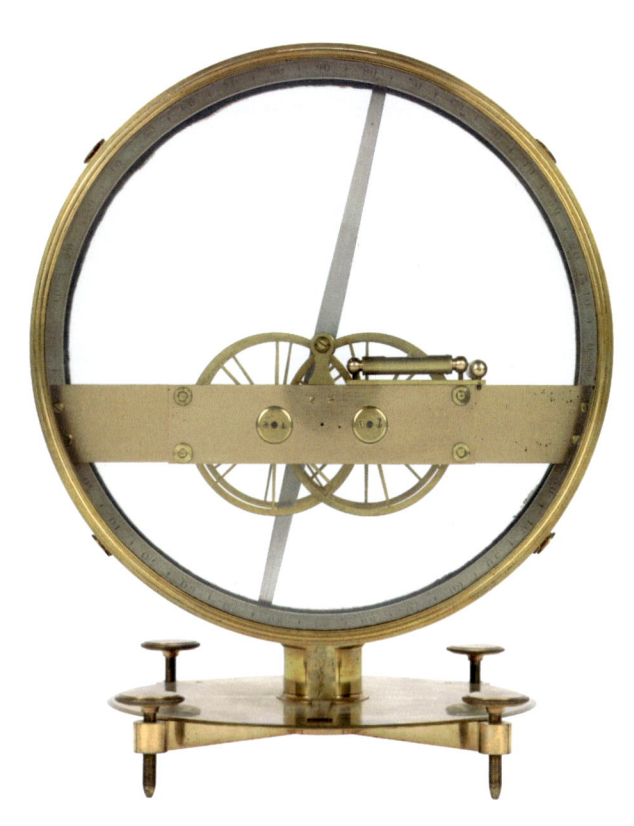

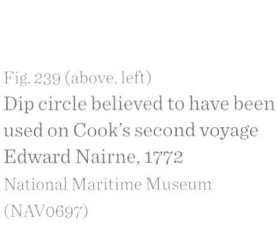

Fig. 239 (above, left)
Dip circle believed to have been used on Cook's second voyage
Edward Nairne, 1772
National Maritime Museum
(NAV0697)

Fig. 240 (above, right)
Compass
G. Graydon, *c*.1824
National Maritime Museum
(NAV0353)

At one time this compass was thought to have belonged to Captain Cook.

Figs 241, 242 (opposite)
Title page and engraving of Nathaniel Portlock, from
A Voyage round the world
1789
National Maritime Museum
(PBC4229)

228   CAPTAIN COOK AND THE PACIFIC

even before the opening of the London-to-Greenwich railway – the world's first suburban line – in 1836 made it even more so by the mid-nineteenth century; surviving the conversion of the Hospital buildings into the Royal Naval College in 1872; and the collection remained open in the Hall until it was transferred over the road and into the long-term care of the NMM in 1936.

Nathaniel Dance's portrait of Cook had originally been commissioned by Sir Joseph Banks to hang in his house in Soho Square, as is discussed in the introduction to this book. It was acquired by the persuasive Locker from Banks's executor in 1829 and was first hung, as can be seen in the watercolour by John Burnet (fig. 237), in the vestibule of the Painted Hall alongside Turner's *Trafalgar* of 1822–24, by far the largest canvas he painted, as well as one of the most controversial at the time. The Turnerelli bust of Banks (fig. 238) also arrived at the Painted Hall by a different route: it was donated by Captain Sir Everard Home in the 1850s. Zoffany's *Death of Captain Cook*, an unfinished work with considerable overpainting, had been given a less prominent place in the Painted Hall and hung high on a wall, where its poor condition was less obvious in the gloom.[8] Locker acquired it from Elizabeth Cook's estate, although how she had acquired it, and who had originally commissioned the work and when; why it remained unfinished; when it was overpainted; and even whether the artist was indeed Zoffany remained unanswered questions for many years.[9] It has recently been established that it was bought in 1827 by a London dealer from the estate of an ex-East India Company Officer, Alexander Kyd, who had inherited it from his uncle, Robert Kyd, also an EIC officer and founder of the Calcutta Botanical Gardens. The dealer then sold or presented it (more probably the former) to Cook's widow, and her executor gave it to Greenwich Hospital on her death, which may have been her unfulfilled personal intention, since it is hardly a picture of domestic scale, let alone one for even a devoted widow to live with. This provenance still leaves questions: for example, it is still not clear why or when Robert Kyd would have commissioned such a painting. Zoffany himself said that he had been inspired to begin the painting after seeing the popular pantomime on the death of Cook at Covent Garden on his return from India in 1789, but the seeds may have been sown a few years earlier, during his tour of India. When Zoffany arrived in Calcutta in September of that year, he was invited to the house of Warren Hastings where he met John Williamson, commander of the frigate *Crocodile*. The two men appear to have got on well and even went on an elephant ride together.[10] Williamson is best known today, as he was then, for having been the officer in charge of the boats when Cook was killed at Kealakekua Bay. Zoffany also socialized with Robert Kyd during his time in Calcutta, and had been Joseph Banks's choice

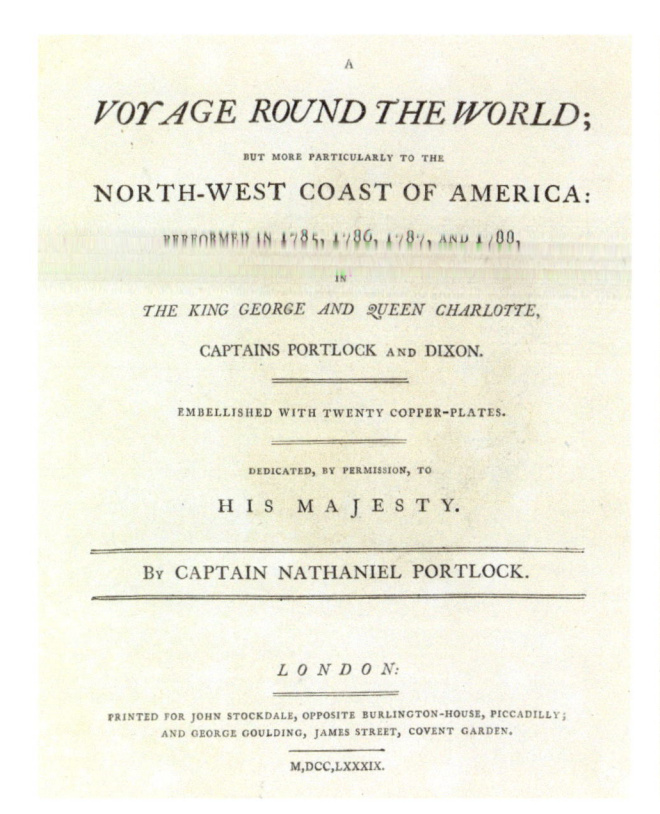

COOK ON DISPLAY 229

Fig. 243
Captain the Honourable John Byron, 1723–86
Joshua Reynolds, 1759
National Maritime Museum, Caird Collection (BHC2592)

Known as 'Foul-weather Jack', from the conditions he endured on his Pacific voyages. One grandson, Admiral George Anson Byron, also made a Pacific voyage in the 1820s and became 7th Lord Byron on the death of another – the more famous poet, George Byron, who was the 6th Lord.

as artist for Cook's second voyage, although, like his patron, he withdrew in the wake of arguments about the proposed accommodation for the scientific party.

The Naval Gallery collection also provided the anonymous portrait of Nathaniel Portlock (see fig. 160), who sailed on Cook's third voyage, took a commercial expedition to the American north-west coast and finally commanded the *Assistant* on Bligh's second – and successful – breadfruit voyage (figs 241, 242). The Native American to the left of the portrait with the distinctive face painting is taken from one of John Webber's engravings of Nootka Sound on Cook's third voyage. The painting was presented in 1902 by one of Portlock's descendants.

These paintings would still be key works in any Cook exhibition today, as indeed they are. More problematically, Callender drew on the old Royal Naval Museum's collection (the Navy's museum having been incorporated into the Royal Naval College, Greenwich, in the 1870s) for a small display of navigational instruments. A compass attributed to Cook was described as 'made by Thomas Graydon [*sic*], an officer of the Royal Engineers in 1768' (fig. 240); two fine telescopes by Dollond were 'contemporary' to Cook; and a dip circle was made by Edward Nairne of Holborn 'to the order of the Board of Longitude' (fig. 239).[11] This last imposing object was one of a small number commissioned by the Board for observations of the transit of Venus in 1769: if it was not actually the one taken by Cook – and the catalogue acknowledged this doubt – it was identical to the one he did take. Although the telescopes were contemporary as stated on the label, the same cannot be said of the compass, which, although a handsome object, was eventually discovered to have been made some 40 years after Cook's death. It has to be understood that these were early days for the Museum, and the pressures of a large collection and initially limited expertise meant that curatorial staff were perforce relying on what were often long-standing but untested attributions.

Equally important to the Cook gallery were five paintings by William Hodges owned by the Admiralty (now the Ministry of Defence Art Collection), which lent the Museum two panels, *View of Tahiti* and *A Review of War Boats of Tahiti*, and three oils on canvas: *Tahiti Revisited* (see fig. 131), *A View of the Cape of Good Hope, taken on the Spot on Board the 'Resolution' and 'Adventure', Capt. Cooke* (see fig. 123) and *A View of Maitavie Bay* (see fig. 32).[12] The Admiralty had originally lent ten paintings but, as Callender would ruefully recall, 'they did not remain [at Greenwich for] long. As soon as the Duff Coopers moved into Admiralty House [in 1937], Lady Diana descended on us and carried off all but two or three.'[13]

Two other fine portraits were also on display in the gallery: Joshua Reynolds's half-length of Cook's predecessor in the Pacific, Captain John Byron (more famous now for being the grandfather of the poet than an explorer), purchased by Sir James Caird (fig. 243); and the portrait attributed to John Webber, *Midshipman James Ward* (see fig. 63), a young man on Cook's last voyage whose claim to fame is that he is reputed to have been the first to sight the Hawai'ian Islands in 1778 and was one of the boat party in Kealakekua Bay on that fateful day in 1779. This portrait was presented directly to the Museum by Ward's descendants in 1934. The gallery additionally held some 20 Cook relics: items of domestic furniture and objects once owned by Cook's wife and passed down through the family of his sister to Henry Baron, who left them to the Royal Naval Museum on his death in 1931 and thence to the NMM in 1936.[14] These, like the other main contents of the Naval Museum, were given to the NMM absolutely – unlike the Greenwich Hospital collections, whose status is that of 'permanent loan'. The relics would survive various iterations of the Cook gallery for decades, but they merited no more than a brief listing in the catalogue and added little to the story of Cook. The Museum had inherited a number of relics, particularly through the old Royal Naval Museum as well as in 1963, following the closure of the Royal United Services Institution Museum. While the NMM always tried to acquire objects with a good provenance, this did not always stand close scrutiny. All of the furniture and most of the domestic Cook relics were finally taken off show in the 1980s, as a consequence of changing and largely 'market-driven' display fashions, and of the building renovations facilitating them.

Although the superb collections in the new museum's care were Callender's great strength, they were also his limiting factor.[15] By his aesthetic leanings, by virtue of the objects available to him and the chronological sequence in which he arranged the Museum's galleries, Callender's Cook – in Gallery VI – was presented as a naval hero to be looked on and admired, but with little explanation as to why. After Callender died suddenly from a heart attack at his desk in 1946, Gallery VI appears to have remained relatively little changed for some years, primarily because there were other, more pressing post-war priorities than altering a display that still worked well on its own terms.

A formative moment for the Museum's representation of Cook arrived in 1955 with a letter from the President of the Hakluyt Society and Head of Maps at the British Library, Raleigh Skelton, who pointed out to Callender's successor, Frank Carr, that 'apart from the material which you have put on exhibition in Caird Gallery VI, this country has never held a worthy exhibition devoted to Cook, even for the centenary of his death in 1879', when, ironically, the Admiralty had 'lent many of its treasures to the fine exhibition of the Société de Géographie in Paris' rather than staging something itself. The Hakluyt Society, a learned society that publishes historical works of travel, exploration and discovery, had its own interests in an exhibition, for it was about to publish the first of J.C. Beaglehole's authoritative editions of Cook's journals. These, said Skelton with masterly understatement, had 'brought to light a good deal of new material' and were already raising 'considerable public interest in Cook'. Carr agreed, and it was settled that *Captain Cook*, a 'tribute to this great English seaman', would open in the Special Exhibitions Gallery in the East Wing in the spring of 1956.[16]

Beaglehole advised on content, suggesting possible objects that might be borrowed for the exhibition. The range of institutional, commercial and private lenders included the British Museum, the Public Record Office, the Royal United Services Institution (now Institute), the Astronomer Royal, Lord Hinchingbrooke and the dealers Maggs of London, who had long been strong supporters of the Museum. Reflecting Beaglehole's archival research, the exhibition relied heavily on manuscript material and other works on paper. Its ten sections covered Cook's early career, the three Pacific voyages and his death, his ships (including a model of *Endeavour* made by C. Whitaker), personal records and relics, his letters and other writings not related to the voyages, portraits of Cook and his officers, his posthumous reputation, and finally a small display of ethnographic objects believed to have been brought back from the three voyages. These objects were largely interpreted by a short, factual description, a note of their provenance and an extract from a relevant journal. The label for the 'New Zealand native weapon, Patoo-Patoo' quoted indirectly from Cook, disclosing that 'This hand weapon, referred to as a "Bludgeon", was described by Cook as being worn by the principal people'. A wooden bowl with carved animal head and tail 'may have been traded at Nootka Sound' and been brought home on the third voyage.[17] Beaglehole was reported to be impressed with the exhibition;[18] so, too, was the public, for the exhibition's run was extended for a further six months to February 1957.

Through the Museum's first temporary exhibition on Cook, the explorer undoubtedly became a more important and rounded figure. Freeing him from Callender's teleological naval hagiography, engaging with the latest scholarly research and working with a broader range of material enabled other narratives to develop and other aspects of Cook's personality and professional life to emerge.

Fig. 244 (right)
Entrance to *Captain Cook and the Exploration of the Pacific* c.1970
National Maritime Museum (B2483-022)

Fig. 245 (below)
Model of *Endeavour* on display in *Captain Cook and Mr Hodges*, 1979
National Maritime Museum (B2483-022)

The model was made by the South African maker Robert Lightley. With one side of the ship open and models of all 94 officers, marines and crew on board, it has been one of the Museum's most popular objects since it was acquired in the 1970s. It was once displayed (as here) with part of the 'real' *Endeavour*'s sternpost – a speculative identification, now known to be incorrect.

232   CAPTAIN COOK AND THE PACIFIC

BUILDING COOK

Over the next half-century, Cook's voyages would be studied through a range of historical lenses: art, empire, literature, anthropology, navigation, geography, astronomy, botany, and so on. The Museum's growing collections would establish themselves as an important resource for international scholars through the work of J.C. Beaglehole, Bernard Smith, Glyndwr Williams, Nicholas Thomas and many others. The more immediate internal effect of the 1956–7 Cook exhibition, however, was that its success fed into a period of strategic agonizing about the purpose and approaches of the Museum.

The reconsideration, eventually entitled *A Sense of Direction*,[19] began in 1957 as a request from Frank Carr for 'concrete proposals for modifying the Museum, so as to lay more stress on conditions at sea in the past'.[20] It would evolve into a frank, thoughtful – if occasionally cruelly witty – and wide-ranging debate. Although there was still some loyalty to the 'spaciousness and harmony' of Callender's galleries, there was general support for the NMM 'restating that it was a museum with an art collection rather than an art gallery supported by some rather nice objects'.[21] The Museum's growing collections and research allowed more complex and informative narratives to be told. Displays of 'admirals and their battles' were a given, but it was agreed that the story of maritime discovery and the voyages of Captain Cook had shown the wide public interest that he could generate, and that his gallery should be extended and improved.

Callender's Cook gallery had already undergone some changes prior to *A Sense of Direction*, with display cases installed in its alcoves to put flesh on the bare bones of the existing interpretation. The gallery was one model on which the Museum could build a 'new look'. The 1960s, '70s and '80s ushered in a series of Cook galleries and exhibitions. In 1968 – the 200th anniversary of Cook's departure on his first voyage – a substantial new presentation was opened in Gallery VI: *Captain Cook and the Exploration of the Pacific* (fig. 244). The display of Turnerelli's bust of Banks, and the handsome one of Cook by Lucien Le Vieux (borrowed from the National Portrait Gallery), imparted a certain shrine-like quality to the gallery entrance, but the pair gave equal status to the natural and navigational sciences. The gallery had an ethnographic display with a similar basic structure to that of the 1956–7 show (fig. 246), and the text accompanying a display of weapons (where the Museum's ethnographic collections are particularly strong) read: 'This hand club, the two carved war clubs and the spear are all reputed to have been collected in the South Seas by crew members of Cook's second and third voyages.' However, there was an attempt to place them in a broader context, with the comment that 'these weapons, like the defensive fortifications, were all made at immense cost of labour with the aid of stone and bone tools only'.

Following the suggestions of *A Sense of Direction*, a self-contained story of navigation was planned for the Cook gallery. This would include the four iconic Harrison chronometers, which were transferred on permanent loan from the

Fig. 246
Ethnographic display
in *Captain Cook and the
Exploration of the Pacific*,
Gallery VI, 1979
National Maritime Museum
(C7675-6)

Admiralty in 1934 and were previously displayed in the Navigation Room.[22] Two years later in 1970, the gallery was revamped and reopened once more. This time the gallery's content revealed more obviously the hand of Carr's successor Basil Greenhill, who had arrived as Director in January 1967. This was particularly noticeable in a section on indigenous craft, a special interest of Greenhill's and a subject that Cook had always observed with a professional eye. This emphasis gave a more sympathetic context to the display of ethnographic objects. Greenhill wrote a small booklet on Cook's voyages to accompany the gallery, which featured indigenous Pacific objects and included Cook's drawing of a Polynesian canoe. These are small details, but suggestive of Greenhill's broader fascination with seafaring and seafarers: to him, Cook was 'a seaman, navigator, cartographer, scientist, explorer and humanitarian', a very different and more nuanced figure than Callender's Cook.[23]

The 1970 Cook gallery lasted only three years, before closing in 1973 to make way for building works that would provide a mezzanine between the two existing floors of the West Wing. This imaginative project was complete by 1975, but the new 'upper and lower mezzanines' were initially placed at the disposal of *The Times* and *Sunday Times* for their very successful temporary exhibition of the following year, *1776: The British Story of the American Revolution*, one of a series of early media-sponsored 'blockbusters' that began with *Treasures of Tutankhamun* at the British Museum in 1972.[24] The new Cook gallery was not installed until 1978, and changed within a few months to become *Captain Cook and Mr Hodges*, an exhibition to mark the bicentenary of Cook's death in 1779' (fig. 247). While it was acknowledged that 'our Cook Gallery will have been opened for only just over a year', it was agreed that 'some special manifestation should be made of this occasion'.[25] The potential that the bicentenary held for promotion of the Museum – and of himself, as leading public champion of what he later called the story of man's encounter with the sea – did not escape Greenhill. He lobbied very effectively for a national service of commemoration at Westminster Abbey, at which he gave the formal address. Two senior members of staff published important research papers on Cook's scientific instruments and ships in a special edition of *Mariner's Mirror*, and Surgeon Vice-Admiral Sir James Watt lectured on 'Medical Aspects of Cook's Voyages' at the Annual General Meeting of the Society for Nautical Research, held at the Museum on 6 June 1979.[26] Again one can sense that Greenhill was pulling the strings there, for scholarly research was a fundamental tenet of his more rigorous vision for the Museum, and his appointment as Director (from a previous career in the Foreign and Commonwealth service) had its background in his long membership and activity in the Society, above all in championing the preservation of historical maritime photography and film. Moreover, the very creation of the Museum had itself been one of the original aims of the Society from its own foundation in 1910.

The leap forward from the 1978 gallery's original ambitions had come when the Deputy Director David (Willie) Waters suggested to Greenhill that the anniversary of Cook's death might provide the perfect opportunity to wheedle the ten remaining Hodges paintings out of the Admiralty, 'where they are virtually inaccessible and are, in consequence, unknown'. Waters closed the memo with the stirring call to arms, 'If anyone should display them, we should',[27] and proposed an art exhibition exploring the impact of the Pacific on Western art – a suggestion that was strongly influenced by Bernard Smith's research on art from the Pacific voyages.[28] Teddy Archibald, the gifted but flamboyantly reactionary Head of Oil Paintings, brusquely dismissed the idea of 'an exhibition on Cook's influence on art', partly because 'I don't think he had one', but also because he believed that the Admiralty had only one Hodges left.[29] He was soon shown to be wrong on the latter point, and perhaps on the former as well, but he fought gamely to the last, arguing – on rather firmer ground this time – that 'another Cook exhibition on top of the splendid one we have could merely confuse the public and lead to overkill. I'm against it.'[30] The Museum got its additional Hodges, despite dark threats of 'other requests . . . not all from this country' being received by the Admiralty, and the exhibition went ahead in a modified form by hanging most of them in a central space cleared in the new Cook gallery on the 'upper' mezzanine floor.[31] At the end of the year-long exhibition, the gallery returned to being *Captain Cook and the Exploration of the Pacific*. While all the Cook displays from this period had a similar 'feel', the layout of the showcase of ethnographic objects in *Captain Cook and Mr Hodges*, to give but one example, shows the increasingly professional, and more expensive, standards that were starting to be applied to gallery design (fig. 246).

The complicated saga of the Admiralty's Hodges paintings and their eventual transfer on loan to the Museum has been well told by Geoff Quilley in *Art for the Nation*.[32] There were a total of 24 Hodges paintings in the Ministry of Defence Art Collection and, with the exception of the large *War-boats of Tahiti* (at time of writing, on display in the Old Admiralty Buildings), all are now housed at Greenwich. Two more, *The Landing at Tanna (Tana), one of the New Hebrides* and *Landing at Mallicolo* (see figs 8, 9), which must at one time have been in the Admiralty's collection, had clearly passed into private hands at some stage in their history, for they were bought by the Museum's Caird Fund through the dealers Colnaghi, in 1957. A third was presented to the Museum by Captain A.W.F. Fuller, with the help of the Art Fund. The highest-profile Hodges acquisition, though, was to happen some eight years after *Captain Cook and Mr Hodges*. The 'missing' portrait of Cook (fig. 234), known from prints but last reported – though perhaps wrongly – as being in the ownership of a Mr W.G. Anderson in 1785, was advertised as one of the lots in Sotheby's sale of the contents of the Irish country house, Mount Juliet, Kilkenny, in 1986. The sale catalogue attributed the painting to 'Circle of William Hodges, R.A.', without an illustration, and with only a brief catalogue entry describing it as 'Portrait of Captain Cook, half length, wearing uniform, inscribed with the identity of the sitter, oil on canvas' and an estimated price of £IR 350–450.[33] At least two people at the sale suspected it was rather more than that, for it was finally knocked down to David Posnett of London's Leger Galleries for £29,500. Posnett sold it on to the NMM for £600,000 – the Museum's most expensive acquisition yet. The National Heritage Memorial Fund contributed £250,000, but even so the acquisition took the NMM's entire purchase vote for 1987–8 and all that had been left of 1986–7. Even then, it was still £100,000 short of the agreed price at the end of 1987.[34] That the Museum was ultimately willing and able to make that commitment was due in large part to the drive of the newly appointed Chairman of the Trustees, Admiral of the Fleet Terence Lewin, Baron Lewin of Greenwich, who was himself a keen naval historian and authority on Cook.

It was neither an uncontroversial sale nor purchase. The original owner was, to say the least, disappointed that Sotheby's had failed to spot a 'sleeper', as such discoveries are known, and despite Sotheby's final contribution of a discount, there was considerable criticism of Leger Galleries for what they were asking of the Museum – some 20 times the auction price they had paid. The Museum also attracted unfavourable publicity for paying it, and its position was not helped by the opinion of the respected art dealer Hugh Leggatt, who felt that the painting was more likely to be the life study than the finished portrait (though no other has yet emerged). This, as the Museum's David Cordingly wrote, had all the makings of yet another art-world scandal, laying the Museum open to the charge of squandering the nation's money. However, with the exception of a 'short but scathing' piece in *The Standard*, entitled 'Cook blow for Maritime Museum', the story faded away, in part because there was actually a strong case for the painting's acquisition, which the Museum argued very well, but also because there was a real threat that the portrait might otherwise have been sold abroad.[35] Three years earlier, the Australian businessman Alan Bond had bought one of the Webber portraits of Cook, owned by Trinity House, for £700,000 and there was no reason to believe that wealthy Australian buyers would not also have been in the market for this one.[36]

The Hodges portrait suggests to some people a 'darker' Cook than the assured Enlightenment man of Nathaniel Dance – one who makes it rather easier to understand contemporary descriptions of Cook's fierce rages and brutal punishments, and a figure rather closer to the post-colonial re-evaluations of his character that began to gather force in the 1980s and '90s. To others, it suggests simply a tough, professional seaman. As Beaglehole once said, well before the Hodges reappeared, 'I have gazed long, and with some intensity, at Cook's portraits, and I cannot say I have learnt much.'[37] Whether or not the Hodges reveals hidden

Fig. 247 (right)
The opening of *Captain Cook and Mr Hodges*, 1979
National Maritime Museum
(C3495-5a)

On the right-hand side, conversing, are Sir Hugh Casson, and Teddy Archibald, former curator of oil paintings.

COOK ON DISPLAY   235

aspects of Cook's character, it was a great acquisition. It was hardly surprising in the circumstances that it was used on the cover of the catalogue for the touring exhibition, *Captain Cook, Navigator*, to which Lewin contributed a thoughtful chapter summarizing the man and the achievements of his three voyages. The exhibition was to travel to three venues in Australia and one in New Zealand before returning to display in Greenwich in 1990.

Displaying Cook had a monetary as well as a reputational value, but instead of generating valuable income for the Museum – which, like all national museums in the UK, was required to be more entrepreneurial in the straitened budgets of the 'Thatcherized' 1980s – the exhibition was nearly a financial disaster when the major sponsor went bankrupt just as it was being installed in Perth. The situation was so severe, wrote David Cordingly, Head of Exhibitions, to the High Commissioner of New Zealand, that 'we nearly closed up shop then and there'. It was only with the support of the Director of the State Library of New South Wales, a contribution from the British government and the 'diplomatic negotiations' of the Museum's Director, Richard Ormond, and Lord Lewin as Chairman that the exhibition was kept open in Perth and was able to show in Sydney for two months. The onward tour to New Zealand had to be cancelled and the exhibits returned directly to Greenwich in 1989.[38]

WIDER HORIZONS

In the early planning stages for the Greenwich version, the suggestion was made that the narrative could be broadened to include three significant legacies of Cook: the South Pacific whaling industry and the voyages of both William Bligh and Matthew Flinders. The first idea was not considered appropriate in a predominantly biographical show, although it was thought an interesting subject in its own right, and there were some concerns that the public might notice that the exhibition looked like a reworking of the *Mutiny on the Bounty* show – the Museum's principal exhibition in 1989 (fig. 249). So the project kept its focus on Cook and was finally opened at Greenwich as *Captain Cook, Explorer*. It covered the Pacific before Cook, using maps and globes and portraits of George Anson and John Byron; Cook's early life and career in the Navy; equipping the *Endeavour* at Deptford; the three voyages; Cook as a navigator and explorer; and ended with his death. The star object of the exhibition was an arresting Hawai'ian feather cloak lent by the Marquis of Salisbury. The opportunity this gave to develop a stronger ethnographic display was not taken, however, for the only other indigenous Pacific objects were Maori clubs in the voyages section and a Tongan club in 'The Death of Cook'. This last section also included an audio background that

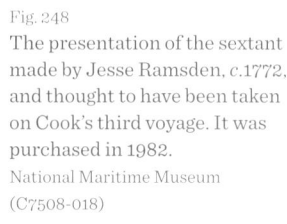

Fig. 248
The presentation of the sextant made by Jesse Ramsden, c.1772, and thought to have been taken on Cook's third voyage. It was purchased in 1982.
National Maritime Museum (C7508-018)

From left: representative of the owners; Alan Stimson, Curator of Navigational Instruments; and Derek Howse, Head of Astronomy and Navigation

236   CAPTAIN COOK AND THE PACIFIC

was intended to convey the confusion of the moment, but which gave the impression that the Hawai'ians were a faceless, baying mob. More recent interpretations of Cook's death, which effectively began in 1985 with Marshall Sahlins's provocative *Islands of History*, were still relatively little known when the gallery was being developed.

It was, however, a good display, using a wide range of what were by now some extraordinarily rich NMM collections: maps and charts, navigational and scientific instruments, ship models and paintings and drawings. The Cook furniture, not shown in a gallery since the 1970s, was not used. The Director, Richard Ormond, was later able to report to the trustees that the exhibition had attracted 90,000 visitors. This was a success for a low-budget, largely home-grown exhibition, but it was slightly less successful than initially hoped. This suggested that in the wake of more than ten galleries and exhibitions since 1956, Teddy Archibald had been right: Cook fatigue had begun to creep in.[39] However, this was probably more apparent among Museum staff than among visitors, for the 1970s, '80s and '90s saw an increasing popular interest in Cook's voyages emerging, with the opening of the Captain Cook Birthplace Museum in Marton, Middlesbrough (1978) and the Captain Cook Memorial Museum in Whitby (1986), while academic research on and public appreciation of the significance of the Pacific voyages was given renewed energy by the publication of Joseph Banks's *Florilegium* during the 1980s. Cook still had a presence at the NMM in the 1990s, but it was lower-key, reflecting broader contexts being explored by scholars, for this was also the period that saw an outpouring of reappraisals (both academic and popular) of Cook and Pacific exploration.

Cook briefly formed the 'Sea' section of a temporary display entitled *Blood, Sea and Ice*, which made use of a space destined to become part of the Neptune Court development (1996–99) and presented the three main exploration stories told in the Museum: those of Cook, Drake and John Franklin. A life-sized bronze statue of Cook, originally modelled as one of a set of seven Pacific explorers for the 'Hall of Discovery' in the New Zealand pavilion at the Seville Expo of 1992, was presented to the Museum in 1994 and temporarily installed on the lawn outside the Queen's House two years later (figs 250, 251). For various reasons it has enjoyed a somewhat peripatetic existence since then, but should be back in its original position by 2018, the 250th anniversary of Cook's first voyage. Two small, short-term exhibitions, *Cook and the 'Endeavour'* and *Cook and his Scientific Instruments*, were held in the Queen's House and the Royal Observatory respectively, to mark the arrival of the Australian-built replica *Endeavour* at Greenwich in 1997. This occasion was marked academically by the Museum's first conference on Cook, 'Science and Exploration in the Pacific', the proceedings of which were edited by the then Head of Research and later Deputy Director, Margarette Lincoln, and published by Boydell and Brewer. It was a rewarding conference in many ways, but especially for Harold Carter's 'note' that identified the unknown watercolour draughtsman on Cook's first voyage – hitherto known only as 'The Artist of the Chief Mourner' – as the Raiatean priest and navigator Tupaia, a moment that has had a profound effect on subsequent studies of Pacific encounters.

Much of the Museum's energy in the late 1990s was taken up with the construction of Neptune Court. This was an ambitious project to provide a better central circulation for the Museum, as well as 12 new galleries, and to present, once again, 'new approaches', to attract a broader family-visitor base. Dance's Cook became one of the 'faces' used to promote the

Fig. 249
Model of the armed transport, *Bounty*, made for the exhibition *Mutiny on the Bounty*
1989
National Maritime Museum
(D4187)

COOK ON DISPLAY   237

Fig. 250 (above)
The unveiling of Anthony Stones's statue of James Cook, by His Royal Highness The Duke of Edinburgh, Senior Trustee of the Museum, 27 February 1997
National Maritime Museum (D8591_1)

Pictured: HRH the Duke of Edinburgh (left) and Admiral of the Fleet Lord Lewin of Greenwich, Chairman of the Trustees of the National Maritime Museum, 1987–95 (right).

Fig. 251 (above, right)
Cook's statue in position, to the south of the Queen's House colonnades
National Maritime Museum (SCU0137/F1350-4)

The statue, modelled by Anthony Stones, was one of a set of seven Pacific explorers made in lightweight materials for the New Zealand pavilion of the Seville Expo, 1992. This bronze cast was presented to the Museum by Sir Arthur Weller, then a Trustee, in 1994.

project, with his exploration of the Pacific once more intended to have its own gallery. In the end this did not materialize, and Cook ended up in a section of a gallery called 'Trade and Empire', which aimed to address the imperial and colonial legacies of maritime expansion, and used ethnographic objects to demonstrate cultural exchange in Pacific exploration. The gallery was not universally admired, with what was seen as a reductive, 'politically correct' treatment of Cook (and slavery) coming in for particular criticism.

The Museum's displays have largely followed new scholarly research, both external and internal, trends in museology, staff expertise, opportunity and directorial whim, and as such have steadily moved on from Cook-the-naval-hero and Cook-the-seaman to interpret him within much broader historical contexts. Between the early 1990s and the present day, the Museum's 'Cooks' – like the Museum's 'Nelsons' – have increasingly become chapters in larger stories rather than the book itself. The *Captain Cook and Mr Hodges* exhibition in the late 1970s, for example, was an important show for many reasons, but it was essentially a familiar narrative, more generously funded and rather better told than previously. By contrast, in 2004, Cook's second voyage was merely one of many contexts for the well-received art show *William Hodges 1744–1797: The Art of Exploration*. Cook's three voyages also formed a section of *Oceans of Discovery*, a gallery that ranged (widely rather than wisely) from early European and Asian voyages to underwater exploration today. More recently, and with much greater success, Cook's important role in testing chronometers on his voyages formed a strong thematic section in the 2014 exhibition *Ships, Clocks and Stars: The Quest for Longitude*. This was curated by two members of staff, both essentially historians of science rather than of seafaring, engaged in a five-year research project on the history of the Board of Longitude with the University of Cambridge. The exhibition went on to travel to the United States and Australia.

New acquisitions continue to strengthen and generate new approaches to the collections. The high-profile purchase of George Stubbs's *Kangaroo* and *Dingo* in 2013 (fig. 252) – at a sum (generously aided by the Art Fund and Lottery) which makes that paid for the Hodges portrait of Cook seem very reasonable – presented the Museum with a show-stopping centrepiece for *The Art and Science of Exploration*. This was a modest exhibition in the Queen's House in 2014 that was an additional output of the Longitude project. With the recent purchase of a copy of Joseph Banks's *Florilegium*, the Museum is now even better equipped to explore natural history collecting on the voyages and their impact in Britain. The acquisition in 2012 of ethnographic material sent from the Pacific back to Britain by missionaries of the London Missionary Society from the 1800s onwards also offers new ways to explore some of the less well-known legacies of Cook's voyages. None of these acquisitions could be said to be 'traditional' Cook objects, but they have enriched and strengthened beyond measure the story that the Museum can tell.

Cook has been a mainstay of the National Maritime Museum since its opening in 1937. Although it might fairly be said that its Cook displays have, in particular ways and contexts, sometimes been formulaic and unambitious – the interpretation of ethnographic objects is a particular example, but not the only one – it is remarkable how consistently he has been used as an agent of change; as a way to try something different, herald new approaches, reach fresh audiences, generate new research, build reputation, earn income and support different agendas. When the conference 'Science and Exploration' was first discussed in the mid-1990s, a respected Pacific scholar wearily remarked that it would be impossible to find anything new to say about Cook. Twenty years on, Cook has gone from strength to strength: the belief, fondly held by some, that the Museum has always presented him in the same tired ways is, like the Southern Continent, a myth.

Fig. 252 (left)
George Stubbs's *Kangaroo* and *Dingo*, on display in *The Art and Science of Exploration*
2014
National Maritime Museum (L7858_011). Acquired with the assistance of the Heritage Lottery Fund; the Eyal and Marilyn Ofer Foundation; The Art Fund (with a contribution from the Wolfson Foundation); and other donors.

Fig. 253 (overleaf)
Advertisement for the *Captain Cook and Mr Hodges* exhibition, 1979
Private collection

NATIONAL MARITIME MUSEUM Greenwich

CAPTAIN COOK and MR HODGES

Paintings of the Second Great Voyage 1772-1775: a special exhibition in the Cook Gallery April 1979-1980

Notes

Chapter 1

1. On Cook as a British version of Columbus, see Alan Frost, 'New Geographical Perspectives and the Emergence of the Romantic Imagination', in Robin Fisher and Hugh J.M. Johnston (eds), *Captain James Cook and His Times* (Vancouver: Douglas and McIntyre, 1979), pp. 5–19, 19.

2. Joseph Conrad, 'Geography and Some Explorers' (1924), in Joseph Conrad, *Last Essays*, edited by Harold Ray Stevens and J.H. Stape (Cambridge: Cambridge University Press, 2010), pp. 3–17, 4.

3. Quoted in Watkin Tench, *A Narrative of the Expedition to Botany Bay* (London: Debrett, 1789), p. 97.

4. Glyn Williams, *Naturalists at Sea: Scientific Travellers from Dampier to Darwin* (New Haven and London: Yale University Press, 2013), p. 21.

5. Quoted in Glyn Williams, *Voyages of Delusion: The Search for the Northwest Passage in the Age of Reason* (London: HarperCollins, 2003), p. 140.

6. Robert E. Gallagher (ed.), *Byron's Journal of His Circumnavigation, 1764–1766* (London: Hakluyt Society, 1964), p. 3.

7. Johann Reinhold Forster, 'The Translator's Preface', in Lewis [sic] de Bougainville, *A Voyage round the World* (London, 1772), p. v. Quoted in Frost, 'New Geographical Perspectives', in Fisher and Johnston (eds), (*Cook and His Times*), pp. 5–19, 6.

8. Quoted in Brian W. Richardson, *Longitude and Empire: How Captain Cook's Voyages Changed the World* (Vancouver: University of British Columbia Press, 2005), p. 11.

9. Conrad (*Last Essays*), pp. 3–17, 16.

10. Lisant Bolton, 'Brushed with fame: Museological investments in the Cook voyage collections', in Michelle Hetherington and Howard Morphy (eds), *Discovering Cook's Collections* (Canberra: National Museum of Australia Press, 2009), pp. 78–91, 82, 83.

11. John Douglas, 'Introduction', in James Cook and James King, *A Voyage to the Pacific Ocean . . . for Making Discoveries in the Northern Hemisphere, in the Years 1776, 1777, 1778, 1779, and 1780*, 3 vols (London: G. Nicol and T. Cadell, 1784), vol. 1, p. lxix.

12. The written word also played an important role in this, of course. Nearly 30 journals and logs remain from the (third) voyage, and many more were lost. For example, James Burney kept three-and-a-half years of diary entries on sheets of Chinese rice paper and folded the whole into an easily concealed 3½-inch bundle. A note on the wrapper explained that he had taken this action so that 'if bereft of our other journals there might be one saved for the Admiralty'. (Quoted in Edward G. Gray, *The Making of John Ledyard: Empire and Ambition in the Life of an Early American Traveller* [New Haven and London: Yale University Press, 2007], p. 70.) The publication of some of these accounts was hugely important in conveying ideas and impressions about the Pacific and its people to European readers.

13. R.A. Skelton, *Captain James Cook – after two hundred years: A Commemorative address delivered before the Hakluyt Society* (London: British Museum, 1969), p. 30.

14. Quoted in Howard T. Fry, *Alexander Dalrymple (1737–1808) and the Expansion of British Trade* (London: Royal Commonwealth Society, 1970), p. 274.

15. Richardson (*Longitude and Empire*), p. 3.

16. See Igor Kopytoff, 'The Cultural Biography of Things: Commoditization as Process', in Arjun Appadurai (ed.), *The Social Life of Things: Commodities in Cultural Perspective* (Cambridge: Cambridge University Press, 1986), pp. 64–91.

17. Geoff Quilley, *Empire to Nation: Art, History and the Visualization of Maritime Britain, 1768–1829* (New Haven and London: Yale University Press, 2011), p. 212.

18. Ibid., p. 217.

19. Christian Feest, 'Foreword', in Adrienne L. Kaeppler, *Holophusicon: The Leverian Museum. An Eighteenth-Century English Institution of Science, Curiosity and Art* (Altenstadt: ZKF Publishers, 2011), p. viii.

20. William Bligh, *A Voyage to the South Sea, undertaken by command of His Majesty, for the purpose of conveying the bread-fruit tree to the West Indies, in His Majesty's ship the Bounty, commanded by Lieutenant William Bligh* (London: George Nicol, 1792), p. 5.

21. Quoted in Frost, 'New Geographical Perspectives', in Fisher and Johnston (eds), (*Cook and His Times*), pp. 5–19, 6.

22. Douglas, 'Introduction', in Cook and King (*Voyage to the Pacific Ocean*), vol. 1, p. lxix.

23. Williams (*Naturalists at Sea*), p. 21.

24. J.C. Beaglehole (ed.), *The Journals of Captain James Cook on his Voyages of Discovery. Vol. II: The Voyage of the 'Resolution' and 'Adventure', 1772–1775* – hereafter referred to as '*Cook* II' (Cambridge: Cambridge University Press, 1961), p. 98.

25. For further details, see William Hauptman, *Captain Cook's Painter, John Webber, 1751–1793: Pacific Voyager and Landscape Artist* (Bern: Kunstmuseum, 1996), p. 52; Bernard Smith, *European Vision and the South Pacific* (New Haven and London: Yale University Press, 1985), p. 346, n. 11.

26. Edmund Burke to Dr Robertson, 10 June 1777, in Charles William, Earl Fitzwilliam and Sir Richard Bourke (eds), *Correspondence of the Right Honourable Edmund Burke*, 4 vols (London: Francis and John Rivington, 1844), vol. 2, p. 183.

27. For a detailed discussion, see John McAleer, 'Exhibiting exploration: Captain Cook, voyages of exploration and cultures of display', in John McAleer and John M. MacKenzie (eds), *Exhibiting the Empire: Cultures of Display and the British Empire* (Manchester: Manchester University Press, 2015), pp. 42–63, 44–8.

28. Nicholas Thomas, 'The Age of Empire in the Pacific', in David Armitage and Alison Bashford (eds), *Pacific Histories: Ocean, Land, People* (Basingstoke: Palgrave Macmillan, 2014), pp. 75–96, 77.

29. James R. Fichter, *So Great a Proffit: How the East Indies Trade transformed Anglo-American Capitalism* (Cambridge, MA: Harvard University Press, 2010), p. 214.

30. Gray (*John Ledyard*), p. 3.

31. Quoted ibid., p. 97.

32. Walter Besant, *English Men of Action: Captain Cook* [first published 1890] (London: Macmillan and Co., 1904), p. 84.

33. See M.K. Beddie, *Bibliography of Captain James Cook*, second edition (Sydney: State Library of New South Wales, 1970).

34. Quoted in Beth Fowkes Tobin, *The Duchess's Shells: Natural History Collecting in the Age of Cook's Voyages* (New Haven and London: Yale University Press, 2014), p. 130.

35. J.C. Beaglehole (ed.), *The Journals of Captain James Cook on his Voyages of Discovery. Vol. I: The Voyage of the 'Endeavour', 1768–1771* – hereafter referred to as '*Cook* I' (Cambridge: Cambridge University Press, 1955), p. cclxxxii.

36. P.J. Marshall and Glyndwr Williams, *The Great Map of Mankind: British Perceptions of the World in the Age of Enlightenment* (London: Dent, 1982), p. 259.

37. Kathleen Wilson, *The Island Race: Englishness, Empire and Gender in the Eighteenth Century* (London: Routledge, 2003), pp. 54–91.

38. Hampshire County Record Office, Winchester, 38M49/5/61/20, Wickham papers, 'Observations on the Island of Madagascar by Colonel Sir George Young' (n.d., c.1806–07), p. 8, note.

39. National Maritime Museum, Greenwich, P/16/4, Thomas Pennant, *Outlines of the Globe*, pp. 222–3.

Chapter 2

1. Daniel Baugh, *The Global Seven Years War, 1754–1763* (London: Longman, 2011), p. 298.

2. N.A.M. Rodger, *The Wooden World: An Anatomy of the Georgian Navy* (London: HarperCollins, 1986), p. 25.

3. Quoted in J.C. Beaglehole, *The Life of Captain James Cook* (Stanford: Stanford University Press, 1974), p. 25.

4. Quoted in John Robson, *Captain Cook's War and Peace: The Royal Navy Years, 1755–1768* (Barnsley: Seaforth Publishing, 2009), p. 63.

5. John Gascoigne, *Captain Cook: Voyager Between Worlds* (London: Continuum Books, 2007), p. 24, citing James Cook and James King, *Voyage to the Pacific Ocean*, 3 vols (London: Nicol and Cadell, 1784), vol. 3, p. 47.

6. Ibid., p. 23.

7. Stephen J. Hornsby, *Surveyors of Empire: Samuel Holland, J.F.W. des Barres and the Making of the Atlantic Neptune* (Montreal and Kingston: McGill-Queen's University Press), pp. 1–9.

8. The commander-in-chief's position was usually a temporary one, renewed annually. Palliser held it between 1764 and 1766.

9. Hornsby (*Surveyors*), p. 37.

10. Admiralty Instructions to Cook, *Cook* I, p. ccixxx.

11. Robson (*Captain Cook's War*), p. 183.

12. *Cook* I, p. ccixxxiii.

13. Derek Howse, 'The Principal Scientific Instruments taken on Captain Cook's Voyages of Exploration, 1776–80', *Mariner's Mirror*, no. 65, 1979, pp. 119–35; Derek Howse and Beresford Hutchinson, *The Clocks and Watches of Captain James Cook, 1769–1969* (London: Antiquarian Horological Society, 1969). Cited in Wayne Orchiston, 'From the South Seas to the Sun: The Astronomy of Cook's Voyages', in *Science and Exploration in the Pacific: European Voyages to the Southern Oceans in the 18th Century*, edited by Margarette Lincoln (Woodbridge: Boydell Press, 1997), pp. 55–72, 57.

14. Joseph Banks, 31 July 1769, quoted in *Cook* I, n. 3, p. 147.

15. Ibid., p. 154.

16. *Cook* I, p. 442.

17. Anne Salmond, *Two Worlds: First meetings between Maori and Europeans 1642–1772* (London: Viking, 1991), pp. 127–28.

18. *Cook* I, n. 2 and p. 171.

19. Sir James Watt, 'Medical Aspects and Consequences of Cook's Voyages', in *Captain James Cook and His Times* (London: Croom Helm, 1979), pp. 138–40.

20. *Cook* I, p. 475.

21. Christine Holmes (ed.), *Captain Cook's Second Voyage: The Journals of Lieutenants Elliott and Pickersgill* (London: Caliban Books, 1984), pp. xxx–xxxi.

22. Elliott, in Holmes (ed.), (*Second Voyage*), p. 14.

23. Pickersgill, ibid., p. 67.

24. *Cook* II, p. 111.

25. Ibid., p. 110.

26. George Forster, *A Voyage Round the World*, edited by Nicholas Thomas and Oliver Berghof, 2 vols (Honolulu: Hawai'i University Press, 2000), vol. 1, p. 80.

27. See Jonathan Lamb, *Scurvy: The Disease of Discovery* (Princeton: Princeton University Press, 2016).

28. See Pieter van der Merwe, '"Icebergs" and Other Recent Discoveries in Paintings from Cook's Second Voyage by William Hodges', *Journal for Maritime Research*, vol. 8 (1), 2006, pp. 34–45; see also pp. 83–84. National Maritime Museum Collections Online, http://collections.rmg.co.uk/collections/objects/13846.html.

29. *Cook* II, p. lxxi.

30. Ibid., p. 200.

31. Jennifer Newell, *Trading Nature: Tahitians, Europeans and Ecological Exchange* (Honolulu: University of Hawai'i Press, 2010), p. 50.

32. Forster (*Voyage Round the World*), vol. 1, pp. 232, 235.

33. Pickersgill, in Holmes (ed.), (*Second Voyage*), p. 95.

34. Ibid., p. 96.

35. *Cook* II, p. 255.

36. Elliott, in Holmes (ed.), (*Second Voyage*), p. 25.

37. *Cook* II, p. 322.

38. Ibid., p. 369.

39. Cited in Watt (*Cook and His Times*), p. 132.

40. Ibid., pp. 958, 960.

41. J. C. Beaglehole (ed.), *The Journals of Captain James Cook on his Voyages of Discovery. Vol. III: The Voyage of the 'Resolution' and 'Discovery', 1776–1780* – hereafter referred to as '*Cook* III' (Cambridge: Cambridge University Press, 1967, p. lxxvi.

42. James Cook to Lord Sandwich, 26 November 1776, Sandwich Papers, National Maritime Museum, quoted ibid., p. 1520.

43. Ibid., p. 769. Quoted by Jonathan Lamb, 'Inchoate Possession: How Captain Kerguelen claimed an island', *Journal for Maritime Research*, vol. 7 (1), 2005, pp. 1–15, DOI: www.tandfonline.com/10.1080/21533369.2005.9668341, accessed 22 May 2016.

44. See, for example, Anne Salmond, *The Trial of the Cannibal Dog: Captain Cook in the South Seas* (London: Allen Lane Penguin, 2003); Nicholas Thomas, *Cook: The Extraordinary Voyages of Captain James Cook* (New York: Walker & Company, 2003); and Frank McLynn, *Captain Cook: Master of the Seas* (New Haven and London: Yale, 2011).

45. *Cook* II, p. 653.

46. *Cook* III, p. 1313.

47. Ibid., p. 194.

48. Ibid., p. 279.

49. Ibid., p. 371.

50. Rod Edmond, *Representing the South Pacific: Colonial Discourse from Cook to Gauguin* (Cambridge: Cambridge University Press, 1997), p. 25.

51. Cited by Glyn Williams, *The Death of Captain Cook: A Hero Made and Unmade* (London: Profile Books, 2008), p. 26.

52. George Gilbert, *Captain Cook's Final Voyage: The Journal of Midshipman George Gilbert* (Honolulu: University of Hawai'i Press, 2008), p. 157. Also cited in part in Williams (*Death of Captain Cook*), p. 16.

Chapter 3

1. William Bligh, *A Voyage to the South Sea* (London: George Nicol, 1792), p. 5.

2. Charles W.J. Withers, *Placing the Enlightenment: Thinking Geographically about the Age of Reason* (London and Chicago: The University of Chicago Press, 2007), pp. 88, 89.

3. Glyndwr Williams, *The Great South Sea: English Voyages and Encounters 1570–1750* (New Haven and London: Yale University Press), p. 2.

4. J.C. Beaglehole, *The Exploration of the Pacific* [first published 1934] (Stanford: Stanford University Press, 1968), p. 40.

5. Quoted by Kenneth Morgan, *Matthew Flinders: Maritime Explorer of Australia* (London: Bloomsbury Academic, 2016), p. 62, citing Kenneth Morgan (ed.), *Australia Circumnavigated: The Voyage of Matthew Flinders in HMS Investigator, 1801–1803* (Farnham and London: Ashgate for the Hakluyt Society, 2015), pp. 406, 407.

6. John Narborough, *An Account of Late Voyages & Discoveries* (1694), quoted by Glyndwr Williams, *Buccaneers, Explorers and Settlers: British Enterprise and Encounters in the Pacific, 1670–1800* (Aldershot: Ashgate Variorum, 2005), p. 28.

7. William Dampier, *A New Voyage Round the World,* 7th edition (London, 1729), title page.

8. William Dampier, in *The Buccaneer Explorer: William Dampier's Voyages*, edited by Gerald Norris (Woodbridge: Boydell & Brewer, 1994) pp. 124–5.

9. Glyndwr Williams, 'The Achievement of the English Voyages', in Derek Howse (ed.), *Background to Discovery: Pacific Exploration from Dampier to Cook* (Los Angeles: University of California Press, 1990), pp. 56–80, 58.

10. Quoted by Daniel A. Baugh, 'Seapower and Science', in Howse, ibid., p. 21.

11. The Rosselli, Jansson and Hack maps can be accessed on the National Maritime Museum Collections Online website.

12. See Glyn Williams, *Voyages of Delusion: The Search for the Northwest Passage in the Age of Reason* (London: Harper Collins, 2002), particularly chapter 8, 'Maps, Hoaxes and Projects', pp. 239–86.

13. John Green, *Remarks in Support of the New Chart of North and South America* (London, 1753), p. 43. Quoted by Glyndwr Williams (*Great South Sea*), p. 269. Green was criticizing French voyagers, but the same could apply to charting by other nations.

14. *Cook* I, p. 413.

15. Withers (*Placing the Enlightenment*), p. 8.

16. W.F.J. Mörzer Bruyns, *Sextants at Greenwich: A Catalogue of the Mariner's Quadrants, Mariner's Astrolabes, Cross-staffs, Backstaffs, Octants, Sextants, Quintants, Reflecting Circles, and Artificial Horizons in the National Maritime Museum, Greenwich* (Oxford: Oxford University Press, 2009), p. 27.

17. Ibid., p. 32.

18. *Finding Longitude: How ships, clocks and stars helped solve the longitude problem* by Richard Dunn and Rebekah Higgitt (Glasgow: Collins, 2014) is a wide-ranging and relevant publication.

19. *Cook* I, p. 392.

20. Derek Howse, 'The principal scientific instruments taken on Captain Cook's voyages of exploration, 1768–80', *The Mariner's Mirror*, published for the Society for Nautical Research, vol. 65 (2), May 1979, pp. 119–35.

21. *Cook* II, p. 174.

22. *Cook* I, p. 275.

23. *Cook* II, p. 174.

24　Ibid., 21 February 1775, p. 643.

25　*Cook* I, p. 138.

26　J.R. Forster, *Observations Made on a Voyage Round the World*, quoted by Beaglehole, *The Exploration of the Pacific*, n. 2, p. 147.

27　See Anne di Piazza and Erik Pearthree, 'A New Reading of Tupaia's Chart', *Journal of the Polynesian Society*, 116 (3), September 2007.

28　Withers (*Placing the Enlightenment*), p. 88.

29　Elliott, in Holmes (ed.), (*Second Voyage*), p. 30.

30　See chapter 7, n. 1.

31　George Vancouver, *A Voyage of Discovery to the North Pacific Ocean and Round the World, 1791–1795*, edited by W. Kaye Lamb (London: Hakluyt Society, 1984), pp. 363, 366.

32　William Perry, Adm 1/1609, in *Cook* I, pp. 632–4.

33　See, for example, the special issue of the *Journal for Maritime Research*, edited by Robert J. Blyth, vol. 15 (1), May 2013.

34　Withers (*Placing the Enlightenment*), p. 110.

35　Anthony Payne, 'The Publication and Readership of Voyage Journals in the Age of Vancouver', in *Enlightenment and Exploration in the North Pacific, 1741–1805*, edited by Stephen Haycox, James Barnett and Caedmon Liburd (Seattle and London: University of Washington Press, 1997), pp. 176–86, quoting Bernard Smith, 'Coleridge's "Ancient Mariner" and Cook's Second Voyage', in *Imagining the Pacific*, edited by Bernard Smith (New Haven and London: Yale University Press, 1992), pp. 135–72; Rod Edmond, *Representing the Pacific: Colonial Discourse from Cook to Gauguin* (Cambridge: Cambridge University Press, 1997), p. 42.

36　Payne ('Publication and Readership'), pp. 176–86.

37　Lynne Withey, *Voyages of Discovery: Captain Cook and the Exploration of the Pacific* (London: Hutchinson, 1987), p. 405.

38　http://www.pascalbonenfant.com/18c/clubs/jt_theroyalsocietyclub.html, accessed 17 September 2016.

39　Williams (*Death of Captain Cook*), pp. 61–129. See also Withers (*Placing the Enlightenment*), particularly pp. 87–134.

Chapter 4

1　John E. Crowley, *Imperial Landscapes: Britain's Global Visual Culture, 1745–1820* (New Haven and London: Yale University Press, 2011), p. 105.

2　David Armitage and Alison Bashford, 'Introduction: The Pacific and its Histories', in David Armitage and Alison Bashford (eds), *Pacific Histories: Ocean, Land, People* (Basingstoke: Palgrave Macmillan, 2014), pp. 1–28, 1, 5.

3　Damon Salesa, 'The Pacific in Indigenous Time', in Armitage and Bashford (eds) (*Pacific Histories*), pp. 31–52, 32.

4　David Igler, *The Great Ocean: Pacific Worlds from Captain Cook to the Gold Rush* (Oxford: Oxford University Press, 2013), p. 4.

5　John Mack, *The Sea: A Cultural History* (London: Reaktion, 2013), p. 65.

6　Epeli Hau'ofa, 'Our Sea of Islands', *Contemporary Pacific*, vol. 6, 1994, pp. 147–62.

7　Salesa ('The Pacific in Indigenous Time'), p. 44.

8　Quoted in Igler (*The Great Ocean*), p. 5.

9　Bronislaw Malinowski, *Argonauts of the Western Pacific: An Account of Native Enterprise and Adventure in the Archipelagoes of Melanesian New Guinea* (New York: Dutton, 1992).

10　Bronwen Douglas, 'Religion', in Armitage and Bashford (eds) (*Pacific Histories*), pp. 193–215, 194.

11　Igler (*The Great Ocean*), p. 21.

12　Armitage and Bashford, 'Introduction', in Armitage and Bashford (eds) (*Pacific Histories*), p. 14.

13　James Cook and James King, *A Voyage to the Pacific Ocean*, 3 vols (London: Nicol and Cadell, 1784), vol. 2, p. 251. For further details, see Joyce E. Chaplin, 'The Pacific before Empire, c.1500–1800', in Armitage and Bashford (eds) (*Pacific Histories*), pp. 53–74.

14　*Cook* I, p. 117.

15　See Mack (*The Sea*), pp. 116–18, for a discussion of Polynesian navigation and the use of stick maps.

16　Salesa ('The Pacific in Indigenous Time'), pp. 44–5.

17　William Hauptman, *Captain Cook's Painter, John Webber, 1751–1793: Pacific Voyager and Landscape Artist* (Bern: Kunstmuseum, 1996), p. 148.

18　Sydney Parkinson, *A Journal of a Voyage to the South Seas, in His Majesty's Ship, the Endeavour* (London: Stanfield Parkinson, 1773), p. 35.

19　George Forster, *A Voyage round the World* (1778), edited by Nicholas Thomas and Oliver Berghof, 2 vols (Honolulu, HI: University of Hawai'i Press, 2000), vol. 1, p. 250.

20　Ibid.

21　Ibid.

22　Ibid.

23　*Cook* II, p. 268.

24　Ibid., p. 268, n. 2. I am grateful to Clifford Thornton for bringing this to my attention. See *Cook's Log*, vol. 37 (2014), p. 12.

25　Anne Salmond, *Two Worlds: First Meetings between Maori and Europeans, 1642–1772* (Honolulu, HI: University of Hawai'i Press, 1996), p. 87.

26　Igler (*The Great Ocean*), pp. 206–7, n. 60.

27　Hauptman (*Captain Cook's Painter*), p. 158.

28　Quoted in Igler (*The Great Ocean*), p. 35.

29　Ibid., pp. 83–4.

30　Parkinson (*Voyage to the South Seas*), p. 109.

31　Nicholas Thomas, 'The Age of Empire in the Pacific', in Armitage and Bashford (eds) (*Pacific Histories*), pp. 75–96, 80.

32　Crowley (*Imperial Landscapes*), p. 93.

33　For more on the visual representation of different people in the period, see David Bindman, *Ape to Apollo: Aesthetics and the Idea of Race in the 18th Century* (London: Reaktion, 2002).

34　Crowley (*Imperial Landscapes*), p. 87. See also Bernard Smith, 'Captain Cook's Artists and the Portrayal of Pacific Peoples', *Art History*, vol. 7, 1984, pp. 295–313.

35　Parkinson (*Voyage to the South Seas*), plate XVI.

36　Ibid., p. 90.

37　Gray, (*Making of John Ledyard*), p. 129.

38　See Geoff Quilley and John Bonehill (eds), *William Hodges, 1744–1797: The Art of Exploration* (New Haven and London: Yale University Press, 2004), pp. 101–3.

39　See Ter Ellingson, *The Myth of the Noble Savage* (Berkeley, CA: University of California Press, 2001).

40　Parkinson (*Voyage to the South Seas*), p. 70.

41　Ibid., p. 70.

42　Crowley (*Imperial Landscapes*), pp. 95–6.

43　Ibid., p. 101.

44　Hauptman (*Captain Cook's Painter*), p. 158.

45　Crowley (*Imperial Landscapes*), p. 102.

46　*Cook* I, p. 122.

47　On collecting in general, see Nicholas Thomas, *Entangled Objects: Exchange, Material Culture, and Colonialism in the Pacific* (Cambridge, MA: Harvard University Press, 1991), pp. 125–84.

48　Parkinson (*Voyage to the South Seas*), p. 51.

49　Quilley and Bonehill (eds), (*William Hodges*), p. 98.

50　Crowley (*Imperial Landscapes*), p. 91.

51　Forster (*A Voyage round the World*), vol. 1, p. 355.

52　Quilley and Bonehill (eds), (*William Hodges*), p. 129.

53　James Cook, *A Voyage towards the South Pole and Round the World*, 2 vols, 2nd edition (London: Strahan and Cadell, 1777), vol. 1, p. 328.

54　Crowley (*Imperial Landscapes*), p. 87.

55　[John Feltham], *The Picture of London for 1806* (London: s.n., 1806), p. 282.

56　Ibid.

Chapter 5

1　John Hawkesworth, *An Account of the Voyages undertaken by the Order of His Present Majesty for making Discoveries in the Southern Hemisphere and successively performed by Commodore Byron, Captain Wallis, Captain Carteret, and Captain Cook*, 3 vols (London: Strahan and Cadell, 1773), vol. 1, p. vii.

2　British Library, Add. MS 29142, John MacPherson to Warren Hastings, 31 December 1778, quoted in Finbarr Barry Flood, 'Correct Delineations and Promiscuous Outlines: Envisioning India at the Trial of Warren Hastings', *Art History*, vol. 29, 2006, pp. 47–78, 59.

3　James Cook and James King, *A Voyage to the Pacific Ocean*, 3 vols (London: Nicol and Cadell, 1784), vol. 1, p. 5.

4   William Dampier, *A Voyage to New Holland* (London: James Knapton, 1703), preface.

5   Ibid.

6   Richard Walter, *A Voyage round the World* (London: John and Paul Knapton, 1748), p. ii.

7   John E. Crowley, *Imperial Landscapes: Britain's Global Visual Culture, 1745–1820* (New Haven and London: Yale University Press, 2010), p. 79.

8   Walter (*Voyage round the World*), p. ii.

9   Crowley (*Imperial Landscapes*), p. 97.

10 Ibid., p. 76.

11 Williams, (*Naturalists at Sea*), p. 74.

12 See A.M. Lysaght, 'Banks's Artists and his *Endeavour* Collections', in *Captain Cook and the South Pacific*, The British Museum Yearbook 3 (London: British Museum, 1979), pp. 9–80.

13 Sydney Parkinson, *A Journal of a Voyage to the South Seas, in His Majesty's Ship, the Endeavour* (London: Stanfield Parkinson, 1773), p. 38.

14 Walter (*Voyage round the World*), p. iii.

15 Ibid., p. viii.

16 See J.R.H. Spencer, 'Coastal Profiles and Landscapes', in D.J. Carr (ed.), *Sydney Parkinson: Artist of Cook's 'Endeavour' Voyage* (London: British Museum [Natural History], 1983), pp. 260–74.

17 Parkinson (*Journal of a Voyage to the South Seas*), p. 123.

18 Ibid., p. 99.

19 Ibid., p. 117.

20 Quoted in Crowley (*Imperial Landscapes*), p. 88.

21 Ibid., p. 89.

22 Quoted in Bernard Smith, 'William Hodges and English *Plein-air* Painting', *Art History*, vol. 6, 1983, pp. 143–52, 145.

23 The NMM holds 26 oils relating to the voyage, of which 24 were either painted for or acquired by the Admiralty, and the other two probably escaped from it by means now unknown. Apart from these two, all were transferred to the Museum from the Ministry of Defence Art Collection in 2017.

24 Johann Reinhold Forster, *Observations made during a Voyage round the World* (1778), edited by Nicholas Thomas, Harriet Guest and Michael Dettelbach (Honolulu: University of Hawai'i Press, 1996), p. 86.

25 Bernard Smith, *European Vision and the South Pacific, 1768–1850* (New Haven and London: Yale University Press, 1985), pp. 57–9.

26 Quilley and Bonehill (eds), (*William Hodges, 1744–1797: The Art of Exploration*), p. 78.

27 Cook II, p. 51.

28 Smith (*European Vision and the South Pacific*), p. 58.

29 Quoted in Rüdiger Joppien and Bernard Smith, *The Art of Captain Cook's Voyages*, 3 vols (New Haven and London: Yale University Press, 1985–88), vol. 2, p. 11.

30 See Pieter van der Merwe, '"Icebergs" and other Recent Discoveries in Paintings from Cook's Second Voyage by William Hodges', *Journal for Maritime Research*, vol. 8, 2006, pp. 34–45, 40.

31 George Forster, *A Voyage round the World in His Britannic Majesty's Sloop Resolution commanded by Capt. James Cook, during the years 1772, 3, 4, and 5*, 2 vols (London: P. Elmsley, 1777), vol. 1, pp. 110–11.

32 *A View of Cape Stephens in Cook's Straits with Waterspout* (BHC1906), which is signed and dated 1776, is one of four of the same size done for the Admiralty. The others, also in the collection of the NMM, are BHC1932, BHC2371 and BHC2377.

33 Quilley and Bonehill (eds), (*William Hodges*), p. 114.

34 Forster (*Observations*), p. 51.

35 Forster (*A Voyage round the World*), vol. 2, p. 369.

36 Smith (*European Vision and the South Pacific*), p. 64.

37 Quilley and Bonehill (eds), (*William Hodges*), p. 124.

38 Ibid., p. 107.

39 Ron Radford, 'William Westall and the Landscape Tradition', in Sarah Thomas (ed.), *The Encounter, 1802: Art of the Flinders and Baudin Voyages* (Adelaide: Art Gallery of South Australia, 2002), pp. 102–17, 104.

40 John J. Bradley and Amanda Kearney, '"He painted the law": William Westall, "Stone Monuments" and Remembrance of Things Past in the Sir Edward Pellew Islands', *Journal of Material Culture*, vol. 16, 2011, pp. 25–45, 30–1.

41 Smith (*European Vision and the South Pacific*), p. 196.

42 Reginald Heber and Amelia Heber, *Narrative of a Journey through the Upper Provinces of India from Calcutta to Bombay, 1824–1825, with Notes upon Ceylon written by Mrs Heber*, 2 vols (London: John Murray, 1828), vol. 2, p. 244. Quoted in Sujit Sivasundaram, *Islanded: Britain, Sri Lanka, and the Bounds of an Indian Ocean Colony* (New Delhi: Oxford University Press, 2013), p. 72.

43 British Library, Add. MS 29142, f. 276, John MacPherson to Warren Hastings, quoted in Natasha Eaton, *Mimesis across Empires: Artworks and Networks in India, 1765–1860* (Durham, NC, and London: Duke University Press, 2013), p. 269, n. 62.

44 Quoted in Flood, 'Correct Delineations', p. 61.

Chapter 6

1   Quoted in Bernard Smith, *European Vision and the South Pacific* (New Haven and London: Yale University Press, 1985), p. 114.

2   Edmund Burke to Dr Robertson, 10 June 1777, in Charles William, Earl Fitzwilliam and Sir Richard Bourke (eds), *Correspondence of the Right Honourable Edmund Burke*, 4 vols (London: Francis and John Rivington, 1844), vol. 2, p. 163.

3   Quoted Wilson, (*The Island Race*), p. 59.

4   For further information, see Katie Taylor, 'Pocket-sized Globes', in *Explore Whipple Collections* (Whipple Museum of the History of Science, University of Cambridge, 2009); http://www.hps.cam.ac.uk/whipple/explore/globes/pocketsizedglobes/, accessed 5 July 2013.

5   Charles L. Batten, *Pleasurable Instruction: Form and Convention in Eighteenth-Century Travel Literature* (Berkeley: University of California Press, 1978).

6   Quoted in Brian W. Richardson, *Longitude and Empire: How Captain Cook's Voyages Changed the World* (Vancouver: University of British Colombia Press, 2005), p. 146.

7   NMM, BHC3159.

8   Williams, (*Naturalists at Sea*), p. 28.

9   O.H.K. Spate, 'Seamen and Scientists: The Literature of the Pacific, 1697–1798', in Roy MacLeod and Philip F. Rehbock (eds), *Nature in its Greatest Extent: Western Science in the Pacific* (Honolulu: University of Hawai'i Press, 1988), pp. 13–26, 14, 15.

10 Glyn Williams, *Voyages of Delusion: The Search for the Northwest Passage in the Age of Reason* (London: HarperCollins, 2003), p. 147.

11 Bernard Smith, *Imagining the Pacific: In the Wake of the Cook Voyages* (Carlton, Victoria: Melbourne University Press, 1992), p. 53.

12 See Thomas Richards, *The Imperial Archive: Knowledge and the Fantasy of Empire* (London: Verso, 1993).

13 For further discussion of explorers' claims for the truthfulness of their accounts, see Dorinda Outram, 'On Being Perseus: New Knowledge, Dislocation, and Enlightenment Exploration', in David N. Livingstone and Charles W.J. Withers (eds), *Geography and Enlightenment* (Chicago: University of Chicago Press, 1999), pp. 281–94, esp. p. 283. See also Barbara Maria Stafford, *Voyage into Substance: Art, Science, Nature, and the Illustrated Travel Account, 1760–1840* (Cambridge, MA: MIT Press, 1984).

14 Williams (*Voyages of Delusion*), p. 354.

15 Ibid., p. 247.

16 Robin Inglis, 'Successors and Rivals to Cook: The French and the Spaniards', in Glyndwr Williams (ed.), *Captain Cook: Explorations and Reassessments* (Woodbridge: Boydell, 2004), pp. 161–78, 177.

17 Alan Frost, 'New Geographical Perspectives and the Emergence of the Romantic Imagination', in Robin Fisher and Hugh J.M. Johnston (eds), *Captain James Cook and His Times* (Vancouver: Douglas and McIntyre, 1979), pp. 5–19, 6–7.

18 Quoted in Helen Wallis, 'Publication of Cook's Journals: Some New Sources and Assessments', *Pacific Studies*, vol. 1, 1978, pp. 163–94, 165.

19 See Nicholas Thomas, *Discoveries: The Voyages of Captain Cook* (London: Penguin, 2004), p. 152.

20 Smith (*European Vision*), p. 46.

21 Paul Kaufman, *Borrowings from the Bristol Library: 1773–1784* (Charlottesville, VA: Bibliographical Society of the University of Virginia, 1960), p. 122.

22 Wilson (*The Island Race*), p. 59.

23 Wallis ('Publication of Cook's Journals'), p. 165.

24 Smith (*European Vision*), p. 62.

25 See Tony Rice, *Voyages of Discovery* (London: Natural History Museum, 2008), p. 176.

26 For a discussion of the impact of Douglas's editorship, see I.S. MacLaren, 'Exploration/Travel Literature and the Evolution of the Author', *International Journal of Canadian Studies*, vol. 5, 1992, pp. 39–68, 43.

27 William Hauptman, *Captain Cook's Painter, John Webber, 1751–1793: Pacific Voyager and Landscape Artist* (Bern: Kunstmuseum, 1996), p. 48.

28 John E. Crowley, *Imperial Landscapes: Britain's Global Visual Culture, 1745–1820* (New Haven and London: Yale University Press, 2011), p. 105.

29 Frost ('New Geographical Perspectives'), pp. 6–7.

30 Quoted in Williams (*Voyages of Delusion*), pp. 336–8.

31 James R. Fichter, *So Great a Profitt: How the East Indies Trade transformed Anglo-American Capitalism* (Cambridge, MA: Harvard University Press, 2010), p. 214.

32 Smith (*Imagining the Pacific*), p. 118.

33 Smith (*European Vision*), p. 62.

34 *London Packet or Lloyd's New Evening Post*, 25 April 1777.

35 Hauptman (*Captain Cook's Painter*), p. 48.

36 Ibid., p. 52.

37 Smith (*European Vision*), p. 346, n. 11.

38 Quoted in Richardson (*Longitude and Empire*), pp. 94–5.

39 Crowley (*Imperial Landscapes*), p. 92.

40 Wallis ('Publication of Cook's Journals'), p. 185.

41 See Rüdiger Joppien, 'The Artistic Bequest of Captain Cook's Voyages: Popular Imagery in European Costume Books of the Late Eighteenth and Early Nineteenth Centuries', in Fisher and Johnston (eds), (*Captain Cook and His Times*), pp. 187–210.

42 For the latest scholarship in this area, see Sadiah Qureshi, *Peoples on Parade: Exhibitions, Empire, and Anthropology in Nineteenth-Century Britain* (Chicago and London: University of Chicago Press, 2011); Pascal Blanchard (ed.), *Human Zoos: Science and Spectacle in the Age of Colonial Empires* (Liverpool: Liverpool University Press, 2008); and the exhibition catalogue *Human Zoos: The Invention of the Savage* (Arles: Actes Sud, 2011).

43 Quoted in Richard Altick, *The Shows of London* (Cambridge, MA: Harvard University Press, 1978), p. 48.

44 Fanny Burney to Samuel Crisp, 1 December 1774, in Frances Burney, *Journals and Letters*, edited by Peter Sabor (Harmondsworth: Penguin, 2001), p. 34.

45 See E.H. McCormick, *Omai: Pacific Envoy* (Auckland: Auckland University Press, 1977).

46 Quoted in Hugh Belsey, 'Some Artists' Studios described in 1785', in *Windows on that World: Essays on British Art presented to Brian Allen* (London: Paul Mellon Centre for British Art, 2012), p. 124.

47 Nick Thomas, 'William Hodges as Anthropologist and Historian', in Ricardo Roque and Kim A. Wagner (eds), *Engaging Colonial Knowledge: Reading European Archives in World History* (Basingstoke: Palgrave, 2011), pp. 235–53, 235.

48 Howard Morphy and Michelle Hetherington, 'Introduction: Encountering Cook's Collections', in Michelle Hetherington and Howard Morphy (eds), *Discovering Cook's Collections* (Canberra: National Museum of Australia Press, 2009), pp. 1–10, 5.

49 Quoted in Lisant Bolton, 'Brushed with Fame: Museological Investments in the Cook Voyage Collections', in Hetherington and Morphy (eds), (*Discovering Cook's Collections*), pp. 78–91, 87.

50 Dr John Douglas, 'Introduction', in James Cook and James King, *A Voyage to the Pacific Ocean*, edited by Dr John Douglas, 3 vols (London: G. Nicol and T. Cadell, 1784), vol. 1, p. lxix.

51 James Cook to Hugh Dalrymple, 22 January 1776, in Howard T. Fry, *Alexander Dalrymple (1737–1808) and the Expansion of British Trade* (London: Royal Commonwealth Society, 1970), p. 279.

52 George Forster, *A Voyage round the World in His Britannic Majesty's Sloop Resolution commanded by Capt. James Cook, during the years 1772, 3, 4, and 5*, 2 vols (London: B. White, 1777), vol. 2, pp. 71, 72, 75.

53 Hauptman (*Captain Cook's Painter*), pp. 80, 109, n. 10.

54 Neil Chambers, *Joseph Banks and the British Museum: The World of Collecting, 1770–1830* (London: Pickering and Chatto, 2007), p. 21.

55 Ibid., p. 9.

56 Smith (*European Vision*), p. 123.

57 British Library, Add. MS 8096, f. 6, Johann Friedrich Blumenbach to Joseph Banks, 30 January 1783, in *The Indian and Pacific Correspondence of Sir Joseph Banks, 1768–1820*, vol. 2; *Letters 1783–1789*, edited by Neil Chambers (London: Pickering & Chatto, 2009), p. 5.

58 Hauptman (*Captain Cook's Painter*), p. 83.

59 Smith (*European Vision*), p. 123.

60 Hauptman (*Captain Cook's Painter*), p. 50.

61 Andrew Kippis, *The Life of Captain James Cook* (London: G. Nicol, 1788), p. 498.

62 Chambers (*Joseph Banks and the British Museum*), p. 11.

63 Kaeppler, (*Holophusicon: The Leverian Museum*), pp. 33, 39.

64 Chambers (*Joseph Banks and the British Museum*), p. 11.

65 Jenny Newell, 'Revisiting Cook at the British Museum', in Jeremy Coote (ed.), *Cook-Voyage Collections of 'Artificial Curiosities' in Britain and Ireland, 1771–2015*, MEG Occasional Paper, no. 5, (Oxford: Museum Ethnographers Group, 2015), pp. 7–32, 12.

66 See Bolton ('Brushed with Fame'), p. 88.

67 Chambers (*Joseph Banks and the British Museum*), p. 12.

68 Newell ('Revisiting Cook at the British Museum'), p.13.

69 Clare Williams (ed.), *Sophie in London, 1786: Being the Diary of Sophie von la Roche* (London: Jonathan Cape, 1933), p. 109.

70 British Museum, *Synopsis of the Contents of the British Museum* (London, 1808), pp. xxiv–xxv, quoted in Chambers (*Joseph Banks and the British Museum*), p. 17.

71 James Peller Malcolm, *Londinium Redivivum, or An Ancient History and Modern Description of London*, 4 vols (London: John Nichols, 1802–7), vol. 2 (1803), p. 520.

72 Forster (*A Voyage round the World*), vol. 2, p. 72.

73 Cook I, p. 638; *Cantabrigia Depicta: A Concise and Accurate Description of the University, Town and County of Cambridge* (Cambridge: J. & J. Merrill, 1790), pp. 89–90. See also Wilfred Shawcross, 'The Cambridge University Collection of Maori Artefacts, made on Captain Cook's First Voyage', *Journal of the Polynesian Society*, vol. 79, 1970, pp. 305–48, 305; and Peter Gathercole, 'Lord Sandwich's Collection of Polynesian Artefacts', in Margarette Lincoln (ed.), *Science and Exploration in the Pacific: European Voyages to the Southern Oceans in the Eighteenth Century* (Woodbridge: Boydell, 2001), pp. 103–15.

74 See Lawrence Keppie, *William Hunter and the Hunterian Museum in Glasgow, 1807–2007* (Edinburgh: Edinburgh University Press, 2007), pp. 24–5, 27. I am grateful to John MacKenzie for this reference. See also Euan W. MacKie, 'William Hunter and Captain Cook: The 18th Century Ethnographical Collection in the Hunterian Museum', *Glasgow Archaeological Journal*, vol. 12, 1985, pp. 1–18.

75 J.D. Freeman, 'Polynesian Collection of Trinity College, Dublin; and the National Museum of Ireland', *Journal of the Polynesian Society*, vol. 58, 1949, pp. 1–18, 3.

76 Kaeppler (*Holophusicon*), p. 265.

77 Charles Longuet Higgins, 'A Few Plain Remarks on Local Museums', *Bedfordshire Architectural and Archaeological Society*, vol. 8, 1865, pp. 321–9, 323.

78 See Altick (*The Shows of London*), pp. 28–9.

79 Kaeppler (*Holophusicon*), p. 10.

80 Susan Burney to Fanny Burney, 16 July 1778, in A.R. Ellis (ed.), *The Early Diary of Frances Burney, 1768–1778*, 2 vols (London: George Bell, 1889), vol. 2, p. 249.

81 Kaeppler (*Holophusicon*), p. 6.

82 Quoted ibid., p. 83.

83 'A Description of the Holophusicon, or, Sir Ashton Lever's Museum', *The European Magazine and London Review*, vol. 1, 1782, pp. 17–21, 17.

84 Ibid., pp. 20–1.

85 Bernard Smith, 'Cook's Posthumous Reputation', in Fisher and Johnston (eds), (*Captain Cook and His Times*), pp. 159–85.

86 [John Feltham], *The Picture of London for 1806* (London: s.n., 1806), p. 282.

Chapter 7

1 William Windham in conversation with James Burney after the news of William Bligh's open boat voyage, cited by J.C. Beaglehole in *Cook* III, p. lxxi.

2 R.J.B. Knight, 'John Lort Stokes and the New Zealand Survey, 1848–1851', in *Pacific Empires: Essays in Honour of Glyndwr Williams*, edited by Alan Frost and Jane Samson (Melbourne: Melbourne University Press, 1999), pp. 87–99.

3   *The Monthly Review and Literary Journal*, no. 80, June 1789, pp. 502–11, 502.

4   James Cook and James King, *A Voyage to the Pacific Ocean undertaken by the command of His Majesty for making discoveries in the Northern Hemisphere*, edited by Dr John Douglas (London: G. Nicol and T. Cadell, 1784), pp. 437, 438.

5   [William Beresford] George Dixon and Nathaniel Portlock, *A Voyage Round the World: But More Particularly to the North-West Coast of America: Performed in 1785, 1786, 1787, and 1788* (London: George Goulding, 1789), p. ix. Confusingly, Dixon and the overall commander of the voyage, Nathaniel Portlock, had two books published under their names with almost exactly the same title. The decision made good sense in one respect, as the two ships split up for the second season in order to cover more of the American coast. One book, which I shall refer to hereafter as 'Dixon', was written by William Beresford, the supercargo on Dixon's *Queen Charlotte*. The other was written by Portlock, although probably with editorial assistance; this will be referred to as 'Portlock'. The books had very different styles and structures – Dixon's being written as a series of letters with relatively little navigational information, while Portlock's was written as a traditional narrative, drawing on his journal. They were clearly meant for different audiences: the popular and the professional. If a review of Dixon's in *The Monthly Review and Literary Journal* is any guide, the experiment was not a resounding success. There was in fact a third book, also with a similar title: *A Voyage Round the World in the Years 1785, 1786, 1787, and 1788, Performed by the King George Commanded by Captain Portlock, and the Queen Charlotte, commanded by Captain Dixon, Under the Direction of the Learned Society for the Advancement of the Fur Trade* (London, for R. Randal, 1789). This rare book is believed to have been largely written by William Colin Lauder, the surgeon on *Queen Charlotte*, who died shortly after the ship left China on the voyage home.

6   David Mackay, *In the Wake of Cook: Exploration, Science and Empire, 1780–1801* (Beckenham: Croom Helm, 1985), pp. 57–82.

7   A snow was a two- rather than three-masted vessel, very similar to a brig.

8   Nathaniel Portlock, *A Voyage Round the World but More Particularly to the North-West Coast of America* (London: John Stockdale and George Goulding, 1789), pp. 5, 6; hereafter referred to as 'Portlock'.

9   Cited in Mackay (*In the Wake of Cook*), p. 21.

10 Ibid., p. 62.

11 Portlock, pp. 6–7.

12 Randolph Cock, 'John Gore', *Oxford Dictionary of National Biography* (Oxford: Oxford University Press, 2002), http://ww.oxforddnb.com/view/article/64857?docPos=3, accessed 13 March 2016. Portlock also names David and Charles Gilmore among the boys taken on the voyage, and this suggests the further involvement of Captain John Gore, for Randolph Cock records him marrying Sarah Gilmore.

13 Portlock, pp. 6, 7.

14 Dixon, p. xi.

15 Mackay (*In the Wake of Cook*), p. 64.

16 Portlock, p. 382.

17 *The Malaspina Expedition, 1789–1794: Journal of the Voyage of Alejandro Malaspina*, 3 vols (London and Madrid: Hakluyt Society and Museo Naval, 2001), vol. 1, p. lxxx.

18 Alain Morgat, 'The Influence of Cook's Voyages on the Lapérouse Expedition, 1785–1788, in *Science and the French and British Navies, 1700–1850* (Greenwich: National Maritime Museum, 2003), pp. 48–61.

19 John Dunmore, *Where Fate Beckons: The life of Jean-François de Galaup de Lapérouse* (Fairbanks: University of Alaska Press, 2007), p. 183.

20 An instrument for copying charts.

21 Dunmore (*Where Fate Beckons*), p. cix.

22 Morgat ('The Influence of Cook's Voyages'), p. 49.

23 La Pérouse, quoted by Dunmore (*Where Fate Beckons*), p. 227.

24 J.-F. de la Pérouse, *The Journal of Jean-François de Galaup de La Pérouse, 1785–1788*, edited by John Dunmore (London: Hakluyt Society, 1994), p. 110.

25 John Dunmore, ibid., p. clxxxiii.

26 As I write this, a British reality-television company is preparing to re-enact Bligh's open-boat voyage.

27 The journal was found in the general naval correspondence of Admiral John Markham while being catalogued by Dr Jane Knight.

28 Quoted by Pieter van der Merwe, in Nigel Rigby, Pieter van der Merwe and Glyn Williams, *Pioneers of the Pacific: Voyages of Exploration, 1787–1810* (Greenwich: National Maritime Museum, 2005), p. 75.

29 See Flinders's obituary in the *Naval Chronicle*, London, 1814.

30 Kenneth Morgan, *Matthew Flinders, Maritime Explorer of Australia* (London: Bloomsbury, 2016), p. 12.

31 William Bligh to Sir Joseph Banks, 9 March 1793 and 30 October 1793, in *The Indian and Pacific. Correspondence of Sir Joseph Banks*, edited by Neil Chambers (London: Pickering and Chatto, 2011), pp. 90, 160.

32 Quoted in a review of Douglas Oliver, *Return to Tahiti: Bligh's Second Breadfruit Voyage* (Melbourne: Miegunyah Press, 1988), Robert Langdon, *Journal of the Polynesian Society*, vol. 100 (1), March 1991, pp. 103–6, 103.

33 John Meares, *A Voyage Made in the Years 1788 and 1789 from China to the North West Coast of America, to which are prefixed an Introductory Narrative of a Voyage Performed in 1786, from Bengal in the Ship Nootka and Observations on the Probable Existence of a North West Passage* (London: J. Walker, 1790), p. viii.

34 For a superb discussion of the cartographical motivations of Vancouver's voyage, see Glyndwr Williams, 'Myth and Reality: The Theoretical Geography of Northwest America from Cook to Vancouver', in *From Maps to Metaphors: The Pacific World of George Vancouver*, edited by Robin Fisher and Hugh Johnston (Vancouver: UBC Press, 1993), pp. 35–50.

35 Admiralty papers, ADM 1/2628, The National Archives. Rob of lemon or orange was produced by boiling down the fruit into a concentrated form. Much of its antiscorbutic value was lost in the process. Cook thought it an efficacious treatment for scurvy, although he did not recommend it, as it was too expensive to produce for widespread use.

36 Hydrographic Office, HO 42 18 fo. 63.

37 Williams ('Myth and Reality'), p. 48.

38 Morgan (*Matthew Flinders*), p. 47.

39 Flinders to Ann Chappelle, 16 March 1799, FLI125; http://flinders.rmg.co.uk/DisplayDocumentaf25.html?ID=93.

40 Matthew Flinders, *A Voyage to Terra Australis Undertaken for the Purpose of Completing the Discovery of that vast Country and Prosecuted in the years 1801, 1802 and 1803* (London: G. & W. Nichol, 1814), book 1, chapter 1, of unpaginated electronic copy at: http://gutenberg.net.au/ebooks/e00049.html, accessed 21 October 2016.

41 Letters to Joseph Banks, quoted in Morgan (*Matthew Flinders*), pp. 56, 63.

42 Greg Keighery and Neil Gibson, 'The Flinders Expedition in Western Australia: Robert Brown, the plants and their influence on W.A. Botany', in Juliet Wege, Alex George, Jan Gathe, Kris Lemson and Kath Napier (eds), *Matthew Flinders and his Scientific Gentlemen* (Welshpool Dc: Western Australia Museum, 2005), pp. 104–13, 105.

43 John Dell, 'The Fauna Encountered in South-Western Australia: its culinary and scientific legacy', in Wege et al. (*Matthew Flinders and his Scientific Gentlemen*), pp. 114–25, 116.

44 Kenneth Morgan, *Australia Circumnavigated: The voyage of Matthew Flinders in HMS Investigator*, 2 vols (London: Ashgate for the Hakluyt Society, 2015), vol. 1, p. 29.

45 Morgan (*Matthew Flinders*), p. 193. Copies of the book were delivered to his London home on 18 July and he died the following day.

46 Hélène Blais, 'Exploration and Colonization in the Pacific: French voyages under the Bourbon Restoration and the July Monarchy', in Peter van der Merwe (ed.), *Science and the French and British Navies, 1700–1850* (Greenwich: National Maritime Museum, 2001), pp. 62–76, 63.

47 Adrian Webb, *The Expansion of British Naval Hydrographic Administration, 1808–1829* (doctoral thesis, University of Exeter, 2010), p. 1.

Chapter 8

1   Howard Morphy and Michelle Hetherington, 'Introduction: Encountering Cook's Collections', in Michelle Hetherington and Howard Morphy (eds), *Discovering Cook's Collections* (Canberra: National Museum of Australia Press, 2009), pp. 1–10, 4.

2   Lisant Bolton, 'Brushed with Fame: Museological Investments in the Cook Voyage Collections', in Hetherington and Morphy (eds), (*Discovering Cook's Collections*), pp. 78–91, 78.

3   Jillian Robertson, *The Captain Cook Myth* (London: Angus & Robertson, 1981).

4   W. Fraser Rae, *The Business of Travel: A Fifty Years' Record of Progress* (London: Thos. Cook and Son, 1891), p. 10.

5   Allan Klenman, *The Faces of Captain Cook: A Record of the Coins and Medals of James Cook* (Victoria, BC: Allan Klenman, 1983), p. 4.

6   Quoted in Alan Frost, 'New Geographical Perspectives and the Emergence of the Romantic Imagination', in Robin Fisher and Hugh Johnston (eds), *Captain Cook and His Times* (Vancouver: Douglas and McIntyre, 1979), pp. 5–19, 6.

7   Arthur Young, *The Autobiography of Arthur Young*, edited by M. Betham-Edwards (London: Smith, Elder and Co., 1898), pp. 69–70. See Mervyn Busteed, *Castle Caldwell, County Fermanagh: Life on a West Ulster Estate, 1750–1800* (Dublin: Four Courts Press, 2006), p. 53.

8   Clare O'Halloran, *Golden Ages and Barbarous Nations: Antiquarian Debate and Cultural Politics in Ireland, c.1750–1800* (Cork: Cork University Press, 2004), p. 98.

9   Quoted in Wilson, (*The Island Race*), p. 86.

10 Harriet Guest, 'Commemorating Captain Cook in the Country Estate', in Gillian Perry, Kate Retford, Jordan Vibert and Hannah Lyons (eds), *Placing Faces: The Portrait and the Country House, 1650–1850* (Manchester: Manchester University Press), pp. 192–218, 209. For an image of the object, see http://acms.sl.nsw.gov.au/album/albumView.aspx?acmsID=421788&itemID=823904, accessed 4 July 2014.

11 According to Royal Society records, 13 medals were struck in gold, 289 in silver and 500 in bronze, with the surplus of the proceeds used to strike some extra gold medals, one of which was presented to Cook's widow. See I. Kaye, 'Captain James Cook and the Royal Society', *Notes and Records of the Royal Society*, vol. 24, 1969, pp. 7–18.

12 British Library, Add. MS 8095, f. 244, Louis Antoine de Bougainville to Joseph Banks, 25 June 1784, in *The Indian and Pacific Correspondence of Sir Joseph Banks, 1768–1820*, vol. 2: *Letters 1783–1789*, edited by Neil Chambers (London: Pickering & Chatto, 2009), p. 61.

13 Sir Benjamin Thompson to Joseph Banks, 24 July 1784, in *The Indian and Pacific Correspondence of Sir Joseph Banks*, vol. 2, p. 63. At least one medal was sent. See British Library, Add. MS 8096, ff. 177–8, Rumford to Banks, 14 June 1785.

14 See Andrew Kippis, *The Life of Captain James Cook* (London: G. Nicol, 1788), pp. 511–13.

15 Klenman (*The Faces of Captain Cook*), p. 6.

16 Guest ('Commemorating Captain Cook in the Country Estate') pp. 194–8.

17 Sujit Sivasundaram, 'Redeeming Memory: The Martyrdoms of Captain James Cook and Reverend John Williams', in Glyndwr Williams (ed.), *Captain Cook: Explorations and Reassessments* (Woodbridge: Boydell, 2004), pp. 201–29.

18 Natasha Eaton, *Mimesis across Empires: Artworks and Networks in India, 1765–1860* (Durham, NC, and London: Duke University Press, 2013), p. 97.

19 Greg Dening, 'Looking across the Beach – Both Ways', in Hetherington and Morphy (eds), *Discovering Cook's Collections*, pp. 11–24, 13.

20 See Harriet Guest, *Small Change: Women, Learning and Patriotism, 1750–1810* (Chicago: University of Chicago Press, 2000), pp. 252–67.

21 Guest ('Commemorating Captain Cook in the Country Estate') p. 200. For an image of the proposal monument, see http://www.britishmuseum.org/research/collection_online/collection_object_details.aspx?objectId=1543888&partId=1&people=128629&peoA=128629-1-9&page=1, accessed 4 July 2014.

22 Ibid., p. 199.

23 Klenman (*The Faces of Captain Cook*), p. v.

24 R.A. Skelton, *Captain James Cook after two hundred years: A Commemorative address delivered before the Hakluyt Society* (London: British Museum, 1969), p. 9. The eulogy inscribed on monument also appeared, attributed to Captain Lord Mulgrave, in the preliminaries of *A Voyage to the Pacific Ocean* (1784). Skelton (*Captain Cook*), p. 31, n. 4.

25 Joan Coutu, *Persuasion and Propaganda: Monuments and the Eighteenth-Century British Empire* (Montreal and Kingston, ON: McGill-Queen's University Press, 2006), p. 20.

26 The arrow was tested in 2004 and turned out to be an antler from the north-west coast of America. See Bolton ('Brushed with Fame'), pp. 78–9.

27 Quoted in Williams, (*Naturalists at Sea*), p. 234.

28 Simon Werrett, 'Russian Responses to the Voyages of Captain Cook', in Williams (ed.), (*Captain Cook*), p. 195.

29 Quoted in Elizabeth C. Childs, 'Second Encounters in the South Seas: Revisiting the Shores of Cook and Bougainville in the Art of Gauguin, La Farge and Barnfield', in Tricia Cusack (ed.), *Framing the Ocean, 1700 to the Present: Envisaging the Sea as a Social Space* (Farnham: Ashgate, 2014), pp. 55–67, 58.

30 Ibid., p. 60.

31 Quoted in James McCarthy, *That Curious Fellow: Captain Basil Hall, R.N.* (Caithness: Whittles Publishing, 2011), p. 137.

32 Ibid., pp. 136–7.

33 Ibid., p. 137.

34 Walter Besant, *English Men of Action: Captain Cook* (London: Macmillan and Co., 1904), p. 66.

35 Quoted in Clare Pettitt, 'Exploration in Print: From the Miscellany to the Newspaper', in Dane Kennedy (ed.), *Reinterpreting Exploration: The West in the World* (New York: Oxford University Press, 2014), pp. 80–108, 84.

36 Besant (*English Men of Action*), pp. 66–7.

37 Matthew Flinders to Charles Baudin, quoted in Marina Carter, *Companions of Misfortune: Flinders and Friends at the Isle of France, 1803–1810* (London: Pink Pigeon Press, 2003), p. 5.

38 Nicholas Thomas, 'The Uses of Captain Cook: Early Exploration in the Public History of Aotearoa New Zealand and Australia', in Annie Coombes (ed.), *Rethinking Settler Colonialism: History and Memory* (Manchester: Manchester University Press, 2006), pp. 140–55, 140.

39 See Tom Griffiths, *Hunters and Collectors: The Antiquarian Imagination in Australia* (Cambridge: Cambridge University Press, 1996), pp. 157–8.

40 Quoted in Graeme Davison, 'Centennial Celebrations', in Graeme Davison, John McCarty and Ailsa McLeary (eds), *Australians 1888* (Sydney: Fairfax, Syme & Weldon, 1987), p. 20.

41 Thomas ('The Uses of Captain Cook'), p. 144. This image is now known only through reproduction, as the original has been missing from the Royal Society of Victoria for more than 50 years.

42 Quoted in Jason Edwards, 'Postcards from the edge? Thomas Woolner's *Captain Cook* for Sydney', *Sculpture Journal*, vol. 23 (2), 2014, pp. 209–20, 210.

43 Quoted ibid., p. 213.

44 See http://christchurchcitylibraries.com/Heritage/Places/Memorials/Statues/, accessed 8 August 2014.

45 'Captain Cook', [Wellington] *Evening Post*, 11 August 1932, p. 9.

46 'Greatest of Navigators', [Wellington] *Evening Post*, 13 August 1932, p. 8.

47 Ibid.

48 Ibid.

49 Richard Altick, *The Shows of London* (Cambridge, MA: Harvard University Press, 1978), p. 300.

50 'The Royal Naval Exhibition: *Pall Mall Gazette* Extra No. 56', June 1891, p. 16.

51 Besant (*English Men of Action*), p. 191.

52 For further discussion of the prosopographical, hagiographical nature of the National Maritime Museum's representation of British exploration, see Claire Warrior, 'On Thin Ice: The Polar Displays at the National Maritime Museum, Greenwich', *Museum History Journal*, vol. 6 (1), 2013, pp. 59–73.

53 *National Maritime Museum Catalogue* (Greenwich: Trustees of the National Maritime Museum, 1937), pp. 154–9.

54 Ibid., p. 165.

55 Ibid., p. 166.

Chapter 9

1   See the brief but informative entry on Callender by Michael Lewis in the *Oxford Dictionary of National Biography* (revised by H.C.G. Matthew), http://www.oxforddnb.com/view/article/32249. See also Kevin Littlewood and Beverley Butler, *Of Ships and Stars: Maritime Heritage and the Founding of the National Maritime Museum, Greenwich* (London: Athlone Press and National Maritime Museum, 1998), to which this chapter is much indebted.

2   *Catalogue of the National Maritime Museum* (London: HMSO, 1937), p. 164.

3   *A Sense of Direction*, a generally undated and unpaginated NMM strategic discussion document, c.1957.

4   John Munday, in *A Sense of Direction*, ibid.

5   George Naish, ibid.

6   Pieter van der Merwe, '"A proud monument of the glory of England": The Greenwich Hospital Collection',

NOTES   247

7. in Geoff Quilley (ed.), *Art for the Nation: The Oil Paintings Collection of the National Maritime Museum* (Greenwich: National Maritime Museum, 2006), pp. 19–37, 24, 25.

7. H.A. Locker, *Memoirs of the Distinguished Naval Commanders whose Portraits are Exhibited in the Naval Gallery of Greenwich* (Greenwich, 1842), advertisement. Quoted by Pieter van der Merwe, in Quilley (ed.), (*Art for the Nation*), pp. 19–38, 25.

8. Charles Mitchell, 'Zoffany's Death of Cook', *The Burlington Magazine for Connoisseurs*, vol. 84, no. 492, March 1944, pp. 56–63.

9. This information has been uncovered recently in the doctoral research of Dr Cicely Robinson.

10. See Lance Bertelsen's interesting essay, 'Patronage and the Pariah of Captain Cook's Third Voyage: Captain John Williamson, Sir William Jones and the Duchess of Devonshire', *Journal for Eighteenth-Century Studies*, vol. 38 (1), March 2015, pp. 29–45.

11. *Catalogue of the National Maritime Museum*, pp. 165–6.

12. The complicated history of the Admiralty collection of Hodges paintings has been described by Geoff Quilley, in '"A tiresome issue": The artistic heritage of empire and the Ministry of Defence Collection', in Quilley (ed.), (*Art for the Nation*), pp. 99–109.

13. Callender to Sir Vincent Baddeley, 26 January 1940, NMM, curatorial records. Baddeley was Principal Assistant Secretary at the Admiralty until 1935, and in Callender's view not an entirely trustworthy supporter of the Museum. The flamboyant politician and socialite Alfred Duff Cooper had been appointed First Lord of the Admiralty in 1937.

14. Sir Vincent Baddeley to Frank Carr, 26 September 1955, citing a letter to Major Hugh Carrington, author of *The Life of Captain James Cook* (London: Sidgwick and Jackson, 1939), originally sent 20 January 1940. Curatorial records, Weapons and Antiquities Dept.

15. E.H.H. Archibald, in *A Sense of Direction*.

16. R.A. Skelton to Frank Carr, 30 August 1955, NMM archives, C42, 'Cook Exhibition'. A full list of temporary exhibitions from 1937 to 1967 is included in Littlewood and Butler (*Of Ships and Stars*), pp. 222–4.

17. *Captain Cook – Special Exhibition held at the National Maritime Museum, 1956–57*, copy of typescript (London, 1957), British Library, General Reference Collection, 08809, identifier 002237920.

18. Dorothy Walker, Whitby Literary and Philosophical Society, to George Naish, 22 September 1956, NMM C42.

19. Littlewood and Butler (*Of Ships and Stars*); see particularly the chapter 'Sea Changes', pp. 158–93.

20. W.E. May, Deputy Director, introductory comments to *A Sense of Direction*.

21. Michael Robinson, ibid.

22. Fighting the science corner were W.E. May, Deputy Director until his retirement in 1968, and the man who eventually followed him in the post, D.W. Waters. Both were ex-naval officers with considerable practical experience of navigation as well as being respected historians, and they firmly established the navigational sciences as an inseparable part of the Museum's 'Cook'.

23. The title of Greenhill's oration for Cook, Westminster Abbey, 11 February 1979, reproduced in *The Mariner's Mirror*, vol. 65, no. 2, May 1979, pp. 101–8. Greenhill may not have been a popular figure in the Museum, but his vision – if very different from Callender's – was equally strong and he carried it through remorselessly. See Basil Greenhill, 'The Last Fifteen Years at the National Maritime Museum', *Museums Journal*, vol. 82, no. 4, March 1983, pp. 213–15.

24. While 1776 was a ticketed show, during this period all NMM temporary exhibitions were free to enter, not least since government funding arrangements discouraged proactive money-making: in effect, any money made was simply deducted from the next year's grant-in-aid.

25. D.W. Waters (Deputy Director) to Director, 26 May 1977, NMM archive G5/4/058.

26. See A.P. McGowan, 'Captain Cook's Ships' and 'The Principal Scientific Instruments Taken on Captain Cook's Voyages of Exploration', *The Mariner's Mirror*, vol. 65, no. 2, May 1979, pp. 109–18, 119–36.

27. Waters to John Munday and Westby Percival-Prescott, 24 November 1977, G5/4/058.

28. Bernard Smith had been publishing on European representations of the Pacific since the late 1950s, with his *European Vision and the South Pacific*, published in 1959, and the four-volume collaboration with Rüdiger Joppien, *The Art of Captain Cook's Voyages*, completed in 1986.

29. E.H.H. Archibald to Percival-Prescott, 22 December 1977, G5/4/058.

30. E.H.H. Archibald to Waters and Percival-Prescott, 18 January 1978, G5/04/058.

31. Westby Percival-Prescott, 3 February 1978, G5/04/058.

32. Quilley (*Art for the Nation*), pp. 99–110.

33. Sotheby's catalogue, *Sale at Mount Juliet, Co. Kilkenny, Wednesday 24th September, 1986*, lot 340.

34. David Cordingly, 'Captain James Cook by William Hodges', *Apollo*, November 1987, pp. 318–22, 318.

35. Ibid., p. 318.

36. The Webber came on the market again some years later after the collapse of the Bond empire, but because it had been granted an export licence in 1983 the Museum was unable to prevent its sale abroad once more, and in 2000 it was bought by the National Portrait Gallery, Canberra, for AUD 5.13m. Britain's National Portrait Gallery had not opposed the original export licence or informed the NMM of the application, and this caused a certain coolness between two national institutions that normally enjoy harmonious relations.

37. J.C. Beaglehole, 'On the character of Captain James Cook', *The Geographical Journal*, vol. CXXII, part 4, December 1956, http://nzetc.victoria.ac.nz/tm/scholarly/tei-BeaChar-t1-body1.html, accessed July 2015.

38. David Cordingly, then Head of Exhibitions, explained to Bryce Harland, the New Zealand High Commissioner in London, that the NMM had been unable to honour the planned opening of *Captain Cook, Navigator* (under which title it toured) because the 'financial straits' in which the Museum found itself after the bankruptcy of the sponsor. David Cordingly to Bryce Harland, October 1989, 2 G5/4/96.

39. Neither the *Bounty* nor the Cook shows reached the Museum's financial targets, which were probably over-optimistically coloured by the huge success of its 1988 Armada exhibition, itself only rivalled by *Henry VIII at Greenwich* in 1991. The contrast between the market appeal of genuinely 'popular', if hackneyed topics and more specialized ones is self-evident and not one that encourages risk-taking innovation.

Index

References to illustrations are shown in *italics* with caption page in brackets if not on the same page as the illustration, e.g. *44-45* (42).

Aberli, Johann Ludwig 137
Aborigines (Australia) 51, *51*, 69, 194, 218
Adams, George 173
Adams, John (also known as Alexander Smith) 184
Admiralty *see* Royal Navy
Adventure
   Adventure at anchor in Table Bay (William Hodges) *41*
   formerly *Marquis of Rockingham* 39
   see also second voyage (*Resolution* and *Adventure*)
Agnese, Battista, map of Magellan's voyage showing the global wind directions *64*
Ahu-toru (Polynesian) 31
Aleutian Islands 55, 60
America, north-west coast
   commemoration of Cook's landings (2028) 7
   Cook's map (Cook, King and Douglas's *A Voyage to the Pacific Ocean*) 79 *(78)*
   Cook's third voyage 50, 55, 60
   La Pérouse's journey 173, 176
   Meares's map 186–7, *187*
   post-Cook knowledge gaps 80, 169
   Vancouver's voyage *see* Vancouver, George
   see also Dixon, George; North-West Passage; Portland, Nathaniel
American Revolution, *1776: The British Story of the American Revolution* (*The Times* and *Sunday Times*, National Maritime Museum, 1975) 234
Anderson, W. G. 235
Anderson, William 41, 51, 120, 137, 152
Anson, George 21, 71, 119, 123, 201, 226, 236
   A Voyage Round the World 147, *147*, 152
Antarctic Circle 41
antiquarian ethnography 202
Archibald, Teddy 234, *235*, 237
art, landscape and exploration
   artist's role and mission 17, 19, 80, 117, 119–20
   Cook's first voyage 120–24
   Cook's second voyage 124–36
   Cook's third voyage, John Webber 137
   legacies 137, 141, 143
   portraits of Pacific Islanders 94, *96–7* (95)
   see also charts and maps; visual images
The Art and Science of Exploration exhibition (National Maritime Museum, 2004) 229, 230
art exhibitions 19, 145, 153
   see also Royal Academy; Society of Artists (later Free Society of Artists)
Art Fund 234, 239
artefacts *see* material culture
'The Artist of the Chief Mourner' 87, 237
Assistant 185, *186*, 230
Astrolabe 173, 176
   Naufrage de l'Astrolabe sur les recifs de l'île de Vanikoro (Louis Le Breton) *178–9* (177)
Astronomer Royal 71, 194, 231
Attahha (or Otago) 88–9, *88* (89)
Australia (New Holland)
   Aborigines 51, *51*, 69, 194, 218
   Captain Cook, Navigator (touring exhibition, National Maritime Museum) 236
   Chart of New South Wales 36–7
   Cook commemorative medal 219, *223*
   Cook's coast mapping 72
   Cook's first voyage 35–6, *36–7*
   Cook's legacy as benevolent 216, 218
   Cook's relics 218
   Cook's statue (by Thomas Woolner) 216, 218
   Dutch ships' tracing of north, south and west coasts 69
   first circumnavigation of 68
   flora and fauna 7, 37, 194

   in Henry Roberts' map 80
   Matthew Flinders's voyage 185, 192–6, *193* (192), *196* (197), 198
   Port Jackson penal colony 176, 192
   red sandstone (northern Queensland) 211, *211*
   tree fragment (northern Queensland) 211, *211*
   William Westall's paintings *195*, 196, *197*
Australian Hydrographic Office, charts depicting course of Cook's first voyage 219, *222*
Australian Museum (Sydney), arrow made of Cook's leg bone 211

backstaff, by Will Garner *70*
Bankes, Thomas, *Universal Geography* 145
Banks, Dorothea, Lady 212
Banks, Joseph, 1st Baronet
   and Bligh's *Bounty* voyage 180
   and Bligh's *Providence* voyage 185
   British Museum donations from 164
   bust of by Turnerelli 226, *227*, 229, 233
   collections of objects 163
   commemorative medals for Cook's second voyage 49
   commission of Cook's portrait by Nathaniel Dance 7, 9, 15, 50, 160, 229
   commissioning ceremony for *King George* and *Queen Charlotte* 170
   commissions from George Stubbs 38, 158, 160
   Cook's debt to 80
   Cook's first voyage 32, 33, 34, 37, 38, 120–21, 147
   Cook's second voyage, withdrawal from 39, 124–5
   death 198
   and Falconer 163
   and Flinders's voyage 192, 193–4
   Florilegium 237, 239
   in Hawkesworth's *An Account of the Voyages* 50
   James King on 170
   and La Pérouse 173
   and Menzies 172, 189
   and Monneron 173
   and Omai (Ma'i) 50, 160
   portrait (engraving after Benjamin West) *10* (11)
   portrait in William Parry's group portrait 160
   Royal Society Dining Club President 79
   and Stanfield Parkinson, legal dispute with 122
   Terra Australis theory 192
   and Tupaia 73, 87
   and Vancouver's voyage 189, 191
Banks, Thomas 210
barometers 71, *72* (73)
Baron, H. A. 219, 233
Barry, James, *Commerce, or the Triumph of the Thames* 210
bartering 90, 112
Bass, George 31, 79, 80, 192
Baudin, Nicolas 194, 196–7
Bauer, Ferdinand 194, 196
Baugh, Daniel 25
Bayly, William 39, 51, 73
   Chart of New Zealand explored in 1767 and 1770 by Lieut. J. Cook 74 *(75)*
Beaglehole, J. C. 11, 51, 60, 65, 231, 233, 235
beaker, used by William Bligh *184*
Beaufort, Rear-Admiral Sir Francis 197, *199* (198)
Beautemps-Beaupré, Charles-François 192
Beechey, Captain Frederick 184
Belcher, George 216
Bellerophon 192
Beranger, Gabriel 202
Bering, Vitus 148
Besant, Walter 21, 214, 216
   Captain Cook title page *215*
Bledisloe, Charles Bathurst, 1st Viscount Bledisloe 218
Bligh, William
   Bounty voyage 19, 94–5, 180, *181* (180), *182*, 183–5, *183*, *184*, 185
   bowl, beaker and bullet as weight used by *184*
   with Cook on third voyage 19, 51, 72

   copy of chart of Bligh's Islands [Fiji] by William Bligh (Matthew Flinders) *186*
   and Hawkesworth's *Voyages Undertaken* 77
   portrait (engraving after John Russell) *180*
   Providence voyage (with *Assistant*) 19, 185–6, *186*, 230
   remembered for his failings 169
   Royal Society Dining Club guest 79
   and scurvy 76–7
   A Voyage to the South Sea title page *182*
   on voyages during George III's reign 16
   see also Mutiny on the Bounty exhibition (National Maritime Museum, 1989) 236, *237*
Blood, Sea and Ice exhibition (National Maritime Museum,1996-99) 237
Blumenbach, Johann Friedrich 163
Board of Longitude 71, 173, 197, 230, 239
boats *see* canoes; maritime technology; ships
Bolton, Lisant 12
Bond, Alan 235
Bougainville, Louis-Antoine de 30–31, 55, 189, 203
Boulton, Matthew 158
Bounty
   model *237*
   see also Bligh, William
Boussole (French ship) 173, 176, *176*
Bouvet de Lozier, Jean-Baptiste Charles 41, 47
Bouvet Island 41
bowls
   Bligh's coconut bowl *184*
   Cook's punch bowl 218
   seal-shaped feast bowl 106, *108*
Boydell, John, *Views in the South Seas* 157
breadfruit 180, 185
   A branch of the bread-fruit tree with fruit (engraving after Sydney Parkinson) *181* (180)
   Tahitian breadfruit-pounder 105–6, *108*
Bridge, John 211
Bristol Library 148–9
British Museum
   and *Captain Cook* exhibition (National Maritime Museum, 1956-57) 231
   Cook-related collections 164–5, 218
   donations to 163–4
   Sandwich Room 113
   South Seas Room 164
   Treasures of Tutankhamun exhibition (1972) 234
Broughton, William 79, 80, 169, 189
Brown, Robert 194, 196
Bruny d'Entrecasteaux, Antoine de 176, 192, 194
Buchan, Alexander 32, 120, 121
bullet (as weight), used by William Bligh *184*
burial places and priests 101–2, *102–4*
Burke, Edmund 19, 145
Burnet, John, sketch of Painted Hall vestibule (Greenwich Hospital Naval Gallery) 227, *229*
Burney, Charles 41, 148
Burney, Fanny 41, 160, 163, 167
Burney, James 41, 75, 77, 79, 163
Burney, Susan 167
Byron, John
   1764-66 circumnavigation 11, 19, 55, 64
   in *Captain Cook, Explorer* exhibition 236
   in Hawkesworth's *An Account of the Voyages* 38, 50, 148–9
   Reynolds's portrait of *230* (231)

Cabot, John 210
Caird, James, Sir 225, 226, *226*, 231
Caldwell, James, 4th Baronet 202
Callender, Geoffrey, Sir 225–6, 228, 230, 231, 233, 234
   Catalogue of the National Maritime Museum 226
   The Naval Side of British History 226
   royal opening of the National Maritime Museum 226
   Sea Kings of Britain 225, 226
Camelford, Lord *see* Pitt, Thomas, 2nd Baron Camelford
Campbell, John 11
Campion, Robert 210
candle snuffers, belonging to the Cooks 219, *220*

249

candlesticks, belonging to Elizabeth Cook 219, *220*
cannibalism 185
canoes 87–8
　see also maritime technology
Captain Cook and Mr Hodges exhibition (National Maritime Museum, 1978) 234, 238
　opening *235*
　poster *240* (239)
Captain Cook and the Exploration of the Pacific gallery (National Maritime Museum, 1970-73)
　entrance to *232*, 233
　ethnographic display 233, *233*, 234
　Navigation Room 233–4
　Pacific objects 234
Captain Cook Birthplace Museum (Marton, Middlesbrough) 237
Captain Cook exhibition (National Maritime Museum, 1956-57) 231, 233
Captain Cook, Explorer exhibition (National Maritime Museum, 1990) 236–7
Captain Cook Memorial Museum (Whitby) 237
Captain Cook, Navigator (touring exhibition, National Maritime Museum, 1988) 237
Captain Cook's voyages of discovery, title page 215
Carr, Frank 231, 233, 234
Carter, George, *The Death of Captain James Cook* (engraving) 206, *208* (209), 209
Carter, Harold 237
Carteret, Lieutenant Philip 11, 38, 65, 77, 148–9, 184
　octant owned by *70*
Cary, John, the Elder 146, *162*
Cary, William *162*
Casson, Hugh, Sir *235*
Catherine the Great, Empress of Russia 203, 206
Cavendish, Thomas 68
charts and maps
　Australia with S.E. Asia and Oceania special edition (commemoration of bicentenary of Cook's first voyage) 219, *222*
　Chart of New South Wales 36–7
　Chart of New Zealand explored in 1767 and 1770 by Lieut. J. Cook (William Bayly) *74* (75)
　Chart of Terra Australis (Flinders' *A voyage to Terra Australis*) 198
　Chart of the Friendly Islands (Cook and Douglas's *A Voyage to the South Pole and Around the World*) *78*
　'A Chart of the Interior Part of North America' (Meares's*Voyages Made in the Years 1788 and 1789*) 186–7, *187*
　Chart of the Island Otaheite (James Cook) *33* (32)
　Chart of the N W Coast of America and N E Coast of Asia (Cook, King and Douglas's *A Voyage to the Pacific Ocean*) *79* (78)
　A chart of the Pacific Ocean (c.1744) 16–17
　A chart of the southern hemisphere (Cook and Douglas' *A Voyage Towards the South Pole*) *81* (80)
　A Chart representing the isles of the South Sea, according to the notions. Chiefly collected from the accounts of Tupaya (J. R. Forster) *86* (87)
　A chart showing part of the coast of N.W. America (Vancouver's *A voyage of discovery*) *190*
　copy of chart of Bligh's Islands [Fiji] by William Bligh (Matthew Flinders) *186*
　A General Chart Exhibiting the Discoveries made by Captn James Cook (Henry Roberts) 79–80, *82–3* (80)
　from John Hawkesworth's *An Account of the Voyages undertaken* 149 (148)
　map of Magellan's voyage showing the global wind directions (Battista Agnese) *64*
　post-Magellan map: *South America and Magellanica* (Joannes A. Doetechum) *66–7*
　pre-Magellan world map (Francesco Rosselli) *65* (64)
　Spanish chart of the Straits of Magellan and Le Maire (William Hack) *62* (63), 64, *68*, 69
　see also art, landscape and exploration; navigators' journals; running surveys
Charvet, Jean-Gabriel 158
Chatham 189

Chatham, Lord *see* Pitt, John, 2nd Earl of Chatham
Chatham Islands 189
Christian, Fletcher 77, 95, 180, 183, 184
chronometers (marine timekeepers)
　Cook's chronometer 218
　on Cook's voyages 71
　Harrison's four chronometers, exhibition of 233–4
　Harrison's H4 marine timekeeper 73, *73*
　Kendall's K1 (copy of Harrison's 'H4') 7, 39, *39* (38), 73
　Kendall's K2 marine timekeeper *184*
　Kendall's K3 chronometer 189, *189* (188)
　Ships, Clocks and Stars. The Quest for Longitude exhibition (National Maritime Museum, 2014) 239
chronometric method 32, 72
Clark, Frederick, Reverend 202
Clerke, Charles
　Cook's third voyage 50, 53, 55, 60, 99, 187
　death 51, 53, 60, 170
　grave 176
　Royal Society Dining Club guest 79
Cleveley, James 137
Cleveley, John, the Younger 137, 160, 164
　Death of Cook (print) 204–5, 206
　Resolution and *Discovery at Morea*, style of 137, *140–41*
　Resolution and *Discovery in Tahiti*, style of 137, *142*
　The *Resolution* and *Discovery off Huaheine*, style of *58–9* (56)
　View of Huaheine (engraving) 159 (158), 160
Cleveley, Robert, *View of a hut in New South Wales* 146 (147)
coast mapping
　running surveys 72–3, 185
　see also charts and maps
coin, commemorating Cook's discovery of Hawaii 220
Cole, Benjamin, octant by *70*
Coleridge, Samuel Taylor, *Rime of the Ancient Mariner* 77
collecting of 'curiosities' 162–3
　see also material culture
Collett, William 164
Colnaghi (art dealers) 234
Colnett, James 169, 172
　A voyage to the South Atlantic and round Cape Horn into the Pacific Ocean title page 170
Colonial Government Museum (Sydney, Australia) 218
Columbus, Christopher 64
commemorative medals 49, 89, 90, 176, 203, *203* (202), 206, 219, *221*, *223*
commerce, and Pacific world 169–70
compasses
　azimuth compass *173*
　Captain Cook's compass (Thomas Graydon) 219, *228*, 230
Conrad, Joseph 9, 11
Constable, Archibald 212
Conyngham, William Burton 202
Cook, Elizabeth (Mrs Cook) 167, 202, 203, 212, 219, *220*, 229
Cook, Captain James
　biographical details
　　'British Columbus' label 9
　　'celebrity' status 11, 21, 23, 79, 201
　　commemoration service (Westminster Abbey, 1979) 234
　　Copley Medal 16, 50, 76, 201–2
　　humble origins 15
　　memorial obelisk (Easby Moor, North Yorkshire) 210
　　methods of command 47
　　possessions exhibited in museums 218–19, *220*
　　Royal Society Dining Club guest 79
　biographies 11, 203
　death 33, 53, 60, 77, 79, 162, 167, 169, 202–3
　　The Apotheosis of Captain Cook (P.J. de Loutherbourg after John Webber) 209–10, *209*
　　'The Apotheosis of Cook' (Pennant's *Outlines of the Globe*) 23, *23* (22)

　　in *Captain Cook, Explorer* exhibition 236–7
　　Death of Captain Cook (after John Webber) 79, 206, *206*, 207
　　The Death of Captain James Cook (after George Carter) 206, *208* (209), 209
　　The Death of Captain James Cook (Johann Zoffany) 206–7, *207* (206), 209, 219, 226, 229–30
　　Death of Cook (after John Cleveley, the Younger) 204–5, 206
　Naval career 25, 26, 28
　portraits and other art works
　　bust by Lucien Le Vieux *232*, 233
　　Captain Cook's Meeting with the Chukchi (John Webber) 102, *106* (107)
　　in *Commerce, or the Triumph of the Thames* (James Barry) 210
　　Cook riding with Omai in Tahiti (from John Rickman's journal) *54*
　　in *A human sacrifice, in a morai, in Otaheite* (John Webber) 105, *107*
　　Landing of Captain Cook at Middleburg, Friendly Islands (William Hodges) 89, *89*
　　portrait by John Webber (bought by Alan Bon) 235
　　portrait by John Webber (engraving after) *61*
　　portrait by Nathaniel Dance (1776) 7, *8* (9), 9, 13–15, 50, 63, 160, 212, 219, *219*, 226, 229, 237–8
　　portrait by William Hodges 211, *224* (225), 234–6, 239
　　Staffordshire earthenware figure 212, *213*
　　statue by Anthony Stones 237, *238*
　　statue by Thomas Woolner (Sydney's Hyde Park) 216, 218
　　statue by William Trethewey (New Zeeland) 218
　　A Striking Likeness of the Late Captain James Cook (engraving), after William Hodges *144* (145), 152
　　Wedgwood blue jasper portrait medallion 219
　publications and journals
　　Chart of the Island Otaheite 33 (32)
　　Endeavour, 1768–71 (journal) *14*
　　journals edited by Beaglehole 231
　　manuscript journals of first and second voyages 225
　　observations of Newfoundland eclipse 28
　　Voyage to the Pacific Ocean (third voyage, with Douglas and King) 19, 78, 79, 85, 149, 152, *152*, 170, 173, 189
　　A Voyage towards the South Pole, and round the World (second voyage, with Douglas) 19, 50, 52, 77, 149, 173, 189
　　see also Cook's afterlives; first voyage (*Endeavour*); second voyage (*Resolution* and *Adventure*); third voyage (*Resolution* and *Discovery*)
Cook and his Scientific Instruments exhibition (National Maritime Museum, 1997) 237
Cook and the 'Endeavour' exhibition (National Maritime Museum) 237
Cook on display *see* museums; National Maritime Museum
Cook's afterlives
　'celebrity' status and death 201
　in Cook's lifetime 201–2
　in death
　　commemorative medal 203, *203* (202), 206
　　crew's relic for Cook's widow 202–3
　　memorial obelisk 210
　　monuments in private estates 210–11
　　objects 210, 211, *211*
　　paintings of Cook's death 204–9, 206–7, 209–10
　　Royal Society Dining Club guest 79
　　writings about Cook 203
　post-death
　　commemorations and medals 219, *221*
　　Cook-related objects 212, *212–13*, 218–19
　　Cook's legacy as benevolent (Australia and New Zeeland) 216, 218
　　impact on other explorers/travellers 212

museums and objects 218–19
part of popular culture 214, 216
Copley Medal 16, 50, 76, 201–2
copperplate method 158
Cordingly, David 235, 236
costume books 158
Council for World Mission 7
Coxe, William, *Account of the Russian Discoveries between Asia and America* 156 (157)
Crocodile 229
Crosley, John 194
'curiosities' 162–3
   see also material culture

Da Fonte, Bartholomew 69
Dalrymple, Alexander 29–30, 39, 169, 187, 192
Dampier, William 11, 19, 69, 117, 119
   New Voyage Round the World 68–9, 147
Dance, George, the Younger, portrait of Captain Hugh Palliser 26
Dance, Nathaniel
   Captain James Cook 7, *8* (9), 9, 13–15, 50, 63, 160, 212, 219, *219*, 226, 229, 237–8
   Full-length portrait of a native, possibly Omai (engraving) 101, *101*
Defoe, Daniel 69, 71, 146–7
d'Entrecasteaux *see* Bruny d'Entrecasteaux, Antoine de
Deptford Royal Dockyard 170, *172*, 189
Des Barres, J. F. W., *Two Views of Cape Breton Island* 27
Dick, John, Sir 170
Dickens, Charles 77, 214, 216
Dillon, Peter 176
dingos, *Portrait of a Large Dog* (George Stubbs) 7, 38, 158, *159*, 160, 239, *239*
dip circle, by Edward Nairne 219, *228*, 230
Discovery (Cook's third voyage)
   refitting and command 50–51
   see also Resolution, HM ship; third voyage (*Resolution* and *Discovery*)
Discovery (Vancouver's ship) 187, 189
   sketch and modification plan *188*
Dixon, George
   American north-west coast expedition (with Nathaniel Portlock)
      accounts of voyage 171–2, 189
      commerce and sea-otter trade 169–70
      King George's Sound Company expedition 19, 21, 170–71
   comparison with John Meares 186–7
   'Dixon-Meares Controversy' 75
   A Voyage Round the World but More Particularly to the North-West Coast of America 169, 170
Dobbs, Arthur 11
Dodd, D. P. 206
Dodd, Robert, *The Mutineers turning Lieut Bligh and part of the Officers and Crew adrift from His Majesty's Ship the Bounty* 183
Doetechum, Joannes A., *South America and Magellanica* (post-Magellan map) 66–7
Dolphin 29, *29*, 30, 32, 39
   model of Dolphin-class frigate *30* (31)
Douglas, John
   on curiosities from Cook's voyages 13
   on material culture from the Pacific 17, 163
   Voyage to the Pacific Ocean (Cook's third-voyage, with Cook and King) 78, 79, 149, 152, *152*, 170, 173, 189
   A Voyage towards the South Pole, and round the World (Cook's second voyage, with Cook) 19, 50, *52*, 77, 149, 173, 189
Drake, Francis 68, 69, 210, 226, 237
du Fresne, Marc-Joseph Marion 50
Duff Cooper, Alfred and Diana 231
Dufour, John 158
Dughet, Gaspard 133
Duke of Edinburgh *see* Philip, Prince, Duke of Edinburgh
Dumont d'Urville, Jules 212
Duncan, Adam 226
Dunmore, John 173, 176
Dutch East India Company 68

Dutch expeditions 68, 69

Eagle, HM ship 25
Earl of Pembroke 12–13, 31, *31*
Easby Moor (North Yorkshire), Cook memorial obelisk 210
East India Company 29, 145, 170, 171, 187
   see also Dutch East India Company
Easter Island 47–8, 68, *117*
   Man of Easter Island (William Hodges) 95, *99*
   A View of the Monuments of Easter Island [Rapanui] (William Hodges) 47, 128, 135–6, *138–9* (136)
Edmond, Rod 77
Edwards, Captain Edward 183, 184
Elcano, Juan Sebastián de 64
Elizabeth, Princess (later Elizabeth II, Queen of the United Kingdom) *226*
Elliott, John 41, 47, 75
Ellis, William 78, 203
Ellison, Thomas 95
empire
   and commemoration of Cook 201
   and Enlightenment 145
   and science 21–3
Encyclopaedia Britannica 145
Endeavour, HM ship
   conversion from *Earl of Pembroke* 31
   conversion plans *12–13*, 31
   Cook and the 'Endeavour' exhibition (National Maritime Museum) 237
   ingot (1976) *221*
   model by C. Whitaker 231
   model by Robert A. Lightley *30* (31), *232*
   renamed *Lord Sandwich* and last years 38
   see also first voyage (*Endeavour*, HM ship)
Endeavour Reef 36
Endeavour River 36–7, *36–7*
English expeditions 68–9
Enlightenment 9, 15, 16, 17, 22, 63, 71, 73, 75, 77, 145, 160
   see also science and knowledge
Espérance (French ship) 192
Etches, Richard Cadmon 170, 172
Euclid 26
The European Magazine and London Review 167

Falconer, Thomas 163
Fanshawe, Admiral Edward Gennys, *Susan Young, The only surviving Tahitian woman, Pitcairn's* 185, *185*
Feest, Christian 15
Feltham, John, *Picture of London for 1806* 113
Field, Judge Barron 216
Fiji islands
   copy of chart of Bligh's Islands [Fiji] by William Bligh (Matthew Flinders) *186*
   reputation for cannibalism 185
first voyage (*Endeavour*, HM ship)
   250th anniversary of voyage (2018) 7
   Admiralty and Royal Society's Venus transit project 29, 39
   art and exploration
      Joseph Banks's employment of artists 32, 120–21
      Sydney Parkinson's contribution 120–24
   charts depicting course of voyage (Australian Hydrographic Office) 219, *222*
   conversion of *Earl of Pembroke* into *Endeavour* *30* (31), 31–2, *31*
   Cook and the 'Endeavour' exhibition (National Maritime Museum) 237
   Cook's appointment and instructions 29–31
   duration of voyage 32
   Joseph Banks's role and contribution 32, 33, 34, 37–8
   Journal of Captain Cook *14*
   journal of Sydney Parkinson *18*
   loss of men 32, 37
   New Holland (Australia) 35–6, *36–7*
   New Zealand and Maori 34–5
   paintings and publications 38

publication of accounts 148–9
reference to Anson's *Voyage* in account 147, 152
return to London and Cook's promotion 37
return to London and *Endeavour*'s last years 38
scurvy 32, 37
ship repairs 36–7
Society Islands 33, 34
'Southern Continent' conundrum 30, 34
Tahiti 32–3
Tupaia and Polynesians' skills 33–4
Van Diemen's Land (Tasmania) 35
Venus transit observations 33, 230
see also Endeavour, HM ship
Flaxman, John 219
Flinders, Matthew
   artist William Westall on Australian expedition 137
   circumnavigation of Australia
      charting of coast 185
      circumnavigation of Australia 192–6
      Investigator ship *193* (192), 194, 196
      Lady Nelson cutter 194–5, *196* (197)
      previous surveys of Australia 192
      prison and return home 196
   copy of chart of Bligh's Islands [Fiji] by William Bligh *186*
   importance of his voyages 19
   on the 'labours of Newton and Cook' 216
   portrait *192*
   on *Providence* voyage with Bligh 185
   Royal Society Dining Club guest 79
   on running surveys 72
   and scurvy 76–7
   second generation explorer after Cook 169
   on Tasman's charts 68
   Van Diemen's Land (Tasmania) voyage 51, 80
   A Voyage to Terra Australis 185, 194, 196
      Chart of Terra Australis 198
Flinders, Samuel 194
fly-whisks 105, 113
   Tongan *fue* 105, *107*
Folger, Captain Mayhew 184
Forster, George (or Georg)
   abilities and character 39
   on cascade in Tuauru Valley, Tahiti 134
   on collecting of 'curiosities' by ship crew 163
   Cook's debt to 80
   on Cook's illness on ship 47–8
   his account of Cook's second voyage 149
   on islanders' maritime prowess 113
   on Otago 88–9
   on Tongan culture 46–7
   on waterspout incident 129
Forster, Johann Reinhold
   abilities and character 39
   on Cascade Cove, New Zeeland 133
   A Chart representing the Isles of the South Sea, according to the notions. Chiefly collected from the accounts of Tupaya (J. R. Forster) *86* (87)
   on circumnavigations being the universal topics of companies 11
   and Cook's account of his second voyage 50
   Cook's debt to 80
   donations to University of Oxford 165
   on Easter Island 136
   John Elliott on 41
   'Physical Geography, Natural History and Ethnic Philosophy' (Cook's second voyage) 149
   on scurvy 48
   study of indigenous practices and rituals 105
   study of Pacific societies 47
   table of linguistic differences *35* (34)
   translation of John Rickman's account 78
   on tropical sun 126
   on Tupaia's chart 75
   on Tupaia's navigation skills 73
Fox, Emmanuel Phillips, *The Landing of Captain Cook at Botany Bay* 218
Franklin, John 237
Free Society of Artists *see* Society of Artists (later Free Society of Artists)

INDEX   251

Freycinet, Louis-Claude de 197
Friendly Islands *see* Tongan islands (Friendly Islands)
Fuller, A. W. F. 234
Furneaux, Tobias 39, 41, 46, 47, 79, 160, 163
fur-trading 21, 60, 75, 173, 176
    see also sea-otter trade

Gainsborough, Thomas, portrait of 4th Earl of Sandwich 228
Garner, Will, backstaff *70*
Garrick, David 148
Géographe (French ship) 194
George III, King of the United Kingdom 16, 29, 46, 51, 160
George IV, King of the United Kingdom 228
George VI, King of the United Kingdom *14*, 225, *226*
gift-giving 90, *90*, 91 (90)
Gilfillan, John Alexander 216
Gillray, James
    The Caneing in Conduit Street 191, *191* (190)
    Very Slippy Weather 157
Glorious First of June, Battle of (1794) 192
globes *34*, 145–6, *162*, 211, *212* (213)
Good, Peter 196
Goodrich Court, Herefordshire, South Sea Room 166
Gopi Mohen Baboo *see* Tagore, Gopi Mohen Baboo
Gore, John 51, 60, 171
Gore, John (boy) 171
Grasset de Saint Sauveur, Jacques, *Tableau des Decouvertes du Capne Cook, & de la Perouse* (engraving) *174–5*
Graydon, Thomas, Captain Cook's compass 219, *228*, 230
Great Northern Expedition (or Second Kamchatka) 148
Green, Charles 33, 37, 72
Greenhill, Basil 234
Greenwich Hospital 9, 50, 171
    Naval Gallery 11, 15, *226*, 225, *227*, 228–30
    see also Royal Naval College
Griffin, William 106, *108*

Hack, William, Spanish chart of the Straits of Magellan and Le Maire *62* (63), 64, *68*, 69
Hadley, John, first double-reflecting octant *71*
Hakluyt Society 231
Hall, Basil 212
Hall, John, *Surrender of the Island of Otaheite to Captain Wallis* 28
Halley, Edmund 29
Hamar, Captain Joseph 25
Harrison, John 80
    exhibition of his four chronometers 233–4
    H4 marine timekeeper 73, *73*
    Kendall's copy of his 'H4' chronometer 7, 39, *39* (38), 73
Hastings, Warren 117, 125, 143, 229
Hau'ofa, Epeli 85
Hawai'i 7, 55, 80, *200*, 202, 219, *220*, 223
Hawkesworth, John
    An Account of the Voyages undertaken
        best-seller commissioned by Admiralty 38, 148–9, 186
        Cook's fury at sensationalist account 50, 149
        Hawkesworth on importance of 'views and figures' 117
        Pitcairn Island's incorrect position 184
        title page *148*
        used by Bligh and Christian 77
        used by La Pérouse 173
    charts and illustrations in *An Account*
        The attack of Captain Wallis (Edward Rooker) *29* (28)
        Chart of the Island Otaheite (James Cook) *33* (32)
        kangaroo image 160
        map *149* (148)
        Surrender of the Island of Otaheite to Captain Wallis (John Hall) 28
    on Cook 13
Hawkins, John 226

Heber, Reginald 141
Heddington, Captain Thomas, *Karakakoa Bay, Owhyee (Hawai'i)* (engraving) *200*, 202
Heritage Lottery Fund (HLF) 7, 239
Heywood, Peter 95
Hibernian Antiquarian Society 202
Hicks, Lieutenant Zachary 37
Hinchingbrooke, Alexander Victor Montagu, Viscount 231
Hitihiti (O-Hedidee) 46, 48, 80
    O-Hedidee, native from the Pacific (engraving, after William Hodges) *47* (46)
Hodges, William
    artist on Cook's second voyage 39, 50, 75, 85, 117, 120, 125–36, 153
    Captain Cook and Mr Hodges exhibition (National Maritime Museum, 1978) 234, *235*, 238, *240* (239)
    collection of his works at National Maritime Museum 7, 15, 219, 230–31
    copperplate reproductions of his paintings 158
    exhibited in influential art exhibitions 19, 145
    paintings of and Admiralty 230–31, 234
    prints in Cook's *Voyage towards the South Pole* 149
    Reynolds on 143
    travel to India 125, 143
    William Hodges 1744-1797: The Art of Exploration (National Maritime Museum, 2004) 239
    works
        Adventure at anchor in Table Bay 41
        Afia-too-ca, a Burying Place in the Isle of Amsterdam (engraving) 102, *104* (105)
        Boats of the Friendly Isles (engraving) *111* (110)
        Captain James Cook 211, *224* (225), 234–6, 239
        A Cascade in the Tuauru Valley, Tahiti 116 (117), 134, *135* (134)
        A Draught, Plan and Section of the Britannia Otahite War Canoe (engraving) 110
        A Draught Plan & Section of an Amsterdam Canoe, seen in the South Seas 110
        Dusky Bay [Cascade Cove] 41, *132–3*, 133
        Easter Island (after Hodges) 117
        Family in Dusky Bay, New Zealand (engraving) *95*, 95
        The Landing at Erramanga 48, *49*
        The Landing at Erramanga (engraving) *151* (150)
        The Landing at Mallicolo 20 (21), 234
        The Landing at Mallicolo (engraving) *151* (150)
        The Landing at Tanna 20 (21), 234
        Landing of Captain Cook at Middleburg, Friendly Islands 89, *89*
        Man of Easter Island 95, *99*
        Native head-dress, necklace and other items (after William Hodges) *114* (115)
        O-Hedidee, native from the Pacific (engraving) *47* (46)
        Omai, native of the South Pacific (engraving) *161* (160)
        Otago, native from the Pacific (engraving) *88*, 88 (89), 89
        The Resolution in the Marquesas 46, 110
        Review of the War Galleys at Tahiti 48, *48* (49)
        A Striking Likeness of the Late Captain James Cook (engraving), after William Hodges *144* (145), 152
        Tahiti, Bearing South-East *92–3*, 113
        Tahiti Revisited 135–6, *136*
        A Toupapow with a Corpse on it, Attended by the Chief Mourner in his Habit of Ceremony (engraving) 102, *104* (105)
        Various articles belonging to South Pacific natives, incl. spears, bow and arrow (after William Hodges) 115
        Various articles belonging to South Pacific natives, incl. weapons with fancy carving (after William Hodges) 165
        View in Dusky Bay with a Maori Canoe *128–9*, 128
        View in Pickersgill Harbour and underpainting of icebergs 41, *42–3*

        View in the Island of New Caledonia (engraving) 118, *150* (151)
        View in the Island of Tanna [Tana] (engraving) 118
        A View of Cape Stephens in Cook's Straits with Waterspout 129, *130–31*, 133
        A view of Maitavie Bay, in the island of Otaheite *44–5* (42)
        View of part of Owharre [Fare] *Harbour, Island of Huahine* 124 (125), 126
        A View of Point Venus and Matavai Bay, looking east 76
        View of Resolution [Vaitahu] *Bay in the Marquesas* 125, *126*
        A View of the Cape of Good Hope 126, *127*
        A View of the Island of New Caledonia 153, *153* (152)
        View of the Islands of Otaha [Taaha] *and Bola Bola* [Bora Bora] *with Part of the Island of Ulietea* [Raiatea] 77 (76)
        A View of the Monuments of Easter Island [Rapanui] 47, 128, 135–6, *138–9* (136)
        View of the Province of Oparree [Pare]*, Island of Otaheite, with part of the Island of Eimeo* [Moorea] 127–8, *127* (126)
        The War-Boats of the Island of Otaheite and the Society Isles 48, *84* (85), 110, *112* (113), 113, 125, 234
        Waterfall in Dusky Bay with a Maori Canoe 128–9, *129* (128)
        A Waterfall in Tahiti 134, *134*
        Woman of New Caledonia 95, *99*
Hogg, Alexander 164
Holland, Samuel 26, 28
Holophusicon 13, 166–7
Home, Captain Sir Everard 229
Hood, Alexander 41, 75
Howe, Richard 226
Howse, Derek *236*
Hudson's Bay Company 170
human sacrifice 105, *107*
Humphrey, George 163
Hunter, John 192
Hunter, William 165–6
Hydrographic Office (Admiralty, Royal Navy) 197–8

indigenous people *see* Pacific Islanders; Polynesians
instrument-makers
    Cook on 72
    see also scientific instruments
Investigator 194, 196
    plans *193* (192)

James II, King of England, Scotland and Ireland 69
Johnson, Dr Samuel 25

Kahura 53
Kalani'op'u 60
kangaroos
    The Kongouro from New Holland (George Stubbs) 7, *22*, 38, 158, 160, 239, *239*
    Sydney Parkinson's pencil sketches 160
Kendall, Larcum
    K1 (copy of Harrison's 'H4' chronometer) 7, 39, *39* (38), 73
    K2 marine timekeeper 184–5, *184*
    K3 chronometer 189, *189* (188)
Kerguelen Land 51
    A view of Christmas Harbour, in Kerguelen's Land (engraving, after John Webber) *50* (51)
Kerguelen-Trémarec, Yves-Joseph de 50
King, Captain James
    donations to Holophusicon 167
    donations to Trinity College, Dublin 166
    on Joseph Banks 170
    on rage for sea-otter pelts 170
    roles on Cook's third voyage 51, 60
    Royal Society Dining Club guest 79
    talking to Fanny Burney about objects from Cook's voyage 163

Voyage to the Pacific Ocean (Cook's third-voyage, with Cook and Douglas) 19, 78, 79, 85, 149, 152, *152*, 170, 178, 189
'we discoverers' quote 75
King, Phillip Parker 196
King George 170
King George's Sound Company 21, 170
Kippis, Andrew 11, 163, 203
   A narrative of the voyages round the world: performed by Captain James Cook title page *214* (215)
Knatchbull, Edward, 9th Baronet 15
Knight, Roger 169
knowledge
   knowledge-harvesting and dissemination 145
   knowledge-transfer and Enlightenment 73
   see also science and knowledge
Krusenstern, Adam Johann von 212
Kyd, Alexander 229
Kyd, Robert 229

La Boudeuse (French ship) 30–31
La Farge, John 212
La Pérouse, Jean-François de Galaup de
   1785-88 voyage 172–3, 176, *177*, 178–9 (177), 192
   on Cook's achievements 11
   medal of Lapérouse's departure 176
   Monument to Monr de la Perouse and his companions, erected at Botany Bay (engraving) *177*
   Naufrage de l'Astrolabe sur les recifs de l'ile de Vanikoro (Louis Le Breton) 178–9 (177)
   Tableau des Decouvertes du Capne Cook, & de la Perouse (engraving) 174–5
la Roche, Sophie von 164
Lady Nelson 194–5
   model *196* (197)
Langle, Fleuriot de 176
languages, table of linguistic differences *35* (34)
Le Breton, Louis, *Naufrage de l'Astrolabe sur les recifs de l'ile de Vanikoro* 178–9 (177)
Le Maire, Isaac 68
Le Vieux, Lucien, bust of James Cook *232*, 233
Ledyard, John 21, 78, 203
Lee, James 121
Leger Galleries 235
Leggatt, Hugh 235
Lesseps, Ferdinand de 176
Lever, Ashton, Sir 13, 166–7
Lewin, Admiral of the Fleet Terence 235, 236
Lightley, Robert A., *Endeavour* model *30* (31), *232*
Lincoln, Margarette 237
Locker, Edward Hawke 15, 228, 229
London Missionary Society 7, 239
London Packet 153
longitude at sea
   Board of Longitude 71, 173, 197, 230, 239
   chronometric method 32, 72
   lunar method 32, 71, 72
   Ships, Clocks and Stars: The Quest for Longitude exhibition (National Maritime Museum, 2014) 239
Lord Sandwich 38
Louis XVI, King of France 173
Louisbourg, fall of (1758) 26
Loutherbourg, Philippe-Jacques de
   The Apotheosis of Captain Cook (after John Webber) 209–10, *209*
   Omai, or, A Trip Round the World scenery and costumes 137
lunar method 32, 71, 72

Mackay, David 170, 171
Mackenzie, Alexander 146
MacLean, Alistair 60
McLynn, Frank 60
MacPherson, John 117
Madagascar 23
Magellan, Ferdinand 64–5, 75, 203
   Americae Retectio: Ferdinand Magellan (engraving after Johannes Stradanus) *63*, 64

map of Magellan's voyage showing the global wind directions (Battista Agnese) *64*
Maggs of London 231
Mahine 94
Ma'i *see* Omai (Ma'i)
Malaspina, Alessandro 148, 172, 189
Malcolm, James 164–5
Maori 34–5, 41, 47, 53, 90, 94, 219, 236
maps *see* charts and maps
Maquinna 223
mar del sur ('South Sea') 85
marine timekeepers *see* chronometers (marine timekeepers)
Mariner's Mirror 234
maritime technology 87–8, 110, *110–12*, 113
   see also scientific instruments
Marquesas Islands 48
   The Resolution in the Marquesas (William Hodges) *46*, 110
   View of Resolution [Vaitahu] *Bay in the Marquesas* (William Hodges) *125*, 126
Marquis of Granby 38, 39
Marquis of Rockingham 39
Marshall, P. J. 22
Mary, Queen, the Queen Mother 226
Maskelyne, Nevil 71, 72
material culture
   discovery 105–6, 113
   examples *107*, *108*, *109* (108), *114* (115), *115*, *164* (165), *165*
   exhibition 162–7
   exhibition in *Captain Cook and the Exploration of the Pacific* gallery (National Maritime Museum) 234
   London Missionary Society's ethnographic material 239
Mayer, Thomas 71
Meares, John
   'Dixon-Meares Controversy' 75
   Nootka Crisis 187
   Voyages Made in the Years 1788 and 1789 189
      'A Chart of the Interior Part of North America' 186–7, *187*
medals *see* commemorative medals
Melanesia 48, 85
Melville, Robert Dundas, 2nd Viscount Melville 212
memorial obelisk (Easby Moor, North Yorkshire) 210
'men of Captain Cook' 21, 75, 169
   see also Pacific voyages (1785-1803)
Mendaña, Alvaro de 48, 65
Mendes da Costa, Emmanuel 22
Menzies, Archibald 75, 172, 189, 191
Mexicana (Spanish vessel) 189
Michell, Samuel Rush 198
Micronesia 85
Miller, John, pocket globes *160*, *163*
Millwall, John 95
Monneron, Paul-Antoine de 173
monuments to Cook 210–11
Moorea (Eimeo) 48, 53, 55
Morgan, Kenneth 196
Morrison, William 184
Mortimer, George 157–8
Mosley, Charles, *The Sailor's Return 24* (25)
Mouat, Captain Patrick 55
Mulgrave, Lord *see* Phipps, Constantine, 2nd Baron Mulgrave
Murray, Lieutenant John 195
Museum Academicum (Göttingen) 163
museums
   Captain Cook Birthplace Museum (Marton, Middlesbrough) 237
   Captain Cook Memorial Museum (Whitby) 237
   Cook's cameo appearance 218
   National Maritime Museum's contextual approach 218–19, 238
   see also National Maritime Museum
Mutiny on the Bounty exhibition (National Maritime Museum, 1989) 236

Nairne, Edward, dip circle by 219, *228*, 230

Nairne and Black 173
Narborough, John 68
National Heritage Memorial Fund 235
national identity
   and Cook's 'celebrity' status 21, 23
   and Naval Gallery at Greenwich 15
National Maritime Museum (NMM)
   1776: The British Story of the American Revolution (*The Times* and *Sunday Times*, 1975) 234
   Caird Fund 234
   Caird Galleries 225–6
      Caird Gallery VI 219, *227*, 231, 233, *233*
   Caird Library and Archive 7
   Catalogue of the National Maritime Museum 226
   contextual approach to Cook exhibits 218–19, 238
   Cook bronze statue by Anthony Stones *237*, 238
   Cook collections
      Cook relics 231
      Dance's portrait of Cook 226, 229
      Hodges's paintings 7, 15, 219, 230–31
      Hodges's portrait of Cook *224* (225), 234–6, 239
      manuscript journals of first and second voyages 225
      Naval Gallery / Greenwich Hospital Collection 228–30
      Royal Naval Museum navigational instruments 230
      Zoffany's *Death of Captain Cook* 226, 229–30
   Cook galleries and exhibitions (by name)
      The Art and Science of Exploration exhibition (2014) 239, *239*
      Blood, Sea and Ice exhibition (1996-99) 237
      Captain Cook and Mr Hodges exhibition (1978) 234, *235*, 238, *240* (239)
      Captain Cook and the Exploration of the Pacific gallery (1970-73) *232*, 233–4, *233*
      Captain Cook exhibition (1956-57) 231, 233
      Captain Cook, Explorer exhibition 236–7
      Captain Cook, Navigator (touring exhibition) 236
      Cook and his Scientific Instruments exhibition 237
      Cook and the 'Endeavour' exhibition 237
      Oceans of Discovery gallery, section of 239
      Pacific Encounters gallery 7
      'Science and Exploration in the Pacific' conference 239
      Ships, Clocks and Stars: The Quest for Longitude exhibition (2014) 239
      'Trade and Empire' gallery, section of 238
      William Hodges 1744–1797: The Art of Exploration (2004) 239
   Cook galleries and exhibitions (overview)
      beginnings (1947-57) 225–31
      building Cook (1957-2000) 233–6
      wider horizons 236–9
   Mutiny on the Bounty exhibition (1989) 236
   Neptune Court development 237–8
   opening of museum 225, *226*
   scholarly research 234
   'Science and Exploration in the Pacific' conference 237
   A Sense of Direction document 233
   see also museums
National Portrait Gallery 233
Naturaliste (French ship) 194
Nautical Almanac 71, 72
Naval Gallery, Greenwich Hospital *see* Greenwich Hospital
navigators' journals 69
neck-rest, Tongan wooden neck-rest 106–7, *108*
Nelson, Horatio 226
New Caledonia (today Noumea) 48
   View in the Island of New Caledonia (engraving, after William Hodges) *118* (150) (151)
   A View of the Island of New Caledonia (William Hodges) 153, *153* (152)
   Woman of New Caledonia (William Hodges) 95, *99*
New Hebrides (today Vanuatu) 48

INDEX 253

New Holland (Australia) *see* Australia (New Holland)
New Zealand 7, 34–5, 41, *42–3*, 43, 48, 51, 53, 69, 72, 73, *74* (75), 80, 124, *132–3*, 133, 169, 216, 218, 219, 223, 237, *238*
   see also Parkinson, Sydney
Newfoundland 28
'noble savage' concept 101, 160, 202
Nootka Crisis/Convention 187, 189
North Island 35, 137–41
Northumberland 26
North-West Passage 11, 55, 60, 69, 71, 79–80, 173, 176, 187, 189, 191
   see also America, north-west coast
Noumea *see* New Caledonia (today Noumea)
Nuu Chuh Chah people 55

Obeyesekere, Gananath 60
Oceania 85
octants
   by Benjamin Cole *70*
   first double-reflecting octant (John Hadley) 71
O-Hedidee *see* Hitihiti (O-Hedidee)
Omai (Ma'i) 15, 46, 50, 51, 53, 79, 94, 160
   portraits
      Full-length portrait of a native, possibly Omai (engraving, after Nathaniel Dance) *101*, 101
      Omai, native of the South Pacific (engraving, after William Hodges) *161* (160)
      Reynolds's painting of 50, 101, 160
      riding with Cook in Tahiti (from John Rickman's journal) *54*
      in William Parry's group portrait 50, 160
Omai, or, A Trip Round the World (pantomime) 137
Orient Line poster *217*
Orio 55
Ormond, Richard 236, 237, *238*
Otago (or Attahha) 88–9, *88* (89)
Otaheite *see* Tahiti (Otaheite)

Pacific *see* Pacific Islanders; Pacific Ocean; Pacific voyages (1785–1803); Pacific world
Pacific Islanders 11, 15, *18*, 19, 73, 80, 89–90, *90*, *91* (90), 94–5, *96–7* (95), *112*, 145, 160
   see also Pacific world; Polynesians
Pacific Ocean, chart (c.1744) *16–17*
Pacific voyages (1785–1803)
   creation of Admiralty's Hydrographic Office 197–8
   gaps in knowledge and post-Cook explorations 169
   see also Bligh, William; Dixon, George; Flinders, Matthew; La Pérouse, Jean-François de Galaup de; Portlock, Nathaniel; Vancouver, George
Pacific world
   introduction 85, 87–8
   personal encounters 88–90, *90*, *91* (90), 94, *112*
   portraits and other paintings 94–5, *95*, *96–7* (95), *98* (99), 99, *99*, *100* (101), 101, *101*
   practices and customs 101–2, *102–4*, 105–7, *105*, *106*, *107*, *108*, 109 (108), 110, *110–12*, 113, *114–15*
   see also Pacific Islanders; Polynesians
Palliser, Admiral Sir Hugh 25, 28, 210–11
   portrait by George Dance, the Younger *26*
Palmerston, Henry Temple, 2nd Viscount Palmerston 125
Pandora 183
pap-boat, belonging to the Cooks 219, *220*
Paramour 29
Parkinson, Stanfield 122
Parkinson, Sydney
   Cook's first voyage 120–24
   death 37
   in Hawkesworth's *An Account of the Voyages* 148
   Journal of a Voyage to the South Seas 122
      title page and frontispiece *120* (121)
   journal of the artist on the *Endeavour 18*
   kangaroo pencil sketches 160
   on Tahitian people 88
   on Tahitian weapons 106
   and Tupaia 87
   works
      A branch of the bread-fruit tree with fruit (engraving) *181* (180)
      The Head of a Chief of New-Zealand, 97 (95)
      An Heiva, or kind of Priest of Yoolee-Etea, & the Neighbouring Islands 101–2, *103* (102)
      A Morai, or Burial Place, in the Island of Yoolee-Etea 101, *102*
      A Native of Otaheite, in the Dress of his Country 87, 88
      Otegoongoon, Son of a New Zealand Chief 90, 94, *96* (95)
      Venus Fort, Erected by the Endeavour's People 32, *121*, 122
      View of a curious Arched Rock, having a River running under it, in Tolago Bay, on the East Coast of New Zealand 123, 124
      View of the great Peak, & the adjacent Country, on the West Coast of New Zealand 123, 124
      View of the North Side of the Entrance into Poverty Bay, & Morai Island, in New Zealand 122–3, *122*
      A war canoe of New Zealand 111 (110)
Parry, William 50, 160
Pasley, Admiral Thomas 192
Patagonian giants 64
Patten, James 166
Payne, Anthony 77
Peace of Amiens (1802) 194
Pearson, George 202
Pembroke 26
Pennant, Thomas
   Outlines of the Globe
      accounts of an 'armchair traveller' 147, 152
      'The Apotheosis of Cook' 23, *23* (22)
      pages from *146* (147), *156* (157), *166*
      Webber's prints 157, *166*
   prints from Cook's third voyage 158
   Thomas Falconer's friend 163
Perry, William 75
Philip, Prince, Duke of Edinburgh 238
Phillip, Arthur 176, 216
Phillips, Molesworth 60
Philosophical Society of Australasia 216
Phipps, Constantine, 2nd Baron Mulgrave 146, 170, 184
Pickersgill, Lt Richard 41, 43, 46–7, 48
   Narrative account of the voyage of HM bark Resolution (second voyage) *40* (41)
Picture of London for 1806 (John Feltham) 113
Picturesque 124, 141
Pingo, Lewis, medal commemorating Captain James Cook *203* (202), 206
Pitcairn Island 65, 77, 180, 184–5
   Susan Young, The only surviving Tahitian woman, Pitcairn's (E. G. Fanshawe) 185, *185*
Pitt, John, 2nd Earl of Chatham 185, 189
Pitt, Thomas, 2nd Baron Camelford 191
   The Caneing in Conduit Street (James Gillray) *191* (190)
plane tables 26
Poedua (Poetua) 55, 90, 99, *100* (101), 101, 137, 153, 160
Polynesia 85
Polynesians
   Ahu-toru 31
   aiding travels of European ships 15
   boat-building and maritime prowess 33–4, 73, 87
   material culture 113
   relations with Europeans 33
   tattooing 94–5
   see also Pacific Islanders; Pacific world
Portlock, Nathaniel
   American north-west coast expedition (with George Dixon)
      accounts of voyage 171–2, 189
      commerce and sea-otter trade 169–70
      King George's Sound Company expedition 19, 21, 170–1
   Assistant voyage with William Bligh 185
   comparison with John Meares 186–7
   portrait *168* (169), 230
   and scurvy 76–7
   veteran of Cook's voyages 19, 21, 169

A Voyage Round the World but More Particularly to the North-West Coast of America 170–72, *229* (228)
Posnett, David 235
priests and burial places 101–2, *102–4*
Prince of Wales 172
Princess Royal 172
Pringle, John, 1st Baronet 16, 202
prints (Cook's second and third voyages) 157–8, *158*
Providence (Bligh's old ship) 189, 192
Providence (Bligh's voyage) 19, 185–6, *186*, 230
Public Record Office 231

Quadra, Juan Francisco de la Bodega y 55, 189
Queen Charlotte 170
Quilley, Geoff 15
   Art for the Nation 234
Quiros, Pedro Fernandes de 65

Rae, W. Fraser 201
Raleigh, Walter 210
Ramsden, Jesse 80, 173
   sextant *173*, *236*
Ramsey, John 41
Recherche (French ship) 192
red sandstone (northern Queensland) 211, *211*
Reliance, HM ship 192
Resolution
   conversion from *Marquis of Granby* 39
   conversion plan from *Marquis of Granby 38*, 39
   A Party from His Majesty's ships Resolution & Discovery shooting sea-horses [walruses] (John Webber) 137, 153, *154–5*
   Resolution and Discovery at Morea, style of John Cleveley, the Younger 137, *140–41*
   Resolution and Discovery in Ship Cove, Nootka Sound (engraving after John Webber) 2, 55, *55* (54)
   Resolution and Discovery in Tahiti, style of John Cleveley the Younger 137, *142*
   The Resolution and Discovery off Huaheine, style of John Cleveley, the Younger *58–9* (56)
   The Resolution beating through the Ice, with the Discovery in the most imminent Danger in the distance (engraving after John Webber) *56–7*
   The Resolution in the Marquesas (William Hodges) 46, 110
   see also second voyage (*Resolution* and *Adventure*); third voyage (*Resolution* and *Discovery*)
Reynolds, Joshua, Sir 228
   Omai's painting 50, 101, 160
   portrait of John Byron *230* (231)
   on William Hodges 143
Richardson, Brian 14
Richardson, Dorothy 160
Rickman, John *54*, 78, 79, 203
Riou, Edward 79, 169
Roberts, Henry 75, 169, 187, 189
   A General Chart Exhibiting the Discoveries made by Captn James Cook 79–80, *82–3* (80)
Robertson, George 30
Rodger, N. A. M. 25
Rodney, George Brydges 21, 226
Roebuck, HM ship 11
Rogers, Woodes 69, 147
Roggeveen, Jacob 68
Rooker, Edward, *The attack of Captain Wallis 29* (28)
Rosario (Spanish ship) 69
Rose, George 170
Rosselli, Francesco, pre-Magellan world map *65* (64)
Royal Academy
   1774, 1775, 1777, 1784 exhibitions 153
   1776 exhibition 101, 137, 160
   1778, 1788 exhibitions 143
   1780 exhibition 210
   paintings of Cook's death 209
Royal Naval College 225, 229, 230
   see also Greenwich Hospital
Royal Naval exhibition (1891), 'The Cook Gallery – Navigation' gallery 218
Royal Naval Museum 226, 230, 231

254   CAPTAIN COOK AND THE PACIFIC

Royal Navy
  Admiralty 21, 22, 29, 37, 39, 78, 152, 186, 193–4, 195, 230–31, 234
  Cook's enlistment and career 25–8
  Hydrographic Office 197–8
  master's mate rank 25
  number of sailors 25
  popular reputation of 25
  types of ships 25
Royal Numismatic Society of New Zealand, medals 219, 223
Royal Observatory, Greenwich 32, 71, 237
Royal Society 16, 21, 22, 28, 29, 32, 39, 50, 76, 201–2, 203, *203* (202), 206
Royal Society Dining Club 79
Royal Society for the Encouragement of Arts, Manufactures and Commerce 210
Royal United Services Institution (later Institute) 218, 231
running surveys 72–3, 185
Russell, John, Captain Bligh portrait (engraving) *180*
Russia 203, 206, 212

Sahlins, Marshall 11, 60
  Islands of History 237
Salesa, Damon 85
Salisbury, Robert Gascoyne-Cecil, 6th Marquess of Salisbury 236
Samwell, David 13, 60, 163, 203
Sandby, Paul 137
Sandwich, John Montagu, 4th Earl of Sandwich 13, 39, 51, 113, 148, 165, 167, 228
Sandwich Islands
  A man of the Sandwich Islands with his helmet (John Webber) *98* (99), 99
  South Sandwich Islands 48
  see also Hawai'i
Sauvages de la Mer Pacifique (wallpaper) 158
Schank, Lieutenant John 194
Schouten, Willem 68
science
  and empire 21–3
  Enlightenment science 15, 63, 71
  see also science and knowledge
'Science and Exploration in the Pacific' conference (National Maritime Museum) 237, 239
science and knowledge
  chart in Dance's portrait of Cook 63
  early expeditions (16th-17th centuries) 64, 65, 68, 69
  state of knowledge by 1700s 69
  state of knowledge from mid 18th century 70, 71–3, 73–80, 82–3 (80)
  see also charts and maps
'scientific gentlemen' 32, 49, 80
scientific instruments
  Cook's instruments 71–2, 73
  La Pérouse's instruments 173
  see also backstaff (Will Garner); barometers; chronometers (marine timekeepers); compasses; *Cook and his Scientific Instruments* exhibition (National Maritime Museum, 1997); dip circle (Edward Nairne); *Mariner's Mirror*; maritime technology; octants; sextants; telescopes
scurvy 26, 32, 37, 39, 41, 48, 50, 64, 75–7, 189
Sea of Japan 169, 173, 176, 189
sea-otter trade 55, 169–70, 171, 191
  A Sea Otter (engraving, after John Webber) *171* (170)
Second Kamchatka (or Great Northern Expedition) 148
second voyage (*Resolution* and *Adventure*)
  accounts of voyage
    Cook's account (*A voyage towards the South Pole*, with Douglas) 19, 50, *52*, 77, 149, 173, 189
    Forsters' accounts 149
  aim to search for Southern Continent 39, 41, 43, 47, 48, 126
  appointment of J. R. Forster 39
  contribution of William Hodges 7, 15, 39, 125–36
  commemorative medal *49*, *89*, 90, 219, *221*

Cook's discoveries and indigenous people's contributions 73, 75
Cook's testing of K1 chronometer 7, *39* (38)
crew, scientists and artists 39, 41
crossing of the Antarctic Circle 41
decision to use two ships 39
Easter Island 47–8
experimental voyage 32
Marquesas Islands *46*, 48
New Caledonia (today Noumea) 48
New Hebrides (today Vanuatu) 48
New Zealand harbours and bays 41, *42–3*, 43, 48
return home of *Adventure* 47, 48
return home of *Resolution* 47–9
scurvy 41, 48
Society Islands 43, 46, 48
South Georgia 48
South Sandwich Islands 48
Tahiti 43, 46, 48
Tongan islands (Friendly Islands) 46–7
Van Diemen's Land (Tasmania) 43
visual images
  art exhibitions 153
  prints 158
A Sense of Direction (National Maritime Museum) 233
Seven Years War (1756-63) 25, 28, 30, 71, 209
Seville Expo (1992), 'Hall of Discovery' (New Zealand pavilion) 237, *238*
Seward, Anna, 'Elegy to Captain Cook' 210
sextants 71, 72
  by Jesse Ramsden *173*, *236*
Sharpe, Bartholomew 69
Sheffield, William, Reverend 163
Shelvocke, George 69, 147
ships
  Royal Navy types of ships 25
  see also maritime technology
Ships, Clocks and Stars: The Quest for Longitude exhibition (National Maritime Museum, 2014) 239
Simcoe, Captain John 26
Skelton, Raleigh 231
Smith, Alexander *see* Adams, John (also known as Alexander Smith)
Smith, Bernard 60, 77, 233, 234
The Snares 189
Society for Nautical Research 234
Society Islands
  Cook's first voyage 33, 34
  Cook's second voyage 41, 43, 48
  Cook's third voyage 53, *54*, 55
  The War-Boats of the Island of Otaheite and the Society Isles (William Hodges) 48, *84* (85), 110, 112 (113), 113, 125, 234
Society of Artists (anti-Free Society of Artists)
  1765-66 exhibition 120
  1773 exhibition 158
  1774 exhibition 127, 153
Solander, Daniel 22, 79, 137, 160, 164
Solebay 25
Solomon, Isles of 65
Solomon, King of Israel, fabled gold mines of 65
Sotheby 235
South Georgia 48
South Island 35
South Sandwich Islands 48
South Sea Bubble 71
South Sea Company 170
'South Sea' (*mar del sur*) 85
'Southern Continent' conundrum
  and 18th-century science 71
  and Cook's first voyage 30, 34
  and Cook's second voyage 39, 41, 43, 47, 48, 126
  and Dalrymple 187
  and Dutch expeditions 68, 69
  proven as myth by Cook and successors 11, 17, 79, 169
  and Spanish expeditions 65
Southern Whale Fishery 187
Spanish expeditions 64–5
Sparrman, Anders 39, 165

Spöring, Herman 37
Staffordshire earthenware figure of Cook 212, *213*
Staines, Captain Thomas 184
The Standard, 'Cook blow for Maritime Museum' 235
Stanford, Edward, terrestrial floor globe 212, *212* (213)
State Library of New South Wales 236
Stephens, Philip 163
Stimson, Alan *236*
Stones, Anthony, Cook bronze statue *237*, *238*
Stradanus, Johannes, *Americae Retectio: Ferdinand Magellan* (engraving) *63*, 64
Strahan and Cadell (publishers) 148
Strait of Le Maire, Spanish chart of the Straits of Magellan and Le Maire (William Hack) *62* (63), 68, 68, 69
Strange, James 21, 152
Stubbs, George
  The Kongouro from New Holland 7, *22*, 38, 158, 160, *239*, *239*
  Portrait of a Large Dog [Dingo] 7, 38, 158, *159*, 160, *239*, *239*
Sullivan, Daniel Giles 218
Sunday Times, *1776: The British Story of the American Revolution* (*The Times* and *Sunday Times*, National Maritime Museum, 1975) 234
Sutil (Spanish vessel) 189
Swallow 184
Swift, Jonathan 69
Sydney Morning Herald 216

Tagore, Gopi Mohen Baboo 209
Tahiti (Otaheite)
  cascade in Tuauru Valley 134
    A Cascade in the Tuauru Valley, Tahiti (William Hodges) *116* (117), 134, *135* (134)
  Cook's chart *33* (32)
  Cook's first voyage 32–3
  Cook's second voyage 43, 46, 48
  A human sacrifice, in a morai, in Otaheite (engraving, after John Webber) *105*, *107*
  A Native of Otaheite, in the Dress of his Country (Sydney Parkinson) *87*, 88
  Review of the War Galleys at Tahiti (William Hodges) 48, *48* (49)
  Susan Young, The only surviving Tahitian woman, Pitcairn's (E. G. Fanshawe) *185*, 185
  Tahiti, Bearing South-East (William Hodges) *92–3*, 113
  Tahiti Revisited (William Hodges) 135–6, *136*
  Tahitian breadfruit-pounder 105–6, *108*
  A view of Maitavie Bay, in the island of Otaheite (William Hodges) *44–5* (42)
  Wallis's voyage 28, 29, 32, 43
  The War-Boats of the Island of Otaheite and the Society Isles (William Hodges) 48, *84* (85), 110, 112 (113), 113, 125, 234
  A Waterfall in Tahiti (William Hodges) 134, *134*
  weapons 106
  A young woman of Otaheite, bringing a present (engraving, after John Webber) *91* (90)
Taita 33, 37
Tarevatoo 110
Tasman, Abel 29, 34, 46, 68, 69
Tasmania *see* Van Diemen's Land (Tasmania)
tattooing 94–5
telescopes 32, 71
  portable reflecting (mid-18th century) *32*
Temple, Earl 211
Theodor, Karl 203
third voyage (*Resolution* and *Discovery*)
  aims of voyage 50
  Aleutian Islands 55, 60
  animals 51, 55
  artistic contribution from John Webber 19, 51, 85, 95, 117, 137, 153
  Bligh's participation 19, 51, 72
  Clerke's participation 50, 53, 55, 60, 99, 187
  command, scientists and artists 51
  Cook's death and return home 60
  Cook's death on Hawai'i 33, 53, 60, 77, 79, 162, 167,

INDEX 255

169, 202–3
distance between Cook and his men 53
experimental voyage 32
John Ledyard's account 21
Kerguelen Land *50* (51), 51
Moorea (Eimeo) 48, 53, 55
New Zealand 51, 53
north-west American coast 50, 55, 60
North-West Passage 55, 60, 79–80
official and unofficial published accounts 78–9
Portlock's participation 19, 21
Society Islands and Omai's house 53, *54*, 55
Tongan islands (Friendly Islands) 53
Van Diemen's Land (Tasmania) 51
Vancouver Island 55
visual images 153, 157–8
Voyage to the Pacific Ocean (third voyage) (Cook, Douglas and King) 19, 78, 79, 85, 149, 152, *152*, 170, 173, 189
Thomas, Nicholas 60, 94, 233
Thompson, Benjamin, Count von Rumford 203
timekeepers *see* chronometers (marine timekeepers)
The Times, 1776: The British Story of the American Revolution (*The Times* and *Sunday Times*, National Maritime Museum, 1975) 234
Tongan islands (Friendly Islands) 46–7, 53, 78, 105, 106–7, *107*, *108*, *109* (108), 110
Torres, Luis Vaz de 65
Townley, Charles 209
travel literature 146–8
Treasures of Tutankhamun (British Museum, 1972) 234
Treaty of Paris (1763) 28
tree fragment (northern Queensland) 211, *211*
Trethewey, William, statue of Captain Cook (New Zeeland) 218
Trinity College, Cambridge, donation of objects from Cook's first voyage 165
Trinity College, Dublin, donations of objects from Cook's voyages 166
Troughton, Edward 173
Tupaia 15, 33–4, *37*, 73, 75, 76, 80, 87
'The Artist of the Chief Mourner' 87, 237
A Chart representing the isles of the South Sea, according to the notions. Chiefly collected from the accounts of Tupaya (J. R. Forster) *86* (87)
Turner, J. M. W., *Battle of Trafalgar* 228, *229*
Turnerelli, Peter, bust of Joseph Banks *226*, *227*, 229, 233

University of Cambridge
Board of Longitude research project 239
donation of objects from Cook's first voyage to Trinity College 165
University of Oxford, donations of objects from J. R. Forster 165

Van Diemen's Land (Tasmania) 35, 43, 51, *51*, 80, 169, 192
Vancouver, Charles 191
Vancouver, George
America, north-west coast voyage 80, 187, 189–91
character and reputation 191
Discovery ship 187, *188*, 189
importance of his voyages 19
and Menzies 172, 191
and New Zealand harbours and bays 41
and Nootka Crisis/Convention 187, 189
public street brawl with Thomas Pitt 191, *191* (190)
Royal Society Dining Club guest 79
and scurvy 76–7
veteran of Cook's voyages 41, 75, 169, 187
A Voyage of Discovery 186
A chart showing part of the coast of N.W. America 190
Vancouver Island 55, 189
Vanuatu (formerly New Hebrides) 48
venereal disease, Hawai'ian islands 55
Venus, transit of across Sun and *Endeavour* voyage 11, 29–30, 33, 230
Vernon, Edward 21, 226

Vespucci, Amerigo 64
visual images
art exhibitions (Cook's second and third voyages) 153
prints (Cook's second and third voyages) 157–8, *158*
works from non-sailing artists 158, 160
see also art, landscape and exploration
visualization of the Pacific *see* art, landscape and exploration
voyages *see* first voyage (*Endeavour*); Pacific voyages (1785-1803); second voyage (*Resolution* and *Adventure*); third voyage (*Resolution* and *Discovery*)

Wales, William 39, 72–73, 127, 171
walking stick ('made of the spear which killed Captain Cook') 211
Wallis, Captain Samuel
1766-68 circumnavigation 11, 29, 30, 31
in Hawkesworth's *An Account of the Voyages 28, 29* (28), 38, 50, 148–9
importance of his voyages 19
loss of men 32
Tahiti 28, 29, 32, 43
wallpaper (*Sauvages de la Mer Pacifique*) 158
Walpole, Horace 148
Ward, James 80, *80*, 231
Waters, D. W. 234
waterspouts 129
A View of Cape Stephens in Cook's Straits with Waterspout (William Hodges) 129, *130–31*, 133
Watt, Surgeon Vice-Admiral Sir James 234
weapons
Captain Cook and the Exploration of the Pacific gallery (National Maritime Museum) 233, *233*
Tongan wooden spear and war clubs 106–7, *109* (108), 110
Webb, Afrian 197–8
Webber, John
artist on Cook's third voyage 19, 51, 85, 95, 117, 137, 153
collaboration with scientific experts 120
collecting of 'curiosities' 163
donations to Bern library 163, 165
donations to British Museum 164
drawings in European costume books 158
illustrations in *Voyage to the Pacific Ocean* (Cook, King and Douglas) 152
in influential art exhibitions 145
Monneron's portrait commission 173
prints from his drawings/paintings 157–8
prints in Pennant's *Outlines of the Globe* 157, 166
works
The Apotheosis of Captain Cook (P.J. de Loutherbourg after Webber) 209–10, *209*
Captain Cook's Meeting with the Chukchi, 102, 106 (107)
Captain James Cook (engraving) 61
Death of Captain Cook (engraving) 79, 206, *206*, 207
A human sacrifice, in a morai, in Otaheite (engraving) 105, *107*
The Inside of a Winter Habitation, in Kamtschatka (engraving) 102, 105, *105*
A man of the Sandwich Islands with his helmet 98 (99), 99
A Man of Van Diemen's Land (engraving) *51*
Midshipman James Ward 80, 231
A night dance by women, in Hapaee (engraving) 53, *53* (52)
Nootka Sound engravings 230
A Party from His Majesty's ships Resolution & Discovery shooting sea-horses [walruses] 137, 153, *154–5*
Poedua, the Daughter of Orio 55, 99, *100* (101), 101, 137, 153, 160
portrait of Cook (bought by Alan Bond) 235
Resolution and Discovery in Ship Cove, Nootka Sound 2, 55, *55* (54)
The Resolution beating through the Ice, with the Discovery in the most imminent Danger in the distance (engraving) 56–7
A Sea Otter (engraving) *171* (170)
Tereoboo, King of Owyhee, bringing Presents to Captain Cook (engraving) 90
Various articles, at the Sandwich Islands (after John Webber) *164* (165)
A View in Ulietea [Raiatea] *86* (87), 87–8
A view of Christmas Harbour, in Kerguelen's Land (engraving) *50* (51)
A View of Huaheine (engraving) 158, *158*
Views in the South Seas 19, 137
Views in the South Seas … View in Macao 177 (176)
Views in the South Seas (Queen Charlottes Sound, New Zealand, engraving) 157, 158
A young woman of Otaheite, bringing a present (engraving) *91* (90)
Wedgwood, blue jasper portrait medallion of Cook 219
Weller, Arthur, Sir 238
West, Benjamin 228
The Death of General Wolfe (engraving) 207, *208* (209), 209
portrait of Joseph Banks (engraving) *10* (11)
Westall, William
artist on Matthew Flinders's expedition 137
A bay on the south coast of New Holland 195
View of Murray's Islands with the natives offering to barter 112 (113)
View of Port Lincoln, South Coast of Australia 119 (118), 120
View of Sir Edward Pellew's Group, Northern Territory 137, 141, *143* (142)
Wreck Reef Bank, taken at low water, 196, *197*
Westmacott, Captain Robert Marsh, *Monument to Monr de la Perouse and his companions, erected at Botany Bay* (engraving) 177
Westminster Abbey, James Cook Commemoration (1979) 234
Whitaker, C., *Endeavour* model 231
Williams, Glyndwr (or Glyn) 22, 60, 69, 79, 189, 233
Williams, Helen Maria, *The Morai* 210
Williams, John 209
Williamson, John 55, 60, 164, 167, 229
Wilson, Kathleen 23
Wilson, Richard 125, 133
wine coasters, commemorating Cook's voyages *210*, 211
Withers, Charles 63, 75, 77
Withey, Lynne 78
Wolfe, General James, *The Death of General Wolfe* (engraving, after Benjamin West) 207, *208* (209), 209
Woollett, William 125
Woolner, Thomas, statue of Captain Cook (Sydney's Hyde Park) 216, 218
Wright, Joseph (of Derby) 133

Xenophon 193 (192), 194

Yanyuwa people 141
Young, Colonel Sir George 23
Young, George, Reverend, *The Life and Voyages of Captain James Cook*, title page 213
Young, Ned 185
Young, Susan 185
Susan Young, The only surviving Tahitian woman, Pitcairn's 185

Zimmermann, Heinrich 78, 203
Zoffany, Johan 39, 79, 124, 163
The Death of Captain James Cook 206–7, *207* (206), 209, 219, 226, 229–30